Applied Mathematical Sciences

EDITORIAL STATEMENT

The mathematization of all sciences, the fading of traditional scientific boundaries, the impact of computer technology, the growing importance of mathematical-computer modelling and the necessity of scientific planning all create the need both in education and research for books that are introductory to and abreast of these developments.

The purpose of this series is to provide such books, suitable for the user of mathematics, the mathematician interested in applications, and the student scientist. In particular, this series will provide an outlet for material less formally presented and more anticipatory of needs than finished texts or monographs, yet of immediate interest because of the novelty of its treatment of an application or of mathematics being applied or lying close to applications.

The aim of the series is, through rapid publication in an attractive but inexpensive format, to make material of current interest widely accessible. This implies the absence of excessive generality and abstraction, and unrealistic idealization, but with quality of exposition as a goal.

Many of the books will originate out of and will stimulate the development of new undergraduate and graduate courses in the applications of mathematics. Some of the books will present introductions to new areas of research, new applications and act as signposts for new directions in the mathematical sciences. This series will often serve as an intermediate stage of the publication of material which, through exposure here, will be further developed and refined. These will appear in conventional format and in hard cover.

MANUSCRIPTS

The Editors welcome all inquiries regarding the submission of manuscripts for the series. Final preparation of all manuscripts will take place in the editorial offices of the series in the Division of Applied Mathematics, Brown University, Providence, Rhode Island.

SPRINGER-VERLAG NEW YORK INC., 175 Fifth Avenue, New York, N. Y. 10010

Printed in U.S.A.

Applied Mathematical Sciences | Volume 32

Theodor Meis
Ulrich Marcowitz

Numerical Solution of Partial Differential Equations

Springer-Verlag
New York Heidelberg Berlin

Theodor Meis
Mathematisches Institut
der Universität zu Köln
Weyertal 86-90
5000 Köln 41
Federal Republic of Germany

Ulrich Marcowitz
Mathematisches Institut
der Universität zu Köln
Weyertal 86-90
5000 Köln 41
Federal Republic of Germany

Translated by Peter R. Wadsack, University of Wisconsin.

AMS Subject Classifications: 65MXX, 65NXX, 65P05

Library of Congress Cataloging in Publication Data

Meis, Theodor.
Numerical solution of partial differential equations.

(Applied mathematical sciences; 32)
Translation of Numerische Behandlung partieller Differentialgleichungen.
Bibliography: p.
Includes index.
1. Differential equations, Partial—Numerical solutions. I. Marcowitz, Ulrich, joint author. II. Title. III. Series.
QA1.A647 vol. 32 [QA374] 510s [515.3'53]
80-26520

English translation of the original German edition *Numerische Behandlung Partieller Differentialgleichungen* published by Springer-Verlag Heidelberg © 1978.

Printed in the United States of America.

9 8 7 6 5 4 3 2 1

ISBN 0-387-90550-2 Springer-Verlag New York Heidelberg Berlin
ISBN 3-540-90550-2 Springer-Verlag Berlin Heidelberg New York

PREFACE

This book is the result of two courses of lectures given at the University of Cologne in Germany in 1974/75. The majority of the students were not familiar with partial differential equations and functional analysis. This explains why Sections 1, 2, 4 and 12 contain some basic material and results from these areas.

The three parts of the book are largely independent of each other and can be read separately. Their topics are: initial value problems, boundary value problems, solutions of systems of equations. There is much emphasis on theoretical considerations and they are discussed as thoroughly as the algorithms which are presented in full detail and together with the programs. We believe that theoretical and practical applications are equally important for a genuine understanding of numerical mathematics.

When writing this book, we had considerable help and many discussions with H. W. Branca, R. Esser, W. Hackbusch and H. Multhei. H. Lehmann, B. Müller, H. J. Niemeyer, U. Schulte and B. Thomas helped with the completion of the programs and with several numerical calculations.

Springer-Verlag showed a lot of patience and understanding during the course of the production of the book. We would like to use the occasion of this preface to express our thanks to all those who assisted in our sometimes arduous task.

Cologne, Fall 1980

Th. Meis
U. Marcowitz

CONTENTS

PART I.
INITIAL VALUE PROBLEMS FOR HYPERBOLIC AND PARABOLIC DIFFERENTIAL EQUATIONS

1. Properly posed initial value problems

In this introductory chapter we will explain what is meant by the concept of *properly posed* initial value problems. We start with the well-known situation for ordinary differential equations, and develop the definition with the help of explanatory examples. This concept is an important one, for problems which are not properly posed cannot, in general, be attacked reasonably with numerical methods.

Theorem 1.1: Let $f \in C^0([a,b] \times \mathbb{R}, \mathbb{R})$ be a continuous function satisfying a Lipschitz condition for a constant $L \in \mathbb{R}$:

$$|f(x,z) - f(x,w)| \leq L|z - w|, \quad x \in [a,b], \quad z,w \in \mathbb{R}.$$

Then for all $\eta, \tilde{\eta} \in \mathbb{R}$ there exists exactly one function

$$u \in C^1([a,b], \mathbb{R}) \quad \text{with} \quad \begin{cases} u'(x) = f(x,u(x)), & x \in [a,b] \\ u(a) = \eta \end{cases}$$

and exactly one function

$$\tilde{u} \in C^1([a,b], \mathbb{R}) \quad \text{with} \quad \begin{cases} \tilde{u}'(x) = f(x,\tilde{u}(x)), & x \in [a,b] \\ \tilde{u}(a) = \tilde{\eta}. \end{cases}$$

If $\hat{L} = \exp(L|b-a|)$, then for all $x \in [a,b]$

$$|u(x) - \tilde{u}(x)| \leq \exp(L|x-a|)|\eta - \tilde{\eta}| \leq \hat{L}|\eta - \tilde{\eta}|.$$

Theorem 1.1 is proved in the theory of ordinary differential equations (cf. e.g. Stoer-Bulirsch 1980, Theorem 7.1.4). It says that the initial value problem

$$\begin{aligned} u'(x) &= f(x,u(x)), \qquad x \in [a,b] \\ u(a) &= \eta \end{aligned}$$

subject to the above conditions, has the following properties:

(1) There exists at least one solution $u(x;\eta)$.

(2) There exists at most one solution $u(x;\eta)$.

(3) The solution satisfies a Lipschitz condition with respect to η:

$$|u(x;\eta) - u(x;\tilde{\eta})| \leq \hat{L}|\eta - \tilde{\eta}|, \qquad x \in [a,b], \quad \eta,\tilde{\eta} \in \mathbb{R}.$$

This motivates the following definition, which is intentionally general and which should be completed, in each concrete case, by specifying the spaces under consideration and the nature of the solution.

<u>Definition 1.2</u>: An initial value problem is called *properly posed* (or *well posed*) if it satisfies the following conditions:

(1) *Existence*: The set of initial values for which the problem has a solution is dense in the set of all initial values.

(2) *Uniqueness*: For each initial value there exists at most one solution.

(3) *Continuous dependence on the initial values:* The solution satisfies a Lipschitz condition with respect to those

initial values for which the problem is solvable. □

We next consider a series of examples of initial value problems for partial differential equations, and examine whether or not these problems are properly posed. Our primary interest will be in classical solutions. These are characterized by the following properties:

(1) The differential equation need only be satisfied in a region G, i.e., in an *open*, connected set. The solution must be as often continuously differentiable in G as the order of the differential equation demands.

(2) Initial or boundary conditions are imposed on a subset Γ of the boundary of G. The solution must be continuously differentiable on $G \cup \Gamma$ as many times as the order of the initial or boundary conditions demand. If only functional values are given, the solution need only be continuous on $G \cup \Gamma$.

Example 1.3: Let $\phi \in C^1(\mathbb{R},\mathbb{R})$ be a bounded function and let $T > 0$. Then one of the simplest initial value problems is

$$\begin{aligned} u_y(x,y) &= 0 \\ u(x,0) &= \phi(x) \end{aligned} \qquad x \in \mathbb{R}, \quad y \in (0,T).$$

Obviously the only solution to this problem is $u(x,y) = \phi(x)$. Therefore we have

$$\|u(\cdot,\cdot;\phi) - u(\cdot,\cdot;\tilde{\phi})\|_\infty = \|\phi - \tilde{\phi}\|_\infty ,$$

so the problem is properly posed.

This initial value problem can be solved "forwards" as above, and also "backwards" since the same relationships exist for $y \in [-T,0]$. However, this situation is by no means typical for partial differential equations. □

An apparently minor modification of the above differential equation changes its properties completely:

Example 1.4: Let ϕ and T be chosen as in Example 1.3. Then the problem

$$\begin{aligned} u_x(x,y) &= 0 \\ u(x,0) &= \phi(x) \end{aligned} \qquad x \in \mathbb{R}, \quad y \in (0,T)$$

is solvable only if ϕ is constant. In this exceptional case there are nevertheless infinitely many solutions $\phi(x) + \psi(y)$, for functions $\psi \in C^1((0,T), \mathbb{R})$ with $\psi(0) = 0$. The problem therefore is not properly posed. □

The following example contains the two previous ones as special cases and in addition leads heuristically to the concept of *characteristics*, which is so important in partial differential equations.

Example 1.5: Let $\phi \in C^1(\mathbb{R},\mathbb{R})$ be a bounded function and let $A, B, T \in \mathbb{R}$ with $A^2 + B^2 > 0$, $T > 0$. We consider the problem

$$\begin{aligned} Au_x(x,y) &= Bu_y(x,y) \\ u(x,0) &= \phi(x) \end{aligned} \qquad x \in \mathbb{R}, \quad y \in (0,T).$$

For $B = 0$, the problem is not properly posed (cf. Example 1.4). Assume $B \neq 0$. On the lines

$$(x_c(t), y_c(t)) = (c - \tfrac{A}{B}t, t), \quad t \in \mathbb{R} \text{ a parameter}$$

we have for $t \in (0,T)$:

$$\frac{d}{dt}u(x_c(t),y_c(t)) = -\frac{A}{B}u_x(x_c(t),y_c(t)) + u_y(x_c(t),y_c(t)) = 0.$$

This implies that

$$u(x_c(t),y_c(t)) \equiv \phi(c) = \phi(x_c(t) + \frac{A}{B}y_c(t)), \quad t \in [0,T).$$

The problem thus has a uniquely determined solution

$$u(x,y) = \phi(x + \frac{A}{B}y)$$

and is properly posed. The family of lines $Bx + Ay = Bc =$ constant (c the family parameter) are called the *characteristics* of the differential equation. They play a distinguished role in the theory and for more general differential equations consist of curved lines. In the example under consideration we see that the problem is properly posed if the initial values do not lie on a characteristic (cf. Example 1.6). Further we note that the discontinuities in the higher derivatives of ϕ propagate along the characteristics. □

In the following example we consider systems of partial differential equations for the first time, and discover that we can have characteristics of differing directions.

Example 1.6: *Initial value problem for systems of partial differential equations.* Let $n \in \mathbb{N}$, $T > 0$, and $\alpha,\beta \in \mathbb{R}$ with $\alpha^2 + \beta^2 > 0$. Define

$$G = \{(x,y) \in \mathbb{R}^2 \mid \alpha x + \beta y \in (0,T)\}$$
$$\Gamma = \{(x,y) \in \mathbb{R}^2 \mid \alpha x + \beta y = 0\}$$

and suppose $\phi \in C^1(\Gamma,\mathbb{R}^n)$ bounded, $q \in C^1(\overline{G},\mathbb{R}^n)$, and $A \in \mathrm{MAT}(n,n,\mathbb{R})$ real diagonalizable. We would like to find a solution of the problem

$$u_y(x,y) = Au_x(x,y) + q(x,y), \qquad (x,y) \in G,$$
$$u(x,y) = \phi(x,y), \qquad (x,y) \in \Gamma.$$

The given system of differential equations may be uncoupled by means of a regular transformation matrix S which takes A to the diagonal form $S^{-1}AS = \Lambda = \mathrm{diag}(\lambda_i)$. Letting $u = Sv$, $q = Sr$, and $\phi = S\psi$, we obtain the equivalent system

$$v_y(x,y) = \Lambda v_x(x,y) + r(x,y), \qquad (x,y) \in G,$$
$$v(x,y) = \psi(x,y), \qquad (x,y) \in \Gamma.$$

Analogously to Example 1.5, we examine the ith equation on the lines $x + \lambda_i y = c$ with the parametric representation

$$(x_c(t), y_c(t)) = (c-\lambda_i t, t), \quad t \in \mathbb{R}.$$

From the differential equation it follows that for the ith component v_i of v, and for all $t \in \mathbb{R}$ with $\alpha(c-\lambda_i t) + \beta t \in (0,T)$:

$$\frac{d}{dt}\, v_i(x_c(t),y_c(t)) = -\lambda_i \frac{\partial v_i}{\partial x}(x_c(t),y_c(t)) + \frac{\partial v_i}{\partial y}(x_c(t),y_c(t))$$
$$= r_i(x_c(t),y_c(t))$$

therefore

$$v_i(x_c(t),y_c(t)) = \eta_i + \int_0^t r_i(x_c(\tau),y_c(\tau))d\tau,$$
$$\eta_i \in \mathbb{R} \text{ arbitrary.}$$

When considering the initial conditions, we have three possible cases:

Case 1: $\alpha\lambda_i - \beta \neq 0$. The two lines Γ and $x + \lambda_i y = c$ have exactly one intersection point

$$(\hat{x}_c, \hat{y}_c) = \left(\frac{-\beta c}{\alpha\lambda_i - \beta}, \frac{\alpha c}{\alpha\lambda_i - \beta}\right)$$

and

$$\eta_i = \eta_i(c) = \psi_i(\hat{x}_c, \hat{y}_c) - \int_0^{\hat{y}_c} r_i(x_c(\tau), y_c(\tau))\,d\tau.$$

Thus for all v_i we obtain the following representation:

$$v_i(x,y) = \eta_i(c) + \int_0^y r_i(c - \lambda_i\tilde{y}, \tilde{y})\,d\tilde{y}$$
$$c = x + \lambda_i y.$$

One can check by substitution that v_i is actually a solution of the ith equation of the uncoupled system.

Case 2: $\alpha\lambda_i - \beta = 0$ and $c = 0$. The two lines Γ and $x + \lambda_i y = c$ are now identical. The ith equation of the uncoupled system can be solved only if coincidentally

$$\frac{d}{dt}\psi_i(x_c(t), y_c(t)) = r_i(x_c(t), y_c(t)),$$
$$\alpha(c - \lambda_i t) + \beta t \in (0,T).$$

In this exceptional case there are, however, infinitely many solutions.

Case 3: $\alpha\lambda_i - \beta = 0$ and $c \neq 0$. The two lines Γ and $x + \lambda_i y = c$ now have no point of intersection. The ith equation of the uncoupled system has infinitely many solutions.

As in Example 1.5, the family of lines $x + \lambda_i y = c$ (c the family parameter) are called the *characteristics* of the system of partial differential equations. If none of the characteristics coincide with the line Γ on which the initial values are given, then we have case 1 for all i, and

$$u(x,y) = Sv(x,y) = S\begin{pmatrix} v_1(x,y) \\ \cdot \\ \cdot \\ \cdot \\ v_n(x,y) \end{pmatrix}$$

is the uniquely determined solution of the original problem. To check the Lipschitz condition, let $\tilde{\phi} \in C^1(\Gamma,\mathbb{R}^n)$ and $\tilde{\phi} = S\tilde{\psi}$. Then it follows that

$$||u(\cdot,\cdot;\phi) - u(\cdot,\cdot;\tilde{\phi})||_\infty \leq ||S||_\infty ||v(\cdot,\cdot;\psi) - v(\cdot,\cdot;\tilde{\psi})||_\infty$$

$$= ||S||_\infty ||\psi-\tilde{\psi}||_\infty \leq ||S||_\infty ||S^{-1}||_\infty ||\phi-\tilde{\phi}||_\infty.$$

Thus we have shown that the problem is properly posed exactly when Γ is not a characteristic.

Let $\mu = \min\{\lambda_i | i = 1(1)n\}$ and $\nu = \max\{\lambda_i | i = 1(1)n\}$. From the representation of u and v we see that the solution u at the fixed point (x_0,y_0) depends on q and additionally only on the values of $\phi(x,y)$ on the line segment connecting the points

$$\left(\frac{-\beta(x_0+\mu y_0)}{\alpha\mu-\beta}, \frac{\alpha(x_0+\mu y_0)}{\alpha\mu-\beta}\right) \text{ and } \left(\frac{-\beta(x_0+\nu y_0)}{\alpha\nu-\beta}, \frac{\alpha(x_0+\nu y_0)}{\alpha\nu-\beta}\right)$$

This segment is called the *domain of dependence* of the point (x_0,y_0). On the other hand, if $\phi(x,y)$ is known only on the segment connecting the points

$$\left(\frac{-\beta a}{\alpha\mu-\beta}, \frac{\alpha a}{\alpha\mu-\beta}\right) \text{ and } \left(\frac{-\beta b}{\alpha\nu-\beta}, \frac{\alpha b}{\alpha\nu-\beta}\right)$$

then the solution can be computed only in the triangle

$$\alpha x + \beta y \geq 0, \quad x + \mu y \geq a, \quad x + \nu y \leq b.$$

This triangle is called the *domain of determinancy* of this segment. These concepts are clarified further by the

example in Figure 1.7. □

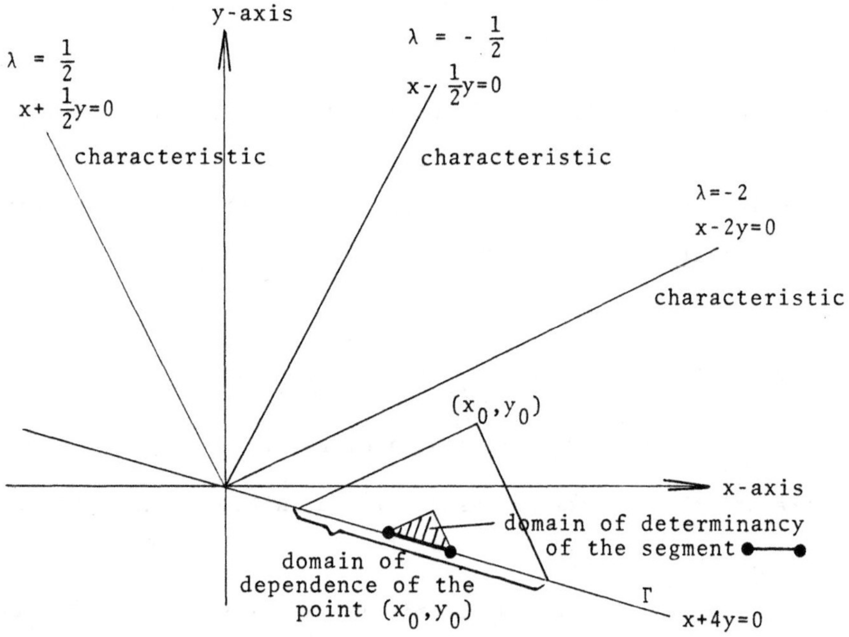

Figure 1.7: Characteristics and domains of determinancy and dependence.

The uncoupling of a system of differential equations as used in the preceding example is also helpful in investigating other problems.

Example 1.8: *Initial boundary value problem for systems of partial differential equations.* Let $n \in \mathbb{N}$, $G = (0,\infty)^2$, $\Gamma = \partial G$, $q \in C^1(\bar{G},\mathbb{R}^n)$ and $\phi, \tilde{\phi} \in C^1([0,\infty), \mathbb{R}^n)$ bounded, and $A \in \mathrm{MAT}(n,n,\mathbb{R})$ real diagonalizable. We seek a solution to the problem

$$
\begin{aligned}
u_y(x,y) &= Au_x(x,y) + q(x,y), \qquad (x,y) \in G \\
u(x,0) &= \phi(x), \qquad x \geq 0 \\
u(0,y) &= \tilde{\phi}(y), \qquad y \geq 0
\end{aligned}
$$

with the compatibility conditions $\phi(0) = \tilde{\phi}(0)$ and $\tilde{\phi}'(0) = A\phi'(0) + q(0,0)$. The system of differential equations can be uncoupled as in Example 1.6. With the notation used there, it follows that for the ith component v_i of v, and for all $t \in \mathbb{R}$ with $t > 0$ and $c - \lambda_i t > 0$:

$$v_i(x_c(t),y_c(t)) = \eta_i + \int_0^t r_i(x_c(\tau),y_c(\tau))d\tau,$$

$$\eta_i \in \mathbb{R} \text{ arbitrary.}$$

There are three possible cases for the initial boundary condition:

Case 1: $\lambda_i < 0$. The characteristic $x + \lambda_i y = c$ has exactly one intersection point with the boundary Γ. With the aid of the values of ϕ or $\tilde{\phi}$, η_i and therefore v_i is uniquely determined.

Case 2: $\lambda_i > 0$. The characteristic $x + \lambda_i y = c$ then has, for $c > 0$, one intersection point with the positive x-axis and one intersection point with the positive y-axis. In this case η_i is overdetermined in general, and the ith equation of the uncoupled system is not solvable.

Case 3: $\lambda_i = 0$. For $c > 0$, η_i is uniquely determined, but the solution $v_i(x_c(t), y_c(t))$ converges continuously to the ith component of $S^{-1}\tilde{\phi}$ for $c \to 0$ only in exceptional cases.

The problem is properly posed if and only if all eigenvalues of A are negative, i.e., when only the first case occurs. □

The following example, the wave equation, is an important special case of Example 1.6.

Example 1.9: *Wave Equation.* Let $T > 0$ and $\phi_1, \phi_2 \in C^1(\mathbb{R}, \mathbb{R})$. With $\alpha = 0$, $\beta = 1$, $A = \begin{pmatrix} 0 & 1 \\ 1 & 0 \end{pmatrix}$, $q \equiv 0$ and the use of componentwise notation, Example 1.6 becomes

$$\left.\begin{aligned} \partial u_1/\partial y &= \partial u_2/\partial x \\ \partial u_2/\partial y &= \partial u_1/\partial x \\ u_1(x,0) &= \phi_1(x) \\ u_2(x,0) &= \phi_2(x) \end{aligned}\right. \qquad x \in \mathbb{R},\quad y \in (0,T).$$

We have $\lambda_1 = 1$, $\lambda_2 = -1$, $S = \begin{pmatrix} 1 & 1 \\ 1 & -1 \end{pmatrix}$ and $S^{-1} = \frac{1}{2}\begin{pmatrix} 1 & 1 \\ 1 & -1 \end{pmatrix}$. Thus the solution of the wave equation becomes

$$\begin{pmatrix} u_1(x,y) \\ u_2(x,y) \end{pmatrix} = S \begin{pmatrix} \psi_1(x+y) \\ \psi_2(x-y) \end{pmatrix} = S \begin{pmatrix} \frac{1}{2}(\phi_1(x+y) + \phi_2(x+y)) \\ \frac{1}{2}(\phi_1(x-y) - \phi_2(x-y)) \end{pmatrix} =$$

$$\frac{1}{2}\begin{pmatrix} \phi_1(x+y)+\phi_2(x+y)+\phi_1(x-y)-\phi_2(x-y) \\ \phi_1(x+y)+\phi_2(x+y)-\phi_1(x-y)+\phi_2(x-y) \end{pmatrix} \qquad x \in \mathbb{R},\quad y \in [0,T).$$

The wave equation can also be written as a partial differential equation of second order by differentiating the first equation with respect to y and the second equation with respect to x, and then subtracting one equation from the other. The initial value for u_2 becomes an initial value for $\partial u_1/\partial y$

with the aid of the first equation. With the change of notation of u for u_1 and ϕ for ϕ_1, we obtain

$$\begin{aligned} u_{yy} - u_{xx} &= 0 \\ u(x,0) &= \phi(x) \qquad\qquad x \in \mathbb{R}, \quad y \in (0,T). \\ u_y(x,0) &= \phi_2'(x) = \psi(x) \end{aligned}$$

Yet another form of the wave equation arises from the coordinate transformation $y = \zeta + \sigma$, $x = \zeta - \sigma$. The differential operators are correspondingly transformed into

$$\frac{\partial}{\partial\zeta} = \frac{\partial}{\partial y} + \frac{\partial}{\partial x}, \qquad \frac{\partial}{\partial\sigma} = \frac{\partial}{\partial y} - \frac{\partial}{\partial x}.$$

It then follows that the given differential equation can also be written as

$$\left(\frac{\partial}{\partial y} + \frac{\partial}{\partial x}\right)\left(\frac{\partial}{\partial y} - \frac{\partial}{\partial x}\right) u(x,y) = 0,$$

i.e.,

$$\frac{\partial^2 u(\zeta-\sigma,\zeta+\sigma)}{\partial\zeta\partial\sigma} = \frac{\partial^2 \tilde{u}(\zeta,\sigma)}{\partial\zeta\partial\sigma} = 0.$$

In order for the wave equation problem in this form to be properly posed, it is necessary that the initial values not be specified for $\zeta \equiv$ constant and $\sigma \equiv$ constant, since these lines are the characteristics. □

Another important type of partial differential equation is exemplified by the heat equation, which is presented in the following examples.

Example 1.10: *Initial value problem for the heat equation.* For $T > 0$, $\alpha > 0$, and $q,\phi \in C^0(\mathbb{R}, \mathbb{R})$ we seek a solution to the following problem:

$$\begin{aligned} u_t(x,t) &= \alpha u_{xx}(x,t) - q(x) \\ u(x,0) &= \phi(x) \end{aligned} \qquad x \in \mathbb{R}, \quad t \in (0,T).$$

However this problem is *not uniquely* solvable and therefore also not properly posed (cf. Hellwig 1977). This shortcoming can be overcome by imposing conditions on q and ϕ and by restricting the concept of a solution. The additional conditions

$$|q(x)| \exp(-|x|) \quad \text{and} \quad |\phi(x)| \exp(-|x|) \quad \text{bounded,}$$

$$|u(x,t)| \exp(-|x|) \quad \text{bounded for} \quad x \in \mathbb{R} \quad \text{and} \quad t \in [0,T)$$

determine linear subspaces of $C^0(\mathbb{R}, \mathbb{R})$ and $C^2(\mathbb{R} \times (0,T), \mathbb{R}) \cap C^0(\mathbb{R} \times [0,T), \mathbb{R})$. For the norm we use

$$||\phi|| = \sup_{x \in \mathbb{R}} \frac{|\phi(x)|}{\exp(|x|)}, \quad ||u|| = \sup_{x \in \mathbb{R},\ t \in [0,T)} \frac{|u(x,t)|}{\exp(|x|)}.$$

One can show that the problem is now properly posed. The solution of the homogeneous equation

$$v_t(x,t) = \alpha v_{xx}(x,t), \quad x \in \mathbb{R}, \quad t \in (0,T)$$
$$v(x,0) = \psi(x), \quad |\psi(x)| \exp(-|x|) \ \text{bounded,} \quad \psi \in C^0(\mathbb{R}, \mathbb{R})$$

can be derived using Fourier transforms (cf. §9):

$$v(x,t) = \begin{cases} (4\pi\alpha t)^{-1/2} \displaystyle\int_{-\infty}^{+\infty} \exp\left(-\frac{(x-\tau)^2}{4\alpha t}\right)\psi(\tau)\,d\tau & \text{for} \quad t \in (0,T) \\ \psi(x) & \text{for} \quad t = 0. \end{cases}$$

To obtain the solution of the inhomogeneous problem, we first consider the equation

$$\alpha w_{xx}(x,t) - q(x) = 0.$$

Obviously

$$w(x,t) = w(x) = \frac{1}{\alpha}\int_0^x \int_0^\xi q(\tau)d\tau d\xi$$

is a particular solution. A straightforward computation shows that $|w(x)|\exp(-|x|)$ is bounded. Now let $v(x,t)$ be the solution of the homogeneous equation with $\psi(x) = \phi(x) - w(x)$. Then $u(x,t) = v(x,t) + w(x)$ is the solution of the original inhomogeneous heat equation.

It is apparent from the integral representation of the solution that there is no finite domain of dependence. Therefore this problem has no practical significance, in contrast to the following initial boundary value problem. □

Example 1.11: *Initial boundary value problem for the heat equation.* Let $T > 0$, $\alpha > 0$; $q,\phi \in C^0([a,b], \mathbb{R})$ and $\psi_a,\psi_b \in C^0([0,T], \mathbb{R})$. We seek a solution of the problem

$$\begin{aligned} &u_t(x,t) = \alpha u_{xx}(x,t) - q(x), && x \in (a,b), \quad t \in (0,T) \\ &u(x,0) = \phi(x), && x \in [a,b] \\ &u(a,t) = \psi_a(t),\ u(b,t) = \psi_b(t), && t \in [0,T). \end{aligned}$$

Since there are two conditions on each of $u(a,0)$ and $u(b,0)$, we also need the following *compatibility conditions*

$$\phi(a) = \psi_a(0), \quad \phi(b) = \psi_b(0).$$

We know from the literature that the problem is properly posed (cf. e.g. Petrovsky 1954, §38). It can be solved, for example, by a Laplace transform (reducing it to an ordinary differential equation) or by difference methods. $u(x,t)$ is dependent on $\phi(\xi)$ for all $\xi \in [a,b]$ and on $\psi_a(s),\psi_b(s)$ for all $s \in [0,t]$.

In contrast to Example 1.3, the problem at hand is not properly posed if one attempts to solve it "backwards" [for $t \in (-T,0)$]--heat conduction processes are not reversible. For the problem then either is not solvable or the solution does not depend continuously on the initial values. This state of affairs is best explained with the following special example:

$a = 0,\ b = \pi,\ \alpha = 1,\ q \equiv 0,\ \psi_a \equiv 0,\ \psi_b \equiv 0,\ \omega \in \mathbb{N},\ \gamma \in \mathbb{R},$
$\phi(x;\gamma,\omega) = \gamma \sin \omega x.$

$$\begin{aligned} u_t(x,t;\gamma,\omega) &= u_{xx}(x,t;\gamma,\omega), && x \in (0,\pi),\quad t \in (-T,0)\\ u(x,0;\gamma,\omega) &= \phi(x;\gamma,\omega), && x \in [0,\pi]\\ u(0,t;\gamma,\omega) &= u(\pi,t;\gamma,\omega) \equiv 0, && t \in (-T,0]. \end{aligned}$$

One obtains the solution

$$u(x,t;\gamma,\omega) = \gamma \exp(-\omega^2 t)\sin \omega x.$$

For the norms we have

$$\|u(\cdot,\cdot;\gamma,\omega) - u(\cdot,\cdot;0,\omega)\|_\infty = \gamma \exp(\omega^2 T)$$

$$\|\phi(\cdot;\gamma,\omega) - \phi(\cdot;0,\omega)\|_\infty = \gamma.$$

The ratio of the norms grows with ω beyond all bounds. Thus there can be no valid Lipschitz condition with respect to dependence on the initial values. □

<u>Example 1.12</u>: *Initial boundary value problem of the third kind for the heat equation. Nonlinear heat equation.* Let $T,\alpha > 0$; $q,\phi \in C^0([a,b],\mathbb{R})$; $\beta_a,\gamma_a,\beta_b,\gamma_b \geq 0$; $\beta_a+\gamma_a > 0$; $\beta_b+\gamma_b > 0$; $\psi_a,\psi_b \in C^0([0,T],\mathbb{R})$. We consider the problem

$$\begin{aligned}
&u_t(x,t) = \alpha u_{xx}(x,t) - q(x), && x \in (a,b), \quad t \in (0,T)\\
&u(x,0) = \phi(x), && x \in [a,b]\\
&\beta_a u(a,t) - \gamma_a u_x(a,t) = \psi_a(t) && \\
&\beta_b u(b,t) + \gamma_b u_x(b,t) = \psi_b(t) && t \in [0,T).
\end{aligned}$$

Compatibility conditions: $\beta_a\phi(a) - \gamma_a\phi'(a) = \psi_a(0)$

$\beta_b\phi(b) + \gamma_b\phi'(b) = \psi_b(0)$.

The boundary conditions imposed here are of great practical significance. They are called *boundary values of the third kind*. The special cases $\gamma_a = \gamma_b = 0$ and $\beta_a = \beta_b = 0$ are called *boundary values of the first and second kinds*, respectively. One can show that the problem is properly posed. The methods of solution are Laplace transforms or finite differences, as in Example 1.11.

The nonlinear heat equation

$$u_t = [\alpha(u)u_x]_x - q(x,u)$$
$$\alpha(z) \geq \varepsilon > 0, \quad z \in \mathbb{R}$$

is frequently rewritten in practice as follows: define a strongly monotone function $f: \mathbb{R} \to \mathbb{R}$ by

$$f(z) = \int_0^z \alpha(\tilde{z})d\tilde{z}$$

and set

$$w(x,t) = f(u(x,t)).$$

It follows that

$$\begin{aligned}
w_x &= \alpha(u)u_x\\
w_t &= \alpha(u)u_t\\
w_t &= \alpha(u)[w_{xx} - q(x,u)]
\end{aligned}$$

With the notation

$$\tilde{\alpha}(z) = \alpha(f^{-1}(z))$$
$$\tilde{q}(x,z) = q(x,f^{-1}(z))$$

one obtains a new differential equation:

$$w_t = \tilde{\alpha}(w)[w_{xx} - \tilde{q}(x,w)].$$

All steady state solutions $(w_t = 0)$ satisfy the simple equation

$$w_{xx} = \tilde{q}(x,w). \qquad \square$$

Example 1.13: *Parabolic differential equation in the sense of Petrovski.* Let $T > 0$, $q \in \mathbb{N}$, $\alpha \in \mathbb{R}$ and $\phi \in C^q(\mathbb{R}, \mathbb{R})$ bounded. We seek a bounded solution for the problem

$$\left.\begin{aligned} u_t(x,t) &= \alpha\left(\frac{\partial}{\partial x}\right)^q u(x,t) \\ u(x,0) &= \phi(x) \end{aligned}\right\} \quad x \in \mathbb{R},\ t \in (0,T).$$

$$q \text{ odd or } (-1)^{q/2}\alpha \le 0.$$

Special cases of this problem are given in Example 1.5 (for $B \neq 0$) and in Example 1.10. The parabolic equations in the sense of Petrovski are *hyperbolic* equations for $q = 1$ and are *parabolic* equations in the ordinary sense for $q = 2$ (see §2). For larger q, the properties of the problem resemble those in the case $q = 2$. One can show that the problem is properly posed. The solution methods are difference methods or Fourier transforms even if $q > 2$ (cf. §9).

The above equation has physical significance even when α is complex. For example, letting $\alpha = \frac{ih}{4\pi m}$ and $q = 2$ yields the *Schrödinger equation* for the motion of a free particle of mass m (h is *Planck's* constant). The condition on q must be modified for complex α, to become

$$\mathrm{Re}(\alpha i^q) \le 0. \qquad \square$$

Example 1.14: *Cauchy-Riemann differential equations.* Let $T > 0$ and $\phi,\psi \in C^0(\mathbb{R},\mathbb{R})$ bounded functions. We consider the problem

$$\begin{aligned} u_y(x,y) &= -v_x(x,y) \\ v_y(x,y) &= u_x(x,y) \qquad\qquad x \in \mathbb{R},\ y \in (0,T) \\ u(x,0) &= \phi(x), \quad v(x,0) = \psi(x). \end{aligned}$$

These two differential equations of first order are frequently combined into one differential equation of second order:

$$u_{xx} + u_{yy} = 0.$$

This equation is called the *potential equation* and is the most studied partial differential equation. The Cauchy-Riemann differential equations are not a special case of the *hyperbolic* system of first order of Example 1.6 since the matrix $A = \begin{pmatrix} 0 & -1 \\ 1 & 0 \end{pmatrix}$ is not real diagonalizable. Rather they are of *elliptic* type (see §2). Although the initial value problem at hand is uniquely solvable for many special cases, there is no continuous dependence on the initial values.

Example: $\gamma, w \in \mathbb{R}$, $\phi(x) = \gamma \sin \omega x$, $\psi(x) = 0$. As a solution of the Cauchy-Riemann differential equations one obtains:

$$\begin{aligned} u(x,y) &= \gamma \sin(\omega x) \cosh(\omega y) \\ v(x,y) &= \gamma \cos(\omega x) \sinh(\omega y). \end{aligned}$$

With $w = (u,v)$ and $\chi = (\phi,\psi)$ this yields:

$$\begin{aligned} \|w\|_\infty &= \gamma \cosh(\omega T) \\ \|\chi\|_\infty &= \gamma. \end{aligned}$$

Thus the solution cannot satisfy a Lipschitz condition with respect to the initial values, and the problem is not properly posed. This property carries over unchanged to the equivalent initial value problem for the potential equation.

In practice, only boundary value problems are considered for *elliptic* differential equations, since the solution does not depend continuously on the initial values. □

2. Types and characteristics

Since initial and boundary value problems in partial differential equations are not always properly posed, it is worthwhile to divide differential equations into various *types*. One speaks of *hyperbolic, elliptic,* and *parabolic* differential equations. Of primary interest are initial value problems for hyperbolic equations, boundary value problems for elliptic equations, and initial boundary value problems for parabolic equations. Typical examples for the three classes of equations are the *wave equation* (hyperbolic, see Example 1.9), the *potential equation* (elliptic, see Example 1.14), and the *heat equation* (parabolic, see Examples 1.11, 1.12). In addition, the concept of the *characteristic* proves to be fundamental for an understanding of the properties of partial differential equations.

In keeping with our textbook approach, we will consider primarily the case of two independent variables in this and the following chapters. We consider first scalar equations of second order, and follow with a discussion of systems of first order. In the general case of m independent variables we restrict ourselves to a few practically important types, since a more complete classification would require the consideration of too many special cases. In particular, for the case $m > 2$ there exist simple equations for which none of the above mentioned problems is properly posed.

Definition 2.1: Let G be a region in $\mathbb{R}^2$ and $a,b,c,f \in C^o(G \times \mathbb{R}^3,\mathbb{R})$ with $a(x,y,z)^2 + b(x,y,z)^2 + c(x,y,z)^2 > 0$ for all $(x,y) \in G$, $z \in \mathbb{R}^3$. The equation

$$a(x,y,p)u_{xx} + 2b(x,y,p)u_{xy} + c(x,y,p)u_{yy} + f(x,y,p) = 0$$

with $p(x,y) = (u,u_x,u_y)$ is called a *quasi-linear second order differential equation*. The quantity

$$au_{xx} + 2bu_{xy} + cu_{yy}$$

is called the *principal part* of the differential equation. The description *quasilinear* is chosen because the derivatives of highest order only occur linearly. The differential equation is called *semilinear* when the coefficients a, b, and c of the principal part are independent of p, and f has the special form

$$f(x,y,p) = d(x,y,u)u_x + e(x,y,u)u_y + g(x,y,u)$$

with functions $d,e,g \in C^o(G \times \mathbb{R},\mathbb{R})$. A semilinear differential equation is called *linear* when the functions d and e are independent of u, and g has the special form

$$g(x,y,u) = r(x,y)u + s(x,y)$$

with functions $r,s \in C^o(G, \mathbb{R})$. A linear equation is called a differential equation with *constant coefficients* when the functions a, b, c, d, e, r, and s are all constant. □

In order to define the various types of second order partial differential equations we need several concepts originating in algebra.

A real polynomial $P(x) = P(x_1,\ldots,x_m)$ is called a (real) *form* of degree k if $P(tx) = t^k P(x)$ holds for all $x \in \mathbb{R}^m$ and all $t \in \mathbb{R}$. A form of degree two is called a *quadratic form*. It may also be represented in matrix form

$$P(x) = x^T Ax, \quad A \in \mathrm{MAT}(m,m,\mathbb{R}), \quad x \in \mathbb{R}^m.$$

Without loss of generality A may be assumed to be symmetric. Then A is uniquely determined by P and vice versa. The usual concepts of symmetric matrices

positive definite	:<=>	all eigenvalues of A greater than zero
negative definite	:<=>	all eigenvalues of A less than zero
definite	:<=>	positive definite or negative definite
positive semidefinite	:<=>	all eigenvalues of A greater than or equal to zero
negative semidefinite	:<=>	all eigenvalues of A less than or equal to zero
semidefinite	:<=>	positive semidefinite or negative semidefinite
indefinite	:<=>	not semidefinite

thus carry over immediately to quadratic forms.

Definition 2.2: To the differential equation of Definition 2.1 assign the quadratic form

$$P(\xi,\eta) = a(x,y,p)\xi^2 + 2b(x,y,p)\xi\eta + c(x,y,p)\eta^2$$

Then the *type* of the differential equation with respect to a *fixed* function $u \in C^2(G, \mathbb{R})$ and a *fixed* point $(x,y) \in G$ is determined by the properties of the associated quadratic form:

Type of d.e.	Properties of $P(\xi,\eta)$
hyperbolic	indefinite (i.e. $ac-b^2 < 0$)
elliptic	definite (i.e. $ac-b^2 > 0$)
parabolic	semidefinite, but not definite (i.e. $ac-b^2 = 0$)

The differential equation is called hyperbolic (elliptic, parabolic) in *all of* G with respect to a *fixed* function, if, with respect to this function, it is hyperbolic (elliptic, parabolic) for all points $(x,y) \in G$. □

The above division of differential equations into various types depends only on the principal part. For semilinear equations, the type at a fixed point $(x,y) \in G$ is the same with respect to all functions $u \in C^2(G, \mathbb{R})$; for constant coefficients a, b, and c, the type does not depend on the point, either.

In many investigations it is not sufficient that the differential equation be hyperbolic or elliptic with respect to a function in all of the region. In such cases one frequently restricts oneself to *uniformly hyperbolic* or *uniformly elliptic* differential equations. By this one means equations for which the coefficients a, b, and c are bounded independent of $(x,y) \in G$, $z \in \mathbb{R}^3$ and for which in addition

$$\left.\begin{array}{ll} ac - b^2 \le -\gamma < 0 & \text{(unif. hyperbolic)} \\ ac - b^2 \ge \gamma > 0 & \text{(unif. elliptic)} \end{array}\right\} \quad (x,y) \in G,\ z \in \mathbb{R}^3$$

where $\gamma \equiv$ const.

Linear second order differential equations with constant coefficients can, by a linear change of coordinates, always be reduced to a form in which the principal part coincides with one of the three *normal forms*

$u_{xx} - u_{yy}$ *hyperbolic normal form*

$u_{xx} + u_{yy}$ *elliptic normal form*

u_{xx} *parabolic normal form.*

Even for more general linear and semilinear equations one can often find a coordinate transformation which achieves similar results. The type of a differential equation is not changed by such transformations whenever these are invertible and twice differentiable in both directions.

For the definition of characteristics we will need several concepts about curves. Let G be a region in $\mathbb{R}^2$ and let I be one of the intervals (a,b), (a,∞), $(-\infty,b)$, or $(-\infty,\infty)$ with $a,b \in \mathbb{R}$. A mapping $\phi \in C^1(I,G)$ is called a *smooth curve* in G if $\phi_1'(t)^2 + \phi_2'(t)^2 > 0$ for all $t \in I$. The vector $(\phi_1'(t),\phi_2'(t))$ is called the *tangent* to the curve ϕ at the point $(\phi_1(t),\phi_2(t))$; the set $\phi(I)$ is called the image of the curve ϕ.

Definition 2.3: We consider the differential equation in Definition 2.1. A vector $\beta = (\beta_1,\beta_2) \in \mathbb{R}^2$, $\beta \neq (0,0)$ is called a *characteristic direction* at the *fixed* point $(x,y) \in G$ with respect to the *fixed* function $u \in C^2(G, \mathbb{R})$ if it is true that

$$a(x,y,p)\beta_2^2 - 2b(x,y,p)\beta_1\beta_2 + c(x,y,p)\beta_1^2 = 0.$$

The image of a smooth curve ϕ in G is called a *characteristic* of the differential equation with respect to u whenever the tangents $(\phi_1'(t),\phi_2'(t))$ are characteristic directions for the differential equation at the points $(\phi_1(t),\phi_2(t))$ with respect to u *for all* $t \in I$. This means that ϕ is a solution of the ordinary differential equation

$$\begin{aligned} &a(\phi_1,\phi_2,p(\phi_1,\phi_2))\phi_2'(t)^2 \\ &-2b(\phi_1,\phi_2,p(\phi_1,\phi_2))\phi_1'(t)\phi_2'(t) \\ &+c(\phi_1,\phi_2,p(\phi_1,\phi_2))\phi_1'(t)^2 = 0, \quad t \in I. \end{aligned} \tag{2.4}$$ □

The condition $a\beta_2^2 - 2b\beta_1\beta_2 + c\beta_1^2 = 0$ can also be put in the form

$$\beta_1 = \frac{b \pm (b^2-ac)^{1/2}}{c}\beta_2 \tag{2.5}$$

when $c \neq 0$. An analogous rearrangement is possible when $a \neq 0$. This implies that a hyperbolic (parabolic, elliptic) differential equation has two (one, no) linearly independent characteristic direction(s) at every point.

Examples: The wave equation $u_{yy}(x,y) - u_{xx}(x,y) = 0$ has characteristic directions $(1,1)$ and $(1,-1)$ at every point. The heat equation $u_t(x,t) = \alpha u_{xx}(x,t) - q(x)$ has characteristic direction $(1,0)$ at every point.

The curve ϕ is not sufficiently determined by the differential equation (2.4). Thus one can impose the normalization condition

$$\phi_1'(t)^2 + \phi_2'(t)^2 = 1, \quad t \in I. \tag{2.6}$$

Subject to the additional condition $a,b,c \in C^1(G \times \mathbb{R}^3, \mathbb{R})$, it can be shown that the initial value problem for arbitrary $t_0 \in I$ for the ordinary differential equations (2.4), (2.6) has exactly two (one) solution(s) with distinct support, if it is the case that the corresponding partial differential equation is hyperbolic (parabolic) in all of G with respect to u. In the hyperbolic case it follows that there are exactly two characteristics through every point $(x,y) \in G$, while in the parabolic case it follows that every point $(x,y) \in G$ is the initial point of exactly one characteristic. The equation has no characteristic when it is of elliptic type.

The differential equation of the characteristic can be simplified when $c(x,y,p(x,y)) \neq 0$ for all $(x,y) \in G$. In lieu of (2.6) we can then impose the normalization condition $\phi_2'(t) = 1$. With (2.5) it follows from (2.4) that

$$\frac{d\psi}{dy} = \frac{b(\psi,y,p(\psi,y)) \pm \sqrt{\Delta(\psi,y,p(\psi,y))}}{c(\psi,y,p(\psi,y))}$$

$$\Delta(\psi,y,p) = b(\psi,y,p)^2 - a(\psi,y,p)c(\psi,y,p).$$

The image set $x = \psi(y)$ is a characteristic. An analogous simplification is possible for $a(x,y,p(x,y)) \neq 0$.

Finally we consider the special case where $c(x,y,p(x,y)) = 0$ at a point. This implies that $(1,0)$ is a characteristic direction. Thus also $a(x,y,p(x,y)) = 0$ implies that $(0,1)$ is a characteristic direction. Since it is possible for hyperbolic equations to have two linearly independent characteristic directions, both cases can occur simultaneously. Indeed, with an affine coordinate transformation one can arrange things so that the characteristic directions at a given point with respect to a fixed function are in arbitrary positions with respect to the coordinate system, so the above representation is possible.

We next consider the type classification and definition of characteristics for systems of first order partial differential equations. Since parabolic differential equations arise in practice almost exclusively as second order equations, we restrict ourselves to the definition of *hyperbolic* and *elliptic*.

Definition 2.7: Let $n \in \mathbb{N}$, G a region in $\mathbb{R}^2$, $h \in C^0(G \times \mathbb{R}^n, \mathbb{R}^n)$ and $A \in C^0(G \times \mathbb{R}^n, MAT(n,n,\mathbb{R}))$. The equation

$$u_y - A(x,y,u)u_x + h(x,y,u) = 0$$

is called a *quasilinear system of first order partial differential equations.* The quantity

$$u_y - Au_x$$

is called the *principal part* of the system. The system is called *semilinear* if A does not depend on u. A semilinear system is called *linear* when h has the special form

$$h(x,y,u) = B(x,y)u + q(x,y)$$

with functions $B \in C^0(G,MAT(n,n,\mathbb{R}))$, $q \in C^0(G,\mathbb{R}^n)$. A linear system is called a system with *constant coefficients* if the functions A, B, and q are all constant. □

Definition 2.8: The system of differential equations in Definition 2.7 is called *hyperbolic (elliptic)* with respect to a *fixed* function $u \in C^1(G, \mathbb{R}^n)$ and a *fixed* point $(x,y) \in G$ if the matrix $A(x,y,u(x,y))$ has n (no) linearly independent real eigenvectors. It is called hyperbolic (elliptic) in *all* G with respect to a *fixed* function if at every point $(x,y) \in G$

it is hyperbolic (elliptic) with respect to this function. □

Definition 2.9: We consider the system of differential equations in Definition 2.7. A vector $\beta = (\beta_1,\beta_2) \in \mathbb{R}^2$, $\beta \neq (0,0)$ is called a *characteristic direction* of the system at a *fixed* point $(x,y) \in G$ with respect to a *fixed* function $u \in C^1(G, \mathbb{R}^n)$ if there exists a $\gamma \in \mathbb{R}$ and an eigenvalue $\lambda = \lambda(x,y,u(x,y))$ of the matrix $A(x,y,u(x,y))$ such that

$$(\beta_1,\beta_2) = \gamma(-\lambda,1).$$

An equivalent condition is $\beta_1 + \lambda\beta_2 = 0$. The image of a smooth curve ϕ in G is called a *characteristic* of the system with respect to u if *for all* $t \in I$ the tangents $(\phi_1'(t),\phi_2'(t))$ are characteristic directions of the system with respect to u at the points $(\phi_1(t),\phi_2(t))$. This means that ϕ is a solution of the following ordinary differential equation:

$$\phi_1'(t) + \lambda(\phi_1,\phi_2,u(\phi_1,\phi_2))\phi_2'(t) = 0, \quad t \in I. \quad \square \quad (2.10)$$

From the above definition and the additional normalization condition $\beta_2 = 1$ it follows at once that a hyperbolic system has as many different characteristic directions at a point $(x,y) \in G$ as the matrix $A(x,y,u(x,y))$ has different eigenvalues. In an elliptic system there is no characteristic direction. The differential equation (2.10) can be simplified since we may impose the additional normalization condition $\phi_2'(t) = 1$. We obtain

$$\psi'(y) + \lambda(\psi,y,u(\psi,y)) = 0.$$

The image set of $x = \psi(y)$ is a characteristic. Consequently the straight lines $y \equiv$ constant are never tangents of a characteristic in this system; this is in contrast to the previously considered second order equations.

Examples: The system $u_y = Au_x + q$ with the presumed real diagonalizable matrix A from Example 1.6 is of hyperbolic type. The characteristics were already given in that example. The Cauchy-Riemann differential equations $u_y = \begin{pmatrix} 0 & -1 \\ 1 & 0 \end{pmatrix} u_x$ of Example 1.14 are a linear first order elliptic system with constant coefficients.

If all the coefficients are not explicitly dependent on u, every quasilinear second order differential equation can be transformed to a 2×2 first order system. This transformation does not change the type of the differential equation. Thus, we consider the following differential equation:

$$a(x,y,u_x,u_y)u_{xx} + 2b(x,y,u_x,u_y)u_{xy}$$
$$+ c(x,y,u_x,u_y)u_{yy} + f(x,y,u_x,u_y) = 0.$$

Setting $v = (u_x,u_y)$ yields the system

$$a(x,y,v)\frac{\partial v_1}{\partial x} + 2b(x,y,v)\frac{\partial v_2}{\partial x} + c(x,y,v)\frac{\partial v_2}{\partial y} + f(x,y,v) = 0$$
$$\frac{\partial v_2}{\partial x} - \frac{\partial v_1}{\partial y} = 0.$$

If $c(x,y,z) \neq 0$ for all $(x,y) \in G$ and all $z \in \mathbb{R}^2$ we can solve for v_y:

$$v_y = \begin{pmatrix} 0 & 1 \\ -a/c & -2b/c \end{pmatrix} v_x - \begin{pmatrix} 0 \\ f \end{pmatrix}.$$

The coefficient matrix has eigenvalues

$$\lambda_{1,2} = \frac{1}{c}\,(-b \pm \sqrt{b^2-ac}).$$

The corresponding eigenvectors are $(1,\lambda_{1,2})$. When $\lambda_1 = \lambda_2$ there is only one eigenvector. Thus the type of this first order system, like the type of the second order differential equation, depends only on the sign of $b^2 - ac$.

We next divide partial differential equations with m independent variables into types, restricting ourselves to the cases of greatest practical importance.

Definition 2.11: Let $m \in \mathbb{N}$, G a region in $\mathbb{R}^m$ and $a_{ik}, f \in C^0(G \times \mathbb{R}^{m+1}, \mathbb{R})$ $(i,k = 1(1)m)$. We use the notation $\partial/\partial x_i = \partial_i$. The equation

$$\sum_{i,k=1}^{m} a_{ik}(x,p(x))\partial_i\partial_k u(x) + f(x,p(x)) = 0$$

with $p(x) = (u(x),\partial_1 u(x),\ldots,\partial_m u(x))$ is called a *quasilinear second order differential equation.* Without loss of generality, we may assume the matrix $A = (a_{ij})$ to be symmetric. Then the *type* of the differential equation with respect to a *fixed* function $u \in C^2(G, \mathbb{R})$ and a *fixed* point $x \in G$ is determined by the following table:

Type of d.e.	Properties of $A(x,p(x))$
hyperbolic	All eigenvalues of $A(x,p(x))$ are different from zero. Exactly $m - 1$ eigenvalues have the same sign.
elliptic	All eigenvalues of $A(x,p(x))$ are different from zero and all have the same sign.
parabolic	Exactly one eigenvalue of $A(x,p(x))$ is equal to zero. All the remaining ones have the same sign.

□

Definition 2.12: Let $m,n \in \mathbb{N}$, G a region in $\mathbb{R}^m$, $h \in C^o(G \times \mathbb{R}^m, \mathbb{R}^n)$ and $A_\mu \in C^o(G \times \mathbb{R}^n, \mathrm{MAT}(n,n,\mathbb{R}))$ for $\mu = 1(1)m-1$. The system

$$\partial_m u(x) - \sum_{\mu=1}^{m-1} A_\mu(x,u(x))\partial_\mu u(x) + h(x,u(x)) = 0$$

is called a *quasilinear first order hyperbolic system* if there exists a $C \in C^1(G \times \mathbb{R}^n, \mathrm{MAT}(n,n,\mathbb{R}))$ with

(1) $C(x,z)$ regular for all $x \in G$, $z \in \mathbb{R}^n$.

(2) $C(x,z)^{-1}A_\mu(x,z)C(x,z)$ symmetric for all $x \in G$, $z \in \mathbb{R}^n$, $\mu = 1(1)m-1$. □

The concepts of *principal part, semilinear, constant coefficients,* and the type with respect to a *fixed* function in *all* of G are defined analogously to Definitions 2.1 and 2.7. The *hyperbolic* type of Definition 2.12 coincides with that of Definition 2.8 in the special case of $m = 2$.

So far we have considered exclusively real solutions of differential equations with real coefficients. The initial and boundary conditions were similarly real. At least insofar as *linear* differential equations are concerned, our investigations in subsequent chapters will often consider *complex* solutions of differential equations with *real* coefficients and *complex* initial or boundary conditions. This has the effect of substantially simplifying the formulation of the theory. It does not create an entirely new situation since we can always split the considerations into real and imaginary parts.

3. Characteristic methods for first order hyperbolic systems

Let $G \subset \mathbb{R}^2$ be a simply connected region and consider the quasilinear hyperbolic system

$$u_y = A(x,y,u)u_x + g(x,y,u). \tag{3.1}$$

Here $A \in C^1(G \times \mathbb{R}^n, \mathrm{MAT}(n,n,\mathbb{R}))$, $g \in C^1(G \times \mathbb{R}^n, \mathbb{R}^n)$, and $u \in C^1(G, \mathbb{R}^n)$ is an arbitrary but fixed solution of the system. For the balance of this chapter we also assume that $A(x,y,z)$ always has n *different real* eigenvalues $\lambda_\mu(x,y,z)$, $\mu = 1(1)n$. Their absolute value shall be bounded independently of x, y, z, and μ. The eigenvalues are to be subscripted so that $\mu < \nu$ implies $\lambda_\mu < \lambda_\nu$.

If the eigenvalues of a matrix are different, they are infinitely differentiable functions of the matrix elements. When multiple eigenvalues occur, this is not necessarily the case. Our above assumption thus guarantees that the $\lambda_\mu(x,y,u(x,y))$ are continuously differentiable (single-valued) functions on G. There are always n linearly independent real eigenvectors. Their Euclidean length can be normalized to 1. They are then uniquely determined by the eigenvalue up to a factor of $+1$ or -1. In the simply connected region $G \times \mathbb{R}^n$ the factor can be chosen so that the eigenvectors are continuously differentiable functions on $G \times \mathbb{R}^n$. Thus there exists an $E \in C^1(G \times \mathbb{R}^n, \mathrm{MAT}(n,n,\mathbb{R}))$ with the following properties:

(1) $E(x,y,z)$ is always regular.

(2) The columns of E are vectors of length 1.

(3) $E(x,y,z)^{-1}A(x,y,z)E(x,y,z) = \mathrm{diag}(\lambda_\mu(x,y,z))$.

Naturally, $-E$ has the same properties as E. Let

$D(x,y,z) = \mathrm{diag}(\lambda_\mu(x,y,z))$. From (3.1) one obtains the *first normal form*

$$E^{-1}u_y = DE^{-1}u_x + E^{-1}g. \tag{3.2}$$

For E, D, and g we suppressed the arguments $(x,y,u(x,y))$. A componentwise notation clarifies the character of this normal form. Let

$$E^{-1} = (e_{\mu\nu}), \quad u = (u_\nu), \quad g = (g_\nu).$$

This implies

$$\sum_{\nu=1}^{n} e_{\mu\nu} \frac{\partial u_\nu}{\partial y} = \lambda_\mu \sum_{\nu=1}^{n} e_{\mu\nu} \frac{\partial u_\nu}{\partial x} + \sum_{\nu=1}^{n} e_{\mu\nu} g_\nu, \quad \mu = 1(1)n \tag{3.3}$$

or

$$\sum_{\nu=1}^{n} e_{\mu\nu}\left[\frac{\partial}{\partial y} - \lambda_\mu \frac{\partial}{\partial x}\right]u_\nu = \sum_{\nu=1}^{n} e_{\mu\nu} g_\nu, \quad \mu = 1(1)n. \tag{3.4}$$

Each equation contains only one differential operator, $\frac{\partial}{\partial y} - \lambda_\mu \frac{\partial}{\partial x}$, which is a directional derivative in a characteristic direction (cf. Example 1.6). However, this does not mean that the system is uncoupled, for in general $e_{\mu\nu}$, λ_μ, and g_ν depend on all the components of u.

For a linear differential equation, it is now natural to substitute

$$v(x,y) = E(x,y)^{-1}u(x,y).$$

This leads to the *second normal form*

$$v_y = Dv_x + \left(\frac{\partial E^{-1}}{\partial y} - D\,\frac{\partial E^{-1}}{\partial x}\right)Ev + E^{-1}g. \tag{3.5}$$

In componentwise notation, these equations become

$$\frac{\partial v_\mu}{\partial y} = \lambda_\mu \frac{\partial v_\mu}{\partial x} + \sum_{\nu=1}^{n} b_{\mu\nu} v_\nu + \tilde{g}_\mu, \quad \mu = 1(1)n \tag{3.6}$$

$$(b_{\mu\nu}) = \left(\frac{\partial E^{-1}}{\partial y} - D\,\frac{\partial E^{-1}}{\partial x}\right)E, \quad (\tilde{g}_\mu) = E^{-1}g.$$

The original hyperbolic system and the normal forms obviously have the same characteristics. They may be represented parametrically as follows:

$$x = \phi(t), \quad y = t$$
$$\phi'(t) + \lambda_\mu(\phi(t),t,u(\phi(t),t)) = 0, \quad \mu = 1(1)n. \qquad (3.7)$$

For each characteristic, μ is fixed. Since the λ_μ are continuously differentiable, they satisfy local Lipschitz conditions. It can be shown that there are exactly n different characteristics through each point of G; thus for each choice of μ there is exactly one. Two characteristics for the same μ cannot intersect each other. No characteristic of our system touches the x-axis. Each characteristic cuts the x-axis at most once.

We will now restrict ourselves to the special case $n = 2$. In this case there are particularly simple numerical methods for handling initial value problems. They are called *characteristic methods*. For simplicity's sake we specify initial values on the set $\Gamma = \{(x,y) \in G | y = 0\}$. We presuppose that:

(1) Γ is a nonempty open interval on the x-axis;

(2) Every characteristic through a point of G intersects Γ.

The second condition can always be satisfied by reducing the size of G. It now follows from the theory that the course of u in all of G depends only on the initial values on Γ. G is the domain of determinancy of Γ. Let Q_1 and Q_2 be two points of Γ. The characteristics through Q_1 and Q_2 then bound the domain of determinancy of the interval $\overline{Q_1Q_2}$.

Since every characteristic intersects the x-axis, one can choose the abscissa of the intersection point (s,0) as a parameter for the characteristics, in addition to μ. A characteristic is uniquely determined when these two are specified. From (3.7) one obtains the parametric representation

$$\frac{\partial}{\partial t}\phi_\mu(s,t) + \lambda_\mu(\phi_\mu,t,u(\phi_\mu,t)) = 0 \qquad (3.8)$$

$$\phi_\mu(s,0) = s, \qquad s \in \Gamma, \quad \mu = 1,2.$$

The solutions are continuously differentiable. Since there are two characteristics through each point (x,y), there are two abscissas, $s_1 = p_1(x,y)$ and $s_2 = p_2(x,y)$, for each point (x,y). It is true for all t that

$$s = p_\mu(\phi_\mu(s,t),t) \qquad s \in \Gamma, \quad \mu = 1,2.$$

Thus p_1 and p_2 are solutions of the initial value problem

$$\frac{\partial p_\mu}{\partial y}(x,y) = \lambda_\mu(x,y,u(x,y))\frac{\partial p_\mu}{\partial x}(x,y), \quad (x,y) \in G, \ \mu = 1,2$$

$$p_\mu(x,0) = x, \qquad x \in \Gamma. \qquad (3.9)$$

To prove this statement one must first show that the initial value problems (3.9) are uniquely solvable and that the solutions are continuously differentiable. For these solutions it is obviously true that

$$p_\mu(\phi_\mu(s,0),0) = p_\mu(s,0) = s, \qquad s \in \Gamma, \ \mu = 1,2.$$

On the other hand, the functions $p_\mu(\phi_\mu(s,t),t)$ do not depend on t, since their derivatives with respect to t are zero:

$$\frac{\partial p_\mu}{\partial x}\frac{\partial \phi_\mu}{\partial t} + \frac{\partial p_\mu}{\partial y} = -\lambda_\mu \frac{\partial p_\mu}{\partial x} + \lambda_\mu \frac{\partial p_\mu}{\partial x} = 0.$$

With the aid of the projections p_1 and p_2 one arrives at

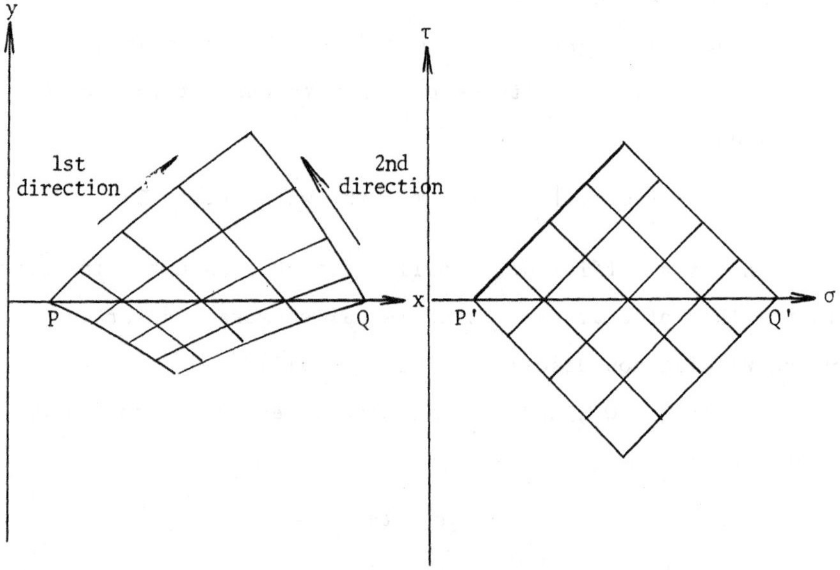

Figure 3.10. The domain of determinancy of the interval $\overline{PQ}$ in the (x,y) plane and in characteristic coordinates (σ,τ).

a new coordinate system in G, called a characteristic coordinate system (cf. Figure 3.10):

$$\sigma = \frac{1}{2}[p_2(x,y) + p_1(x,y)]$$

$$\tau = \frac{1}{2}[p_2(x,y) - p_1(x,y)].$$

By previous remarks, the transformation is one-to-one. On Γ one has $\sigma = x$ and $\tau = y = 0$.

The characteristic methods determine approximations for u, x, and y at the lattice points with characteristic coordinates

$$\{(\sigma,\tau) \mid \sigma = kh,\ \tau = \ell h \text{ with } k,\ell \in \mathbb{Z}\}.$$

Here h is a sufficiently small positive constant. The simplest method of characteristics is called *Massau's method*, which we will now describe in more detail.

Let Q_0, Q_1, and Q_2, in order, denote the points with coordinates

$$\sigma = kh, \qquad \tau = \ell h$$
$$\sigma = (k-1)h, \qquad \tau = (\ell-1)h$$
$$\sigma = (k+1)h, \qquad \tau = (\ell-1)h.$$

Massau's method uses the values of u, x, and y at Q_1 and Q_2 to compute the values at Q_0. Since the initial values for $\tau = 0$ are known, a stepwise computation will yield the values at the levels $\tau = h$, $\tau = 2h$, etc. Here one can obviously restrict oneself to the part of the lattice with $k + 1$ even or $k + 1$ odd. We note that $\sigma - \tau$ is the same at Q_0 and Q_1, as is $\sigma + \tau$ at Q_0 and Q_2. Therefore, Q_0 and Q_1 lie on the characteristic $p_1(x,y) = (k-1)h$ and Q_0 and Q_2 lie on the characteristic $p_2(x,y) = (k+1)h$ (cf.

Figure 3.11). In this coordinate system the characteristics are thus the straight lines with slope +1 and -1.

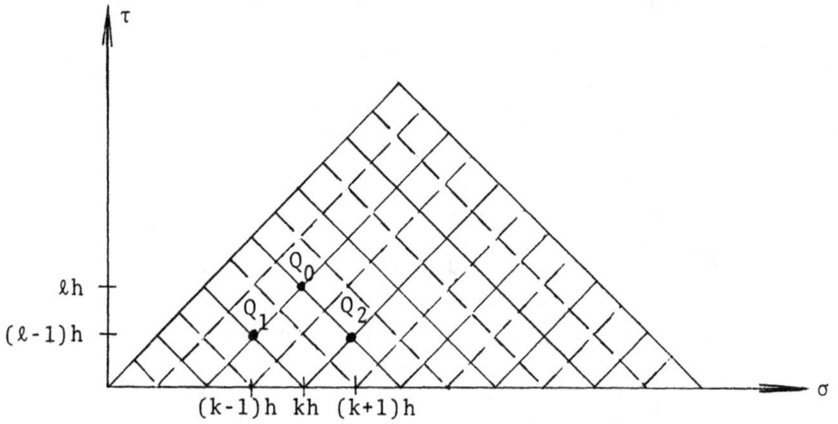

Figure 3.11. Layers in Massau's method.

The numerical method begins with the first normal form (3.4) and the differential equations (3.8) for ϕ_1 and ϕ_2. A, λ, and E^{-1} are regarded as constant on the intervals $\overline{Q_1Q_0}$ and $\overline{Q_2Q_0}$. Their values are fixed at Q_1 and Q_2, respectively. The derivatives along the characteristics are approximated by the simplest difference quotients. We use superscripts $j = 0,1,2$ to denote the approximations for u, x, y, A, λ_μ, $E^{-1} = (e_{\mu\nu})$ and $g = (g_\nu)$ at the points Q_0, Q_1, Q_2. Then we have

$$\frac{x^0-x^j}{y^0-y^j} \approx \frac{\partial\phi_j}{\partial y}(p_j(x^j,y^j),y^j)$$

$$\frac{u_\nu^0-u_\nu^j}{y^0-y^j} \approx (\frac{\partial}{\partial y} - \lambda_j{}^j \frac{\partial}{\partial x})u_\nu(x^j,y^j,u^j). \qquad \nu = 1,2;\ j = 1,2$$

In detail, the computation runs as follows:

(1) Determine A^j, $(E^{-1})^j$, $\lambda_\mu{}^j$ for $j = 1,2$ and $\mu = 1,2$.

(2) Determine x^o and y^o from the system of equations

$$\frac{x^o-x^j}{y^o-y^j} + \lambda_j{}^j = 0, \qquad j = 1,2$$

or

$$(x^o-x^1) + \lambda_1{}^1(y^o-y^1) = 0$$

$$(x^o-x^1) + \lambda_2{}^2(y^o-y^1) = (x^2-x^1) + \lambda_2{}^2(y^2-y^1).$$

(3) Determine u_1^o and u_2^o from the system of equations

$$\sum_{\nu=1}^{2} e_{j\nu}^j \frac{u_\nu^o-u_\nu^j}{y^o-y^j} = \sum_{\nu=1}^{2} e_{j\nu}^j g_\nu^j, \qquad j = 1,2$$

or

$$e_{11}^1(u_1^o-u_1^1) + e_{12}^1(u_2^o-u_2^1) = (y^o-y^1)(e_{11}^1 g_1^1 + e_{12}^1 g_2^1)$$

$$e_{21}^2(u_1^o-u_1^1) + e_{22}^2(u_2^o-u_2^1) = e_{21}^2[(y^o-y^2)g_1^2 + u_1^2 - u_1^1]$$

$$+ e_{22}^2[(y^o-y^2)g_2^2 + u_2^2 - u_2^1].$$

The rewriting of the systems of equations in (2) and (3) is done for reasons of rounding error stability.

When h is sufficiently small, the matrices in both systems of equations are regular. For when h is sufficiently small, we have

$$\lambda_2^2 \approx \lambda_2^1 \neq \lambda_1^1$$

and

$$\begin{pmatrix} e_{11}^1 & e_{12}^1 \\ e_{21}^2 & e_{22}^2 \end{pmatrix} \approx \begin{pmatrix} e_{11}^1 & e_{12}^1 \\ e_{21}^1 & e_{22}^1 \end{pmatrix} = \text{regular matrix.}$$

Massau's method sometimes converges in cases where the initial value problem has no continuous solution. As a rule, it is easily seen numerically that such a case has occurred, for then the same pair (x,y) occurs for different pairs (σ,τ). Then there is no single-valued mapping $(x,y) \to (\sigma,\tau)$. The accuracy of Massau's method for hyperbolic systems is comparable to that of the Euler method for ordinary differential equations. But there also exist numerous characteristic methods of substantially greater accuracy. The extrapolation methods (see Busch-Esser-Hackbusch-Herrmann 1975) have proven themselves particularly useful. For nondifferentiable initial values, implicit characteristic methods with extrapolation are also commendable. All these methods differ from Massau's method in their use of higher order difference quotients. All in all one can say that under the conditions formulated above--two variables, systems of two equations, A has distinct real eigenvalues--the characteristic methods are probably the most productive. There also exist generalizations for more generally posed problems; unfortunately, they are much more complicated and much less useful. For that reason we want to conclude our treatment of characteristic methods at this point and turn to other methods, known as difference methods on rectangular lattices.

The theory of normal forms may be found in Perron (1928), and the convergence proof for characteristic methods in Sauer (1958). A FORTRAN program may be found in Appendix I.

4. Banach spaces

There are many cases in which initial value problems for linear partial differential equations can be reduced to initial value problems for ordinary differential equations. However, in such cases the ordinary differential equations are for maps of a real interval into an appropriate Banach space of non-finite dimension. One result of this reformulation of the problem is that it is easier to make precise the concept of a *properly posed* initial value problem, as discussed in Chapter 1. Lax-Richtmyer theory concerns itself with stability and convergence criteria for difference methods. As it starts with the reformulated problems, a knowledge of these "Banach space methods" is absolutely essential for an understanding of the proofs. The situation is different for practical applications of difference methods. For then one almost always begins with the original formulation as an initial value problem for a hyperbolic or parabolic differential equation. Elliptic equations do not play a role here, since the corresponding initial value problems are not properly posed.

In this section are defined the basic concepts of *Banach space, linear operator, differentiability* and *integral* in a Banach space, etc. Also presented are several important theorems which are necessary for the development of Banach space methods.

Definition 4.1: Let B be a vector space over a field $\mathbb{K} = \mathbb{C}$ $(\mathbb{K} = \mathbb{R})$. B is called a *complex (real) Banach space* whenever the following holds:

1. In B there is a distinguished function $\|\cdot\|$: $B \to [0,\infty)$, called a *norm*, with the following properties:

(a) $\|a\| = 0 \iff a = 0 \qquad a \in B$

(b) $\|\lambda a\| = |\lambda|\,\|a\| \qquad \lambda \in \mathbb{K},\ a \in B$

(c) $\|a+b\| \leq \|a\| + \|b\| \qquad a,b \in B.$

2. The space B is *complete* with respect to the topology induced by $\|\cdot\|$; i.e., every Cauchy sequence $\{a_n\}_{n \in \mathbb{N}}$ of elements of B converges to an element a in B. Recall that $\{a_n\}$ is called a *Cauchy sequence* if for every positive ε there exists an integer n_0 such that $n,m > n_0$ implies $\|a_n - a_m\| < \varepsilon$. The sequence $\{a_n\}$ is said to converge to the element a in B if the sequence $\{\|a - a_n\|\}$ converges to 0. □

Every Banach space thus consists of a vector space together with a defined norm. Thus two Banach spaces with the same underlying vector space are distinct if the norms are different. In particular it is worth noting that an infinite dimensional vector space which is complete with respect to one norm by no means need have this property with respect to any other norm.

In the following, we will speak simply of Banach spaces insofar as it is clear from context whether or not we are dealing with complex or real spaces. Since later developments will make heavy use of Fourier transforms, we will almost exclusively consider complex Banach spaces.

Example 4.2: The vector space $\mathbb{C}^n$ becomes a Banach space with either of the two norms

$$\|x\| = \max_j |x_j|, \quad \|x\| = \Big(\sum_j x_j \bar{x}_j\Big)^{1/2}.$$

The same is true for any other norm on $\mathbb{C}^n$ (cf. Dieudonne 1960, Ch. V). □

Example 4.3: The set of all maps $x : \mathbb{Z} \to \mathbb{C}^n$ for which the infinite series

$$\sum_{j=-\infty}^{+\infty} \sum_{k=1}^{n} x_k(j)\overline{x_k(j)}$$

converges, becomes a vector space over $\mathbb{C}$ with the usual definition of addition and multiplication by a scalar. With the definition

$$||x|| = \Big(\sum_{j=-\infty}^{\infty} \sum_{k=1}^{n} x_k(j)\overline{x_k(j)}\Big)^{1/2}$$

this vector space becomes a Banach space, which we denote by $\ell^2(\mathbb{C}^n)$ (cf. Yosida 1968, Ch. I.9). □

Example 4.4: Let $K \subset \mathbb{R}^m$ be a compact set. The vector space $C^0(K,\mathbb{C}^n)$ is complete with respect to the norm

$$||f||_\infty = \max_j \max_{x \in K} |f_j(x)|$$

and is therefore a Banach space. Here the completeness of the space results from the fact that, with this definition of the norm, every Cauchy sequence represents a uniformly convergent sequence of continuous functions. Such a sequence is well known to converge to a continuous limit function, and thus to an element of the space. The space $C^0(K,\mathbb{C}^n)$ is not complete, however, with respect to the norm

$$||f||_2 = \Big(\int_K \sum_{j=1}^{n} f_j(x)\overline{f_j(x)}\, dx\Big)^{1/2}$$

We can see this from the following counterexample: in $C^0([0,2],\mathbb{C})$ the sequence $\{f_\mu\}_{\mu \in \mathbb{N}}$ where

$$f_\mu(x) = \begin{cases} x^\mu & \text{for } x \in [0,1) \\ 1 & \text{for } x \in [1,2] \end{cases}$$

is a Cauchy sequence. It converges, but not to a continuous limit function.

In the following, whenever we speak of the *Banach space* $C^o(K,\mathbb{C}^n)$, we always mean the vector space of continuous functions $f:K \to \mathbb{C}^n$ together with the norm $\|\cdot\|_\infty$. □

Example 4.5: Let G be a region in $\mathbb{R}^m$ and $\Lambda = \{f:G \to \mathbb{C}^n \mid f$ square-intebrable in $G\}$, where f is called square-integrable in G if the integral

$$\int_G \sum_{j=1}^{n} [f_j(x)\overline{f_j(x)}]\, dx$$

exists as a Lebesgue integral and is finite. Λ becomes a vector space over $\mathbb{C}$ with the usual definition of addition and multiplication by a scalar. The map $|||\cdot|||: \Lambda \to [0,\infty)$ defined by

$$|||f||| = \left(\int_G \sum_j f_j(x)\overline{f_j(x)}\, dx\right)^{1/2}$$

has all the properties of a norm with the exception of 1(a), since $|||f||| = 0$ for all $f \in N$ where

$$N = \{f \in \Lambda \mid \{x \in G \mid f(x) \neq 0\} \text{ has measure zero}\}.$$

One eliminates this deficiency by passing to the quotient space Λ/N. The elements of Λ/N are equivalence classes of maps in Λ, where the elements of a class differ only on sets of measure zero. Λ/N becomes a vector space over $\mathbb{C}$ in a canonical way. With the definition

$$\|\tilde{f}\| = |||f|||, \quad \tilde{f} \in \Lambda/N, \quad f \in \tilde{f}$$

this vector space becomes a Banach space, which we denote by $L^2(G,\mathbb{C}^n)$. Although the vector space and norm properties are easily checked, the proof of completeness turns out to be substantially more difficult (cf. Yosida 1968, Ch. I.9). In order to simplify notation and language, we will not distinguish between the equivalence classes $\tilde{f} \in L^2$ and their representatives $f \in \tilde{f}$ in the sequel, since the appropriate meaning will be clear from context. □

The following definition introduces the important concept of a dense set.

Definition 4.6: Let B be a Banach space and let D_1, D_2 be subsets of B with $D_1 \subset D_2$. Then D_1 is called *dense* in D_2 if for every $a \in D_2$ and for every $\varepsilon > 0$ there exists a $b \in D_1$ such that $\|a - b\| < \varepsilon$. □

In our future considerations those vector subspaces of a Banach space which are dense in the Banach space play a significant role. We first consider several Banach spaces of continuous functions with norm $\|\cdot\|_\infty$. Because of Weierstrass's fundamental theorem, it is possible to display simple dense subspaces.

Theorem 4.7: *Weierstrass Approximation Theorem.* Let $K \subset \mathbb{R}^m$ be a compact set. Then the vector space of polynomials with complex coefficients defined on K is dense in the Banach space $C^0(K,\mathbb{C})$.

A proof may be found in Dieudonne (1960), Ch. VII.4. It follows immediately from this theorem that the spaces $C^k(K,\mathbb{C})$, $k = 1(1)\infty$, and $C^\infty(K,\mathbb{C})$ are dense in $C^0(K,\mathbb{C})$, since they are supersets of the space of polynomials. In addition, we even have:

Theorem 4.8: (1) The vector space

$$V = \{f \in C^\infty([a,b],\mathbb{C}) \mid f^{(\nu)}(a) = f^{(\nu)}(b) = 0,\ \nu = 1(1)\infty\}$$

is dense in the Banach space $C^0([a,b],\mathbb{C})$.

(2) The vector space of bounded functions in $C^\infty(\mathbb{R},\mathbb{C})$ is dense in the Banach space of bounded functions in $C^0(\mathbb{R},\mathbb{C})$.

The proof requires the following lemma.

<u>Lemma 4.9</u>: Let $c_1,c_2,d_1,d_2 \in \mathbb{R}$ with $d_1 < c_1 < c_2 < d_2$. Then there exists a function $h \in C^\infty(\mathbb{R},\mathbb{C})$ with

(1) $h(x) = 1$ for $x \in [c_1,c_2]$

(2) $h(x) = 0$ for $x \in \mathbb{R} - (d_1,d_2)$

(3) $h(x) \in (0,1)$ for $x \in (d_1,d_2) - [c_1,c_2]$.

A proof of this lemma may be found in Friedman (1969), part 1, Lemma 5.1.

Proof of 4.8(1): We first show that the space

$$\tilde{V} = \{f \in V \mid f(a) = f(b) = 0\}$$

is dense in the Banach space $W = \{f \in C^0([a,b],\mathbb{C}) \mid f(a) = f(b) = 0\}$. Now let $f \in W$ and $\varepsilon > 0$ be given. Then there exists a $\delta > 0$ with $|f(x)| < \frac{\varepsilon}{3}$ for $x \in [a,a+\delta) \cup (b-\delta,b]$. Choose $h \in C^\infty(\mathbb{R},\mathbb{C})$ where $d_1 = a$, $c_1 = a+\delta$, $c_2 = b-\delta$ and $d_2 = b$ as in Lemma 4.9. Since $C^\infty([a,b],\mathbb{C})$ is dense in $C^0([a,b],\mathbb{C})$, there exists a function $g \in C^\infty([a,b],\mathbb{C})$ with $\|f-g\|_\infty < \frac{\varepsilon}{3}$. Now let $\tilde{g} = g \cdot h$. Then:

$\tilde{g} \in C^{\infty}([a,b],\mathbb{C})$;

$h^{(\mu)}(a) = h^{(\mu)}(b) = 0, \ \mu = 0(1)\infty$ by 4.9(2);

$\tilde{g}^{(\nu)}(x) = \sum_{\mu=0}^{\nu} \binom{\nu}{\mu} g^{(\nu-\mu)}(x)h^{(\mu)}(x), \quad \nu = 0(1)\infty$;

$\tilde{g}^{(\nu)}(a) = \tilde{g}^{(\nu)}(b) = 0, \quad \nu = 0(1)\infty$;

$|f(x)-\tilde{g}(x)| = |f(x)-g(x)|$ for $x \in [a+\delta,b-\delta]$ by 4.9(1);

$|g(x)| \leq |f(x)-g(x)| + |f(x)| < \frac{2}{3}\varepsilon$ for $x \in [a,a+\delta) \cup (b-\delta,b]$;

$|f(x)-\tilde{g}(x)| \leq |f(x)| + |h(x)||g(x)| < \varepsilon$
for $x \in [a,a+\delta) \cup (b-\delta,b]$;

$||f-g||_{\infty} < \varepsilon$.

And this shows that $\tilde{V}$ is dense in W. To prove contention (1), let an arbitrary $f \in C^{0}([a,b],\mathbb{C})$ be chosen. Find a function $h(x)$ subject to Lemma 4.9 with $d_1 < c_1 < a < c_2 < d_2 < b$. Then

$$p(x) = f(a)h(x) + f(b)[1-h(x)]$$

is obviously a function in V with $p(a) = f(a)$ and $p(b) = f(b)$. But by the previous considerations, the function $f(x) - p(x)$ can be approximated arbitrarily closely by functions in $\tilde{V} \subset V$.

Proof of 4.8(2): Let f be a bounded function in $C^{0}(\mathbb{R},\mathbb{C})$ and let $\varepsilon > 0$. By (1) there exist on each of the intervals $[k,k+1]$ $(k \in \mathbb{Z})$ functions $g_k \in C^{\infty}([k,k+1],\mathbb{C})$ with

$$g_k^{(\nu)}(k) = g_k^{(\nu)}(k+1) = 0, \quad \nu = 1(1)\infty$$

$$|f(x) - g_k(x)| < \varepsilon, \qquad x \in [k,k+1].$$

It follows from the proof of (1) that the functions g_k can be chosen so that additionally

$$g_k(k) = f(k), \quad g_k(k+1) = f(k+1).$$

Thus

$$g(x) = g_k(x), \quad x \in [k,k+1)$$

is a bounded function in $C^\infty(\mathbb{R},\mathbb{C})$ with $\|f - g\|_\infty < \varepsilon$. □

Next we consider two examples of dense subspaces of the Banach space $L^2(G,\mathbb{C})$.

Theorem 4.10: Let G be a region of $\mathbb{R}^m$. Then:

(1) The vector space $C_0^\infty(G,\mathbb{C})$ of infinitely differentiable functions with compact support defined on G is dense in the space $L^2(G,\mathbb{C})$.

(2) The vector space of polynomials with complex coefficients defined on G is dense in the space $L^2(G,\mathbb{C})$, if G is bounded.

Proof: Conclusion (1) is well known. (2) follows from (1) and Theorem 4.7. □

Definition 4.11: Let B_1, B_2 be Banach spaces and let D be a subspace of B_1. A mapping

$$A: D \to B_2$$

is called a *linear operator* if it is true for all $a,b \in D$ and all $\lambda,\mu \in \mathbb{K}$ that

$$A(\lambda a + \mu b) = \lambda A(a) + \mu A(b).$$

A linear operator A is called *bounded* if there exists an $\alpha > 0$ with $\|A(a)\| \leq \alpha\|a\|$ for all $a \in D$. The quantity

$$\|A\| = \inf\{\alpha \in \mathbb{R}_+ \mid \|A(a)\| \leq \alpha\|a\| \text{ for all } a \in D\}$$

is then called the *norm* of the linear bounded operator A. We define

$$L(D,B_2) = \{A: D \to B_2 \mid A \text{ linear and bounded}\}. \quad \square$$

In order to define bounded linear operators from B_1 to B_2 it suffices to define them on a dense subspace of B_1, because of the following theorem.

<u>Theorem 4.12</u>: Let B_1 and B_2 be Banach spaces, let D be a dense vector subspace of B_1, and let $A \in L(D,B_2)$. Then there exists exactly one operator $\hat{A} \in L(B_1,B_2)$ which agrees with A on D. Furthermore, $\|\hat{A}\| = \|A\|$. $\hat{A}$ is called the extension of A on B_1.

Proof: Let $a \in B_1$. Since D is dense in B_1, there exists a sequence $\{a_j\}$ of elements in D converging to a. It follows from

$$\begin{aligned}\|A(a_j)-A(a_k)\| = \|A(a_j-a_k)\| &\leq \|A\| \; \|a_j-a_k\| \\ &\leq \|A\| \; (\|a-a_j\| + \|a-a_k\|)\end{aligned}$$

that $\{A(a_j)\}$ is a Cauchy sequence in B_2. Being a Banach space, B_2 is complete, and so there exists a (uniquely determined) limit for the sequence $\{A(a_j)\}$ in B_2, which we denote by c. The quantity c depends only on a, and not on the particular choice of sequence $\{a_j\}$; for suppose $\{\tilde{a}_j\}$ is another sequence of elements of D converging to a. Then one estimates

$$\begin{aligned}\|c-A(\tilde{a}_j)\| &\leq \|c-A(a_j)\| + \|A(a_j)-A(\tilde{a}_j)\| \\ &\leq \|c-A(a_j)\| + \|A\| \; (\|a-a_j\| + \|a-\tilde{a}_j\|).\end{aligned}$$

Passage to the limit shows that the sequence $\{A(\tilde{a}_j)\}$ also converges to c. We define the mapping $\hat{A}: B_1 \to B_2$ by the rule $\hat{A}(a) = c$. $\hat{A}$ is well defined and agrees with A on D.

To see the linearity of $\hat{A}$, let $a,b \in B_1$ and let $\{a_j\}$ and $\{b_j\}$ be two sequences of elements of D converging to a and b, respectively. Then the sequence $\{\lambda a_j + \mu b_j\}$ converges to the element $\lambda a + \mu b$. It follows that

$$\hat{A}(\lambda a + \mu b) = \lim_{j\to\infty} A(\lambda a_j + \mu b_j)$$

$$= \lambda \lim_{j\to\infty} A(a_j) + \mu \lim_{j\to\infty} A(b_j) = \lambda\hat{A}(a) + \mu\hat{A}(b).$$

To see the boundedness of $\hat{A}$, let a and $\{a_j\}$ be given as above. Then $\|A(a_j)\| \le \|A\|\,\|a_j\|$, and by the continuity of the norm

$$\|\hat{A}(a)\| = \|\lim_{j\to\infty} A(a_j)\| = \lim_{j\to\infty} \|A(a_j)\|$$

$$\le \|A\| \lim_{j\to\infty} \|a_j\| = \|A\|\|\lim_{j\to\infty} a_j\| = \|A\|\,\|a\| .$$

It follows from this that $\|\hat{A}\| = \|A\|$.

To see that $\hat{A}$ is uniquely determined, we proceed indirectly. Let $\hat{\hat{A}}$ be a bounded linear operator from B_1 to B_2 which agrees with A on D. Suppose there is an $a \in B_1 - D$ for which $\hat{\hat{A}}(a) \ne \hat{A}(a)$, and let $\{a_j\}$ be a sequence of elements of D converging to a. Then

$$\|\hat{\hat{A}}(a) - \hat{A}(a)\| \le \|\hat{\hat{A}}(a) - A(a_j)\| + \|\hat{A}(a) - A(a_j)\|$$

$$= \|\hat{\hat{A}}(a) - \hat{\hat{A}}(a_j)\| + \|\hat{A}(a) - \hat{A}(a_j)\| \le \|\hat{\hat{A}}\|\,\|a - a_j\| + \|\hat{A}\|\,\|a - a_j\| .$$

Passage to the limit leads to a contradiction. □

By the theorems of functional analysis (in particular the *Hahn-Banach Theorem)*, a bounded linear operator can be extended to all of B_1, preserving the norm, even when the domain of definition D is not dense in B_1. However, the extension is not unique in that case.

Definition 4.13: Let B_1 and B_2 be Banach spaces and M a set of bounded linear operators mapping B_1 to B_2. M is called *uniformly bounded* if the set $\{\|A\| \mid A \in M\}$ is bounded. This is equivalent to the existence of a constant $\alpha > 0$ such that $\|A(a)\| \leq \alpha\|a\|$ for all $A \in M$ and all $a \in B_1$. □

Theorem 4.14: *Principle of uniform boundedness.* Let B_1 and B_2 be Banach spaces and M a set of bounded linear operators mapping B_1 to B_2. Suppose there exists a function $\beta\colon B_1 \to \mathbb{R}_+$ with $\|A(a)\| \leq \beta(a)$ for all $a \in B_1$ and all $A \in M$. Then the set M is uniformly bounded.

For a proof of Theorem 4.14, see Yosida (1968), Ch. II.1. Observe that the function β need not be continuous or linear.

Definition 4.15: Let B be a Banach space and $[T_1,T_2]$ a real interval. A mapping

$$u\colon [T_1,T_2] \to B$$

is called *differentiable at the point* $t_0 \in [T_1,T_2]$, if there exists an element $a \in B$ such that

$$\lim_{\substack{h\to 0 \\ t_0+h \in [T_1,T_2]}} \frac{\|u(t_0+h)-u(t_0)-h\cdot a\|}{|h|} = 0.$$

The element a is uniquely determined and is called the *derivative* of u *at the point* t_0. It is denoted by $u'(t_0)$ or $\frac{du}{dt}(t_0)$. The mapping u is called *differentiable* if it is differentiable at every point of $[T_1,T_2]$. The mapping u is called *uniformly differentiable* if it is differentiable and if

$$\frac{1}{|h|}\|u(t+h)-u(t)-hu'(t)\|$$

converges uniformly to zero as $h \to 0$ for $t \in [T_1,T_2]$. The mapping u is called *continuously differentiable* if it is differentiable and if the derivative $u'(t)$ is continuous on $[T_1,T_2]$. □

It follows immediately from the above definition that a mapping which is differentiable at a point t_0 is also continuous there. It follows from the generalized Mean Value Theorem (cf. Dieudonne 1960, Theorem (8.6.2)) that for a continuously differentiable function u:

$$\frac{1}{|h|}\|u(t+h)-u(t)-hu'(t)\| \leq \sup_{0<\nu<1} \|u'(t+\nu h)-u'(t)\|.$$

A continuously differentiable function is therefore also uniformly differentiable.

A treatment of inhomogeneous initial value problems requires the concept of an integral for continuous maps of a real interval into a Banach space. We list a number of its properties without proof, referring to Dieudonne (1960), Ch. VIII.

<u>Definition 4.16</u>: Let B be a Banach space, $I = [T_1,T_2]$ a real interval, and $v \in C^0(I,B)$. A mapping $u: I \to B$ is called an *antiderivative* for v on I if u is differentiable on I and

$$u'(t) = v(t), \qquad t \in I. \qquad □$$

The following lemma permits the definition of the (definite) integral.

Lemma 4.17: (1) If u_1, u_2 are two antiderivatives of v in I, then $u_1 - u_2$ is constant on I.

(2) Every continuous function v has (at least) one antiderivative in I.

Definition 4.18: Let B be a Banach space, $I = [T_1, T_2]$ a real interval, $v \in C^o(I,B)$, and u an antiderivative of v in I. Then

$$\int_{T_1}^{T_2} v(t)dt = u(T_2) - u(T_1)$$

is called the *integral* of v from T_1 to T_2. □

It follows from Lemma 4.17 that the above definition is independent of the particular choice of antiderivative. We now list some properties of the integral which will be needed in Section 7.

Theorem 4.19: Let B be a Banach space, $I = [T_1, T_2]$ and $J = [T_3, T_4]$ real intervals, $v \in C^o(I,B)$, $C \in L(B,B)$, and $g, f, f_x \in C^o(I \times J, B)$. Then

$$(1) \quad \int_{T_1}^{T_2} C(v(t))dt = C\left(\int_{T_1}^{T_2} v(t)dt\right)$$

(2) For every $\varepsilon > 0$ there exists a $\delta > 0$ such that for every partition $\{x_k\}$ with

$$T_1 = x_o \le t_o \le x_1 \le \ldots \le x_k \le t_k \le x_{k+1} \le \ldots \le x_n = T_2$$

and

$$x_{k+1} - x_k < \delta, \qquad k = 0(1)n-1$$

it is true that

$$\left\| \int_{T_1}^{T_2} v(t)dt - \sum_{k=0}^{n-1} v(t_k)(x_{k+1} - x_k) \right\| < \varepsilon.$$

(3) The function

$$u(t) = \int_{T_1}^{t} v(\tau)d\tau$$

is a continuously differentiable antiderivative of v on I.

(4) The function

$$w(x) = \int_{T_1}^{T_2} f(t,x)dt$$

is continuously differentiable on J with

$$\frac{d}{dx} w(x) = \int_{T_1}^{T_2} f_x(t,x)dt, \qquad x \in J.$$

(5) $$\int_{T_1}^{T_2} \int_{T_3}^{T_4} g(t,x)dxdt = \int_{T_3}^{T_4} \int_{T_1}^{T_2} g(t,x)dtdx.$$

The solutions of initial value problems in the following chapters will make frequent use of implicit difference equations. That is to say, one has linear mappings $x \to y = T(x)$ of a Banach space B into itself, which are implicitly defined by linear equations $R(y) = S(x)$, where $R,S \in L(B,B)$. When the equation $R(y) = S(x)$ is uniquely solvable for all $x \in B$, we simply write $T = R^{-1}S$. However, we do not mean to imply by this that there exists a linear map R^{-1} defined on all of B. In most concrete applications, the unique solvability of the equation $R(y) = S(x)$ is nevertheless established via the existence of a bounded linear map R^{-1}. The main aid in these proofs is the Banach fixed point theorem, or contraction theorem, which can be formulated as follows for bounded linear operators.

<u>Theorem 4.20</u>: Let $D \in L(B,B)$, $\|D\| < 1$ and $R = I+D$. Then there exists a mapping $R^{-1} \in L(B,B)$ with $R^{-1}R = RR^{-1} = I$.

For a proof, see Yosida (1968), Ch. II.1 (*Neumann* series). The existence of R^{-1} for implicit difference equations derives from so-called stability results, together with the Banach fixed point theorem.

Theorem 4.21: Let $R \in C^{0}([0,1],L(B,B))$, $R(0) = I$ and $\Lambda = \{\lambda \in [0,1] \mid$ there exists an $R^{-1}(\lambda) \in L(B,B)$ with $R^{-1}(\lambda)R(\lambda) = R(\lambda)R^{-1}(\lambda) = I\}$. If the set $\{R^{-1}(\lambda) \mid \lambda \in \Lambda\}$ is uniformly bounded, then $\Lambda = [0,1]$.

Proof: Assume $\|R^{-1}(\lambda)\| < M$ for all $\lambda \in \Lambda$. By the uniform continuity of R, there exists a $\delta > 0$ so that for all $\lambda,\tilde{\lambda} \in [0,1]$ and $|\lambda-\tilde{\lambda}| < \delta$ it is true that

$$\|R(\lambda) - R(\tilde{\lambda})\| < 1/M.$$

Now let $\lambda_0 \in \Lambda$ and $|\lambda-\lambda_0| < \delta$. Then it follows that

$$R(\lambda) = R(\lambda_0) + R(\lambda) - R(\lambda_0) = R(\lambda_0)\{I + R^{-1}(\lambda_0)[R(\lambda) - R(\lambda_0)]\}.$$

Since

$$\|R^{-1}(\lambda_0)\| \, \|R(\lambda) - R(\lambda_0)\| < 1$$

both factors on the right side are invertible. $0 \in \Lambda$ thus implies $[0,\delta) \subset \Lambda$, which implies $[0,2\delta) \subset \Lambda$, etc. □

5. Stability of difference methods

In this section, we first define the fundamental concepts *properly posed initial value problem*, and *consistency*, *stability*, and *convergence* of difference methods in Banach space terminology. Then we prove the essential theorem of Lax-Richtmyer, which states that stability is necessary and sufficient for convergence of a consistent difference method. Additionally, we discuss several examples, which are designed to give substance to the abstract concepts.

Definition 5.1: Let B be a Banach space, D_A a vector subspace of B, $T \in \mathbb{R}_+$, and $A: D_A \to B$ a linear operator. We consider the following *initial value problem* for the *initial value* $c \in D_A$: Find a differentiable mapping $u: [0,T] \to D_A$ (called a *solution*) with

$$\begin{aligned} u'(t) &= A(u(t)), \quad t \in [0,T] \\ u(0) &= c. \end{aligned} \tag{5.2}$$

Henceforth we denote this problem by $P(B,T,A)$. □

Since A is linear, the set of all initial values for which problem $P(B,T,A)$ has a solution is a vector subspace of D_A. When the problem has a unique solution, then for fixed $t_0 \in [0,T]$ we can define a linear operator by the assignment $c \to u_c(t_0)$, where $u_c(t)$ denotes the solution of $P(B,T,A)$ for the initial value c. These linear operators will be used in the definition of a *properly posed* initial value problem.

Definition 5.3: The initial value problem $P(B,T,A)$ is called *properly posed* if there exist a vector subspace D_E such that

$$D_E \subset D_A \subset B,$$

and a collection $M_o = \{E_o(t) \mid t \in [0,T]\}$ of linear operators

$$E_o(t) : D_E \to D_A, \quad t \in [0,T]$$

with the following properties:

(1) D_E (and hence D_A) is dense in B.

(2) For all initial values $c \in D_E$ problem $P(B,T,A)$ has exactly one solution $u_c(t)$, which is represented for all $t \in [0,T]$ by $u_c(t) = E_o(t)(c)$.

(3) The set M_o is a uniformly bounded set of linear operators. We will call the operators $E_o(t)$ in M_o *solution operators* from now on. □

One sees immediately that the above concept represents a sharpening of the corresponding considerations in Section 1. The Lipschitz condition required by Definition 1.2 follows at once from (3):

$$\|u_c(t)-u_{\tilde{c}}(t)\| = \|E_o(t)(c)-E_o(t)(\tilde{c})\| \leq L\|c-\tilde{c}\|, \quad t \in [0,T].$$

Since the linear operators $E_o(t)$ are bounded and since their domain of definition D_E is dense in B, they may be extended uniquely, by Theorem 4.12, to an operator of the same norm defined on all of B.

Definition 5.4: Let $E_o(t)$ be solution operators as in Definition 5.3. Let $E(t)$ denote the extension of $E_o(t)$ to B, and let $M = \{E(t) \mid t \in [0,T]\}$. The map $E(\cdot)(c)\colon [0,T] \to B$ determined by $c \in B$ is called the *generalized solution* of initial value problem $P(B,T,A)$ for the initial value c, and the operators $E(t)$ in M are called *generalized solution operators.*

Generalized solutions are not solutions if they extend beyond D_A. We list some properties of the concepts just introduced.

Theorem 5.5: Let $P(B,T,A)$ be a properly posed initial value problem. Then it is true that:

(1) The set $M = \{E(t) \mid t \in [0,T]\}$ is uniformly bounded, with the same bound as the set $M_0 = \{E_0(t) \mid t \in [0,T]\}$.

(2) Every generalized solution $E(\cdot)(c)$ of $P(B,T,A)$ for initial value $c \in B$ may be approximated uniformly, in the sense of the norm on B, by solutions of $P(B,T,A)$. This means that for every $\varepsilon > 0$ there exists an element $\tilde{c} \in D_E$ such that

$$\|E(t)(c) - E_0(t)(\tilde{c})\| < \varepsilon, \quad t \in [0,T].$$

(3) Every generalized solution $E(\cdot)(c)$ of $P(B,T,A)$ for initial value $c \in B$ belongs to $C^0([0,T],B)$.

(4) The linear operators $E(t) \in C^0(B,B)$ satisfy the semigroup property:

$$E(r+s) = E(r)E(s), \quad r,s,r+s \in [0,T].$$

(5) For all $c \in D_E$

$$E(t)A(c) = AE(t)(c)$$

is satisfied.

(6) For all $c \in D_E$, $u(t) = E_0(t)(c)$ is continuously differentiable.

Proof of (1): Follows immediately from Theorem 4.12.

Of (2): Let $\varepsilon > 0$ be given. Let L denote the bound for the norms of the operators $E(t)$ given by (1). Since D_E is dense in B, there exists a $\tilde{c} \in D_E$ such that

$$\|c-\tilde{c}\| < \frac{\varepsilon}{L}.$$

Since $E(t)(\tilde{c}) = E_0(t)(\tilde{c})$, we estimate as follows:

$$\|E(t)(c)-E_0(t)(\tilde{c})\| = \|E(t)(c)-E(t)(\tilde{c})\| \leq L\|c-\tilde{c}\| < \varepsilon .$$

Of (3): Let $s \in [0,T]$ and $\varepsilon > 0$ be given. By (2), we choose an element $\tilde{c} \in D_E$ for which

$$\|E(t)(c) - E_0(t)(\tilde{c})\| < \frac{\varepsilon}{3}, \quad t \in [0,T],$$

holds. Since $E_0(\cdot)(\tilde{c})$ is a differentiable map and therefore continuous, there exists a $\delta > 0$ such that

$$\|E_0(s+h)(\tilde{c})-E_0(s)(\tilde{c})\| < \frac{\varepsilon}{3}, \quad |h| < \delta, \quad s+h \in [0,T].$$

Altogether, for all h with $|h| < \delta$ and $s+h \in [0,T]$, this yields the estimate

$$\|E(s+h)(c)-E(s)(c)\| \leq \|E(s+h)(c)-E_0(s+h)(\tilde{c})\| +$$
$$\|E_0(s+h)(\tilde{c})-E_0(s)(\tilde{c})\| + \|E_0(s)(\tilde{c})-E(s)(c)\| < \varepsilon.$$

Of (4): Let $r \in (0,T]$. In addition to problem $P(B,T,A)$ we consider $P(B,r,A)$. The quantities from Definitions 5.3 and 5.4 which pertain to $P(B,r,A)$ will be identified by an additional index r. For a solution u of $P(B,T,A)$, v obviously is a solution of $P(B,r,A)$ with $v(t) = u(t+\tilde{r}-r)$, $\tilde{r} \in [r,T]$. We consequently define

$$D_{E,r} = \text{span}\{\bigcup_{\tilde{r}\in[r,T]} E_0(\tilde{r}-r)(D_E)\}.$$

Let r,s be arbitrary, but fixed in $(0,T]$ with $r+s \le T$. Without loss of generality, $0 < r \le s$. Then

$$D_E \subset D_{E,s} \subset D_{E,r} \subset D_A \subset B$$

$$E_{o,r}(t)(c) = E_o(t)(c), \qquad t \in [0,r], \quad c \in D_E$$

$$E_r(t)(c) = E(t)(c), \qquad t \in [0,r], \quad c \in B$$

$$E_o(s)(c) \in D_{E,r}, \qquad c \in D_E.$$

One obtains for $c \in D_E$

$$\begin{aligned} E(r)\circ E(s)(c) &= E_r(r)\circ E_o(s)(c) = E_{o,r}(r)\circ E_o(s)(c) \\ &= E_o(r+s)(c) = E(r+s)(c). \end{aligned}$$

This proves the contention that D_E is dense in B.

Of (5): For $c \in D_E$ and $t \in [0,T]$ one has the following estimate:

$$\begin{aligned} \|E(t)\circ A(c) - A\circ E(t)(c)\| &\le \|E(t)\circ A(c) - \tfrac{1}{h}E(t)[E(h)(c)-c]\| \\ &\quad + \|\tfrac{1}{h} E(t)[E(h)(c)-c] - A\circ E(t)(c)\| \\ &\le \tfrac{1}{|h|}\|E(t)\|\ \|E(h)(c)-c-hA(c)\| \\ &\quad + \tfrac{1}{|h|}\|E(t+h)(c)-E(t)(c)-hA\circ E(t)(c)\| . \end{aligned}$$

Since $E(\cdot)(c) = E_o(\cdot)(c)$ is differentiable on the entire interval $[0,T]$, the conclusion follows by passing to the limit $h \to 0$.

Of (6): Since

$$u'(t) = A(u(t)) = A\circ E(t)(c) = E(t)\circ A(c) = E(t)(A(c))$$

the continuity of u' follows from (3). □

With the next two examples we elucidate the relationship of these considerations to problems of partial differential equations.

Example 5.6: *Initial value problem for a parabolic differential equation.* Let $T > 0$, $\phi\colon \mathbb{R} \to \mathbb{C}$, and $a \in C^{\infty}(\mathbb{R}, \mathbb{R})$ with $a' \in C_0^{\infty}(\mathbb{R}, \mathbb{R})$ and $a(x) > 0$ $(x \in \mathbb{R})$. It follows that

$$0 < K_1 \le a(x) \le K_2, \qquad x \in \mathbb{R}.$$

We consider the problem

$$\begin{aligned} u_t(x,t) &= [a(x)u_x(x,t)]_x \\ u(x,0) &= \phi(x) \end{aligned} \qquad x \in \mathbb{R}, \quad t \in (0,T). \qquad (5.7)$$

The problem is properly posed only if u and ϕ are subject to certain growth conditions (cf. Example 1.10). The choice of Banach space in which to consider this problem depends on the nature of these growth conditions and vice versa. We choose $B = L^2(\mathbb{R},\mathbb{C})$ and

$$D_A = \{f \in B \mid f \in C^1(\mathbb{R},\mathbb{C}),\ af' \text{ absolutely continuous},\ (af')' \in B\}.$$

D_A is a superspace of $C_0^{\infty}(\mathbb{R},\mathbb{C})$ and therefore is dense in B by Theorem 4.10. We define a linear operator $A\colon D_A \to B$ by the assignment $f \to (af')'$. Problem (5.7) is thus transformed into form (5.2). That this is properly posed can be shown with the use of known properties from the theory of partial differential equations. For this one may choose $D_E = C_0^{\infty}(\mathbb{R},\mathbb{C})$, for example. Generalized solutions exist, however, for arbitrary square integrable initial functions, which need not even be continuous. The operators $E_0(t)$ and $E(t)$, which by no means always have a closed representation, can be written as integral operators for $a(x) \equiv$ constant (cf. Example 1.10). □

Example 5.8: *Initial value problem for a hyperbolic differential equation.* Let $T > 0$, $\phi: \mathbb{R} \to \mathbb{C}$ and $a \in C^\infty(\mathbb{R}, \mathbb{R})$ with $|a(x)| \le K$, $x \in \mathbb{R}$. We consider the problem

$$\begin{aligned} u_t(x,t) &= a(x)u_x(x,t) \\ u(x,0) &= \phi(x) \end{aligned} \qquad x \in \mathbb{R}, \quad t \in (0,T). \qquad (5.9)$$

For simplicity's sake we choose the same Banach space B as in Example 5.6 and set

$$D_A = \{f \in B \mid f \text{ absolutely continuous, } af' \in B\}.$$

We define the linear operator A by the assignment $f \to af'$. All other quantities are fixed in analogy with Example 5.6. Once again it can be shown that the problem is properly posed. □

We are now ready to define the concept of a *difference method* for a properly posed problem $P(B,T,A)$ as well as the related properties of *consistency, stability,* and *convergence.*

Definition 5.10: Let $P(B,T,A)$ be a properly posed initial value problem and $M = \{E(t) \mid t \in [0,T]\}$ the corresponding set of generalized solution operators, as given in Definition 5.4, and $h_o \in (0,T]$.

(1) A family $M_D = \{C(h): B \to B \mid h \in (0,h_o]\}$ of bounded linear operators defined on B is called a *difference method* for $P(B,T,A)$ if the function $||C(\cdot)||$ is bounded in every closed interval of $(0,h_o]$.

(2) The difference method M_D is called *consistent* if there exists a dense subspace D_C in B such that, for

all $c \in D_C$, the expression

$$\frac{1}{|h|} \| [C(h) - E(h)](E(t)(c)) \|$$

converges uniformly to zero for $t \in [0,T]$ as $h \to 0$.

(3) The difference method M_D is called *stable* if the set of operators

$$\{C(h)^n \mid h \in (0,h_o], n \in \mathbb{N}, nh \leq T\}$$

is uniformly bounded.

(4) The difference method M_D is called *convergent* if the expression

$$\| C(h_j)^{n_j}(c) - E(t)(c) \|$$

converges to zero for all $c \in B$, for all $t \in [0,T]$, and for all sequences $\{h_j\}$ of real numbers in $(0,h_o]$ converging to 0. Here $\{n_j\}$ is an admissible sequence of natural numbers if $\{n_j h_j\}$ converges to t and $n_j h_j \leq T$. □

The following theorem explains the relationship between the above concepts.

Theorem 5.11: *Lax-Richtmyer.* Let M_D be a consistent difference method for the properly posed initial value problem $P(B,T,A)$. Then the difference method M_D is convergent if and only if it is stable.

Proof:

(a) *Convergence implies stability:* We will proceed indirectly, and thus assume that M_D is convergent but not stable. Then there exists a sequence $\{h_j\}$ of elements in $(0,h_o]$ and a sequence of natural numbers $\{n_j\}$ related by the condition $n_j h_j \in [0,T]$, $j \in \mathbb{N}$ so that the sequence $\{\|C(h_j)^{n_j}\|\}$ is not bounded. Since $[0,T]$ and $[0,h_o]$ are compact, we may assume without loss of generality that the sequences $\{n_j h_j\}$ and $\{h_j\}$ converge to $t \in [0,T]$ and $h \in [0,h_o]$. Assume $h > 0$. From a certain index on, n_j is constant. By Definition 5.10(1), $\|C(\cdot)\|$ is bounded in $[h/2,h_o]$. Consequently $\|C(h_j)^{n_j}\| \le \|C(h_j)\|^{n_j}$ is bounded. This is a contradiction, and $\{h_j\}$ must converge to zero. Since M_D is a convergent difference method, the sequence

$$\{\|C(h_j)^{n_j}(c) - E(t)(c)\|\}$$

also converges to zero, for every $c \in B$. Hence there exists a $j_o(c) \in \mathbb{N}$ so that for all $j > j_o(c)$ it is true that

$$\|C(h_j)^{n_j}(c) - E(t)(c)\| < 1$$
$$\|C(h_j)^{n_j}(c)\| < 1 + \|E(t)(c)\| .$$

We set

$$K(c) = \max_{j \le j_o(c)} \{1 + \|E(t)(c)\| , \|C(h_j)^{n_j}(c)\|\} .$$

It then follows for all $c \in B$ that

$$\|C(h_j)^{n_j}(c)\| \le K(c), \qquad j \in \mathbb{N}.$$

Applying Theorem 4.14 yields that

$$\{C(h_j)^{n_j}\}$$

is a uniformly bounded set of operators. Contradiction!

(b) *Stability implies convergence:* Let $c \in D_C$, $t \in [0,T]$, $\{h_j\}$ a sequence of real numbers in $(0,h_o]$ converging to zero, and $\{n_j\}$ a related sequence of natural numbers, so that $\{n_j h_j\}$ converges to t and $n_j h_j \le T$. For

$$\psi_j(c) = C(h_j)^{n_j}(c) - E(t)(c), \quad j \in \mathbb{N}$$

it is true by Theorem 5.5(4) (*semigroup property* of E) that

$$\psi_j(c) = \sum_{k=0}^{n_j-1} C(h_j)^k[C(h_j)-E(h_j)]E((n_j-1-k)h_j)(c)$$

$$+ E(\rho_j)\ [E(n_jh_j-\rho_j) - E(t-\rho_j)]\,(c), \qquad j \in \mathbb{N},$$

where

$$\rho_j = \min\{t, n_jh_j\}, \qquad j \in \mathbb{N}.$$

Now let $\varepsilon > 0$ be given. Then we obtain the following estimates:

(α) By the stability of M_D there exists a constant K_C such that

$$\|C(h_j)^k\| \le K_C, \quad j \in \mathbb{N}, \quad k = 0(1)n_j.$$

(β) By the consistency of M_D there exists a $j_1 \in \mathbb{N}$ such that

$$\|[C(h_j)-E(h_j)]E((n_j-1-k)h_j)(c)\| < \varepsilon h_j, \quad j > j_1.$$

(γ) By Theorem 5.5(1) there exists a constant K_E such that

$$\|E(\tau)\| \le K_E, \qquad \tau \in [0,T].$$

(δ) By Theorem 5.5(3) there exists a $j_2 \in \mathbb{N}$ such that

$$\|[E(n_jh_j-\rho_j) - E(t-\rho_j)](c)\| < \varepsilon, \qquad j > j_2.$$

Altogether, it follows from (α) - (δ) that

$$\|\psi_j(c)\| < n_jK_C\varepsilon h_j + K_E\varepsilon \le (K_CT + K_E)\varepsilon, \quad j > \max\{j_1, j_2\}.$$

This already proves that the difference method M_D is convergent for all $c \in D_C$. Now for $\tilde{c} \in B$ and $c \in D_C$ we can write

$$\psi_j(\tilde{c}) = C(h_j)^{n_j}(\tilde{c}) - E(t)(\tilde{c})$$

$$= C(h_j)^{n_j}(c) - E(t)(c) + C(h_j)^{n_j}(\tilde{c}-c) - E(t)(\tilde{c}-c),$$

$$\|\psi_j(\tilde{c})\| \le \|\psi_j(c)\| + K_C\|\tilde{c}-c\| + K_E\|\tilde{c}-c\|, \qquad j \in \mathbb{N}.$$

For a given $\eta > 0$ we then choose a $c \in D_C$ so that the last two terms on the right side of the last inequality are less than $2\eta/3$. Considering this together with the previous inequality shows that there exists a $j_0 \in \mathbb{N}$ such that

$$\|\psi_j(\tilde{c})\| < \eta, \qquad j > j_0. \qquad \square$$

Remark on Theorem 5.11: In discussing convergence, we have not spoken of *order of convergence* and *order of consistency* as yet. The precise situation is as follows. In case the expression

$$\frac{1}{|h|}\|[C(h) - E(h)](E(t)(c))\|$$

is of order $O(h^p)$ for all $c \in D_C$ (*order of consistency*), then it follows from the above proof, under the additional condition that $n_j h_j = t$ for all $c \in D_C$, that there is also convergence of order $O(h^p)$, i.e.

$$\|C(h)^n(c) - E(t)(c)\| = O(h^p), \quad nh = t.$$

In case c does not lie in D_C, the order of convergence is substantially worse, as a rule. For c in a subspace of D_C on the other hand, the order of convergence can be even better than p. $\square$

As a rule, the proof that a given initial value problem is properly posed is very tedious. In the literature, one often finds only existence and uniqueness theorems. Condition

(3) of Definition 5.3 is then unsatisfied. However, if there also exists a consistent and stable difference method, then this condition, too, is satisfied.

Theorem 5.12: Let $P(B,T,A)$ be a problem satisfying conditions (1) and (2) of Definition 5.3. Further let there be given a family $M_D = \{C(h) \mid h \in (0,h_o]\}$ of operators $C(h) \in L(B,B)$ with the following properties:

(1) For all $c \in D_E$ the expression

$$\frac{1}{|h|}\|C(h)(E_o(t)(c)) - E_o(t+h)(c)\|$$

converges to zero as $h \to 0$. Convergence is uniform for all $t \in [0,T]$.

(2) The set of operators

$$\{C(h)^n \mid h \in (0,h_o],\ n \in \mathbb{N},\ nh \leq T\}$$

is uniformly bounded.

Then $P(B,T,A)$ is properly posed.

Proof: Assume

$$\|C(h)^n\| \leq L, \quad h \in (0,h_o],\ n \in \mathbb{N},\ nh \leq T.$$

For $t \in (0,T]$, let $h = t/m$ where $m \in \mathbb{N}$. For $c \in D_E$ it follows that

$$\|E_o(mh)(c)\| \leq \|C(h)^m(c)\|$$

$$+ \sum_{\nu=0}^{m-1} \|C(h)^{\nu+1}E_o((m-1-\nu)h)(c) - C(h)^{\nu}E_o((m-\nu)h)(c)\|.$$

For fixed c, we now choose an m so large that

$$\|C(h)E_o((m-1-\nu)h)(c) - E_o((m-\nu)h)(c)\| \leq \|c\|\, h.$$

This yields

$$\|E_0(t)(c)\| \le (L+Lt)\ \|c\| \le L(1+T)\ \|c\|$$

$$\|E_0(t)\| \le L(1+T), \qquad t \in [0,T]. \qquad \square$$

One can now also drop condition (1) of Definition 5.3 from the hypotheses of Theorem 5.12. Then $P(\bar{D}_E,T,\tilde{A})$, where $\tilde{A}$ is the restriction of A to $\bar{D}_E \cap D_A$, is properly posed. The difference method consequently still converges for all $c \in \bar{D}_E$, i.e., for all c for which the existence of a generalized solution is guaranteed.

<u>Theorem 5.13</u>: *Kreiss.* Let $P(B,T,A)$ be a properly posed problem, $M_D = \{C(h) \mid h \in (0,h_0]\}$ a stable difference method for $P(B,T,A)$, and $\{Q(h) \mid h \in (0,h_0]\}$ a uniformly bounded set of linear operators $Q(h) : B \to B$. Then $\{C(h) + hQ(h) \mid h \in (0,h_0]\}$ is also stable.

Proof: By hypothesis there exist constants $K_1 > 0$ and $K_2 > 0$ such that

$$\|C(h)^\mu\| \le K_1, \qquad \mu \in \mathbb{N},\ h \in (0,h_0],\ \mu h \le T$$

$$\|Q(h)\| \le K_2, \qquad h \in (0,h_0].$$

On the other hand, we have a representation

$$[C(h) + hQ(h)]^\mu = \sum_{\lambda=0}^{\mu} h^\lambda \sum_{\kappa=1}^{\binom{\mu}{\lambda}} P_{\lambda,\kappa}$$

with operators $P_{\lambda,\kappa}$ which are products of μ factors. $C(h)$ occurs $(\mu - \lambda)$ times as a factor, and $Q(h)$, λ times. We gather the factors $C(h)$, which are not divisible by $Q(h)$, as powers. Now in $P_{\lambda,\kappa}$ there are at most $\lambda+1$ powers of $C(h)$ gathered in this way, so that we obtain the estimate

$$\|P_{\lambda,\kappa}\| \leq K_1^{\lambda+1}K_2^{\lambda}.$$

Altogether, it follows for $\mu \in \mathbb{N}$ and $h \in (0,h_0]$ with $\mu h \leq T$ that

$$\|[C(h)+hQ(h)]^{\mu}\| \leq \sum_{\lambda=0}^{\mu} h^{\lambda}\binom{\mu}{\lambda} K_1^{\lambda+1}K_2^{\lambda}$$

$$\leq K_1(1+hK_1K_2)^{\mu} \leq K_1 \exp(TK_1K_2).$$

Consequently, $\{C(h) + hQ(h)\}$ is stable. □

Lax-Richtmyer theory (Theorems 5.11, 5.12, 5.13, 7.2, 7.4) is relatively simple and transparent. But one must not overlook the fact that this result was made possible by three restricting hypotheses:

(1) The differential equations $u'(t) = A(u(t))$ under consideration are linear. Moreover, the operators A do not depend on t.

(2) All difference operators are defined on the same Banach space, and map this space into itself.

(3) The difference operators are defined for all step sizes h in an interval $(0,h_0]$.

The generalization of the theory to nonconstant operators A and quasi-linear differential equations presents considerable difficulties. A good treatment of the problems involved may be found in Ansorge-Hass (1970). Hypotheses (2) and (3) also are an idealization relative to the procedures followed in practice. Assume for the moment that the elements of the Banach space B are continuous functions on a real interval. In the numerical computations, we consider instead the restrictions of the functions to the lattice

$$\{x_j \in \mathbb{R} \mid x_j = j\Delta x, \quad N_1 \leq j \leq N_2\},$$

where Δx, N_1, and N_2 all depend on the length of the step h in the t-direction. The restrictions of the functions to the lattice form a finite-dimensional vector space. The significance of that for the practical execution of the numerical computations is naturally decisive. For initial boundary value problems, the definition of a difference operator for all $h \in (0,h_0]$ often presents substantial difficulties. However, it also suffices to have a definition for step widths $h_\nu = K\cdot 2^{-\nu}$, $\nu = 1(1)\infty$, for a fixed constant $K > 0$.

We will now show that, under certain natural conditions, the essential parts of Lax-Richtmyer theory remain correct for these "finite" difference methods. For the rest of this chapter, we make the following general assumptions: $P(B,T,A)$ is a fixed initial value problem in the sense of Definition 5.1. The problem is uniquely solvable for all initial values $c \in D_E \subset B$, where D_E is a vector space containing at least one nonzero element. We do not demand that D_E be dense in B. The solutions of the problem we denote as before by $E_0(t)(c)$. The operators $E_0(t)$ are linear. Their continuity need not be demanded.

<u>Definition 5.14</u>: Let $K \in \mathbb{R}_+$.

(1) The sequence $M_D = \{(B_\nu, r_\nu, C_\nu) \mid \nu \in \mathbb{N}\}$ is called a *strongly finite difference method* if it is true for all $\nu \in \mathbb{N}$ that:

(a) B_ν is a finite dimensional Banach space with norm $\|\cdot\|^{(\nu)}$ (the space of *lattice functions*).

(b) r_ν is a linear mapping (*restriction*) of B to B_ν, and $\lim_{\nu\to\infty} \|r_\nu(c)\|^{(\nu)} = \|c\|$ holds for every fixed $c \in B$.

(c) C_ν is a linear mapping (*difference operator*) of B_ν to itself.

(2) M_D is called *consistent* if there exists a vector space $D_C \subset D_E \subset \bar{D}_C \subset B$ such that

$$\lim_{\nu\to\infty} \frac{1}{h_\nu} \| C_\nu \circ r_\nu \circ E_0(t)(c) - r_\nu \circ E_0(t+h_\nu)(c) \|^{(\nu)} = 0$$

for all $c \in D_C$ with $h_\nu = K\cdot 2^{-\nu}$. Convergence is uniform for all $t \in [0,T]$.

(3) M_D is called *stable* if the set

$$\{\|C_\nu^n\|^{(\nu)} \mid \nu \in \mathbb{N},\ n \in \mathbb{N},\ nK2^{-\nu} \leq T\}$$

is bounded.

(4) M_D is called *convergent* if for all $t = \mu_1 K 2^{-\mu_2} \in [0,T]$ with $\mu_1, \mu_2 \in \mathbb{N}$ and all $c \in D_E$ it is true that

$$\lim_{\substack{\nu\to\infty \\ \nu\geq\mu_2}} \| C_\nu^{n_\nu} \circ r_\nu(c) - r_\nu \circ E_0(t)(c) \|^{(\nu)} = 0,$$

where $n_\nu = t\cdot 2^\nu/K = \mu_1 2^{\nu-\mu_2}$. □

<u>Theorem 5.15</u>: Whenever the strongly finite difference method M_D is consistent and stable, then (1) $P(\bar{D}_E,T,A)$ is properly posed, and (2) M_D is convergent.

Conclusion (1) corresponds to the assertion of Theorem 5.12, and conclusion (2) to one direction of Theorem 5.11. In addition, (1) implies that the operators $E_0(t)$ are continuous for fixed t. It is easily seen that Theorem 5.13 also can be carried over to finite difference methods in a reasonable way.

Proof of Theorem 5.15(1): For fixed $c \in D_C$ and $t = \mu_1 K\cdot 2^{-\mu_2}$ we make the following definitions:

$$h_\nu = K\cdot 2^{-\nu}, \qquad \nu = \mu_2(1)\infty$$

$$n_\nu = \mu_1 2^{\nu-\mu_2}, \qquad \nu = \mu_2(1)\infty$$

$$t_{\nu\kappa} = (n_\nu - \kappa)h_\nu, \ \kappa = 0(1)n_\nu$$

$$d_{\nu\kappa} = \| C_\nu^{\kappa+1}\circ r_\nu\circ E_0(t_{\nu,\kappa+1})(c) - C_\nu^{\kappa}\circ r_\nu\circ E_0(t_{\nu\kappa})(c)\|^{(\nu)}, \qquad \kappa = 0(1)n_\nu-1.$$

In addition, always assume

$$\|C_\nu^n\|^{(\nu)} \le L, \qquad \nu \in \mathbb{N},\ n \in \mathbb{N},\ nK2^{-\nu} \le T.$$

It follows that

$$d_{\nu\kappa} \le L\,\|C_\nu\circ r_\nu\circ E_0(t_{\nu,\kappa+1})(c) - r_\nu\circ E_0(t_{\nu\kappa})(c)\|^{(\nu)}$$

or

$$d_{\nu\kappa} \le L\,\|C_\nu\circ r_\nu\circ E_0(t_{\nu,\kappa+1})(c) - r_\nu\circ E_0(t_{\nu,\kappa+1}+h_\nu)(c)\|^{(\nu)}.$$

By the consistency of M_D, for every $\varepsilon > 0$ there exists a $\nu_0(c,\varepsilon)$ so that $\nu \ge \nu_0(c,\varepsilon)$ implies $d_{\nu\kappa} \le \varepsilon L h_\nu$. We can now estimate $E_0(t)(c)$:

$$\begin{aligned}\|E_0(t)(c)\| &= \lim_{\nu\to\infty} \|r_\nu\circ E_0(t)(c)\|^{(\nu)} \\ &\le \limsup_{\nu\to\infty} \|C_\nu^{n_\nu}\circ r_\nu(c)\|^{(\nu)} + \limsup_{\nu\to\infty} \sum_{\kappa=0}^{n_\nu-1} d_{\nu\kappa} \\ &\le L\,\|c\| + \varepsilon LT.\end{aligned}$$

Since $\varepsilon > 0$ is arbitrary, we also have $\|E_0(t)(c)\| \le L\,\|c\|$. This inequality, however, was derived under the assumptions $t = \mu_1 K\cdot 2^{-\mu_2}$ and $c \in D_C$. Since the function $E_0(\cdot)(c)$ is differentiable, it is also continuous for fixed c. The previously admitted t values are dense in $[0,T]$. Hence the inequality holds for all $t \in [0,T]$. Finally, use the inclusion $D_E \subset \overline{D}_C$. Then it follows for all $c \in D_E$ and $t \in [0,T]$ that

$$\|E_0(t)(c)\| \leq L\,\|c\|\,.$$

This proves conclusion (1) of Theorem 5.15.

Proof of 5.15(2): Again we assume that a fixed $c \in D_C$ and $t = \mu_1 K 2^{-\mu_2}$ have been chosen. Similarly to the above, one can then estimate:

$$\|C_\nu^{n_\nu}\circ r_\nu(c) - r_\nu\circ E_0(t)(c)\|^{(\nu)} \leq \sum_{\kappa=0}^{n_\nu - 1} d_{\nu\kappa} \leq \varepsilon LT, \quad \nu \geq \nu_0(c,\varepsilon).$$

This inequality obviously implies convergence for all $c \in D_C$. Now let $c \in D_E$ be arbitrarily chosen and let $\tilde{c} \in D_C$. This yields

$$\begin{aligned}\|C_\nu^{n_\nu}\circ r_\nu(c) - r_\nu\circ E_0(t)(c)\|^{(\nu)} &\leq \|C_\nu^{n_\nu}\circ r_\nu(c) - C_\nu^{n_\nu}\circ r_\nu(\tilde{c})\|^{(\nu)} \\ &\quad + \|C_\nu^{n_\nu}\circ r_\nu(\tilde{c}) - r_\nu\circ E_0(t)(\tilde{c})\|^{(\nu)} \\ &\quad + \|r_\nu\circ E_0(t)(\tilde{c}) - r_\nu\circ E_0(t)(c)\|^{(\nu)} \\ &\leq L\,\|r_\nu(c-\tilde{c})\|^{(\nu)} + \|r_\nu\circ E_0(t)(c-\tilde{c})\|^{(\nu)} \\ &\quad + \|C_\nu^{n_\nu}\circ r_\nu(\tilde{c}) - r_\nu\circ E_0(t)(\tilde{c})\|^{(\nu)}.\end{aligned}$$

By passing to the limit $\nu \to \infty$ one obtains

$$\limsup_{\nu\to\infty} \|C_\nu^{n_\nu}\circ r_\nu(c) - r_\nu\circ E_0(t)(c)\|^{(\nu)} \leq L\,\|c-\tilde{c}\| + \|E_0(t)(c-\tilde{c})\|.$$

Here $E_0(t)$ is bounded and $\|c - \tilde{c}\|$ can be made arbitrarily small. □

6. Examples of stable difference methods

This chapter is devoted to a presentation of several difference methods, whose stability can be established by elementary methods. We begin with a preparatory lemma and definition.

Lemma 6.1: Let $f \in C^0(\mathbb{R},\mathbb{C})$ be square integrable, $a \in C^0(\mathbb{R}, \mathbb{R}_+)$ bounded, and $\Delta x \in \mathbb{R}$. Then

$$(1)\quad \int_{-\infty}^{+\infty} \overline{f(x)}\{a(x+\Delta x/2)[f(x+\Delta x)-f(x)] - a(x-\Delta x/2)[f(x)-f(x-\Delta x)]\}dx$$

$$= -\int_{-\infty}^{+\infty} a(x)|f(x+\Delta x/2)-f(x-\Delta x/2)|^2dx.$$

$$(2)\quad \int_{-\infty}^{+\infty} |a(x+\Delta x/2)[f(x+\Delta x)-f(x)] - a(x-\Delta x/2)[f(x)-f(x-\Delta x)]|^2dx$$

$$\leq 4\int_{-\infty}^{+\infty} a(x)^2|f(x+\Delta x/2)-f(x-\Delta x/2)|^2dx.$$

Proof of (1): We have

$$\int_{-\infty}^{+\infty} g(x+\Delta x)dx = \int_{-\infty}^{+\infty} g(x)\, dx$$

for all functions g for which the integrals exist. Thus we can rewrite the left hand integral as follows:

$$\int_{-\infty}^{+\infty} \overline{f(x)}\{a(x+\Delta x/2)[f(x+\Delta x)-f(x)]-a(x-\Delta x/2)[f(x)-f(x-\Delta x)]\}dx$$

$$= \int_{-\infty}^{+\infty} [a(x)\overline{f(x-\Delta x/2)}f(x+\Delta x/2)-a(x)\overline{f(x-\Delta x/2)}f(x-\Delta x/2)$$

$$- a(x)\overline{f(x+\Delta x/2)}f(x+\Delta x/2)+a(x)\overline{f(x+\Delta x/2)}f(x-\Delta x/2)]dx$$

$$= -\int_{-\infty}^{+\infty} a(x)|f(x+\Delta x/2)-f(x-\Delta x/2)|^2 dx.$$

Proof of (2): For $\alpha,\beta \in ¢$ we have

$$|\alpha + \beta|^2 \leq 2(|\alpha|^2 + |\beta|^2).$$

Therefore

$$|a(x+\Delta x/2)[f(x+\Delta x)-f(x)]-a(x-\Delta x/2)[f(x)-f(x-\Delta x)]|^2$$

$$\leq 2[a(x+\Delta x/2)^2|f(x+\Delta x)-f(x)|^2+a(x-\Delta x/2)^2|f(x)-f(x-\Delta x)|^2].$$

Each of the summands inside the square brackets, when integrated with the appropriate translation, yields the value

$$\int_{-\infty}^{+\infty} a(x)^2|f(x+\Delta x/2)-f(x-\Delta x/2)|^2 dx.$$

The desired conclusion follows once we add the two together. □

Definition 6.2: Let $\Delta x \in \mathbb{R}$, M an arbitrary set, and $f\colon \mathbb{R} \to M$. We define:

$$T_{\Delta x}(x) = x + \Delta x, \qquad x \in \mathbb{R}$$
$$T_{\Delta x}(f)(x) = f(T_{\Delta x}(x)).$$

$T_{\Delta x}$ is called a *translation operator*.

The translation operator is a bounded linear map of $L^2(\mathbb{R},¢)$ into itself with $\|T_{\Delta x}\| = 1$. The operator is invertible and

$$T_{\Delta x}^{-1} = T_{-\Delta x}.$$

To derive a difference method for the initial value problem in Example 5.6 we discretize the differential equation (5.7),

$$u_t(x,t) = [a(x)u_x(x,t)]_x$$

where $a \in C^\infty(\mathbb{R}, \mathbb{R})$, $a' \in C_0^\infty(\mathbb{R}, \mathbb{R})$, $a(x) > 0$ for $x \in \mathbb{R}$, as follows:

$$D(x,t,\Delta x) = (\Delta x)^{-2}\{a(x+\Delta x/2)[u(x+\Delta x,t)-u(x,t)]$$

$$- a(x-\Delta x/2)[u(x,t)-u(x-\Delta x,t)]\} \approx [a(x)u_x(x,t)]_x$$

$$u(x,t+h)-u(x,t) \approx \alpha hD(x,t,\Delta x) + (1-\alpha)hD(x,t+h,\Delta x)$$

$$\Delta x,\ h \in \mathbb{R}_+,\ \alpha \in [0,1].$$

Using $\lambda = h/(\Delta x)^2$ and the operator

$$H(\Delta x) = a(x-\Delta x/2)T_{\Delta x}^{-1} - [a(x+\Delta x/2)+a(x-\Delta x/2)]I$$

$$+ a(x+\Delta x/2)T_{\Delta x}$$

we obtain

$$\begin{aligned} C(h)-I &= \alpha\lambda H(\Delta x) + (1-\alpha)\lambda H(\Delta x)\circ C(h), \\ [I-(1-\alpha)\lambda H(\Delta x)]\circ C(h) &= I + \alpha\lambda H(\Delta x). \end{aligned} \tag{6.3}$$

The method is *explicit* for $\alpha = 1$ and *implicit* for $\alpha \in [0,1)$. The case $\alpha = 0$ is also called *totally implicit*. For $\alpha = 1/2$ it is called the *Crank-Nicolson method*. The relationships are depicted graphically in Figure 6.4. In the following, we will examine only the cases $\alpha = 1$ and $\alpha = 0$.

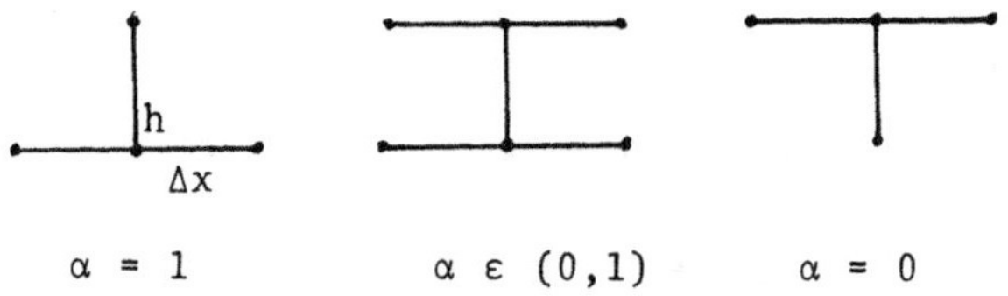

Figure 6.4

The difference method is applied as follows. Begin at a fixed time t_o with a known approximation $w(x,t_o)$ to the exact solution $u(x,t_o)$. Compute a new approximation $w(x,t_o+h)$ to $u(x,t_o+h)$ by the rule

$$w(x,t_o+h) = C(h)[w(x,t_o)].$$

In the implicit case, this means solving the system of equations

$$[I-(1-\alpha)\lambda H(\Delta x)][w(x,t_o+h)] = [I+\alpha\lambda H(\Delta w)][w(x,t_o)].$$

On a computer, $w(x,t_o)$ and $w(x,t_o+h)$ can only be obtained for a discrete set of x-values, so the latter are chosen equidistant so that the translation operator does not lead outside the set.

<u>Theorem 6.5</u>: *Explicit difference method for a parabolic differential equation.* Let $P(B,T,A)$ be the initial value problem of Example 5.6 with $B = L^2(\mathbb{R},\mathbb{C})$. Consider the set $M_D = \{C(h) \mid h \in (0,h_o]\}$ where $C(h) = I + \lambda H(\Delta x)$. Then

(1) M_D is a consistent difference method for $P(B,T,A)$ with order of consistency $O(h)$.

(2) Under the additional hypothesis

$$0 < \lambda \max_{x\in\mathbb{R}} a(x) \leq 1/2 \quad \text{(stability condition)}$$

M_D is stable. In particular, for all $h \in (0,h_o]$ and $n \in \mathbb{N}$ with $nh \leq T$:

$$\|C(h)^n\|_2 \leq 1.$$

Proof of (1): Every operator $C(h)$ maps into B, is linear, and is bounded, since such is the case for the operators I, $T_{\Delta x}$, and $T_{\Delta x}^{-1}$, and since the function $a(x)$ is bounded. To prove consistency, we choose $D_C = C_0^\infty(\mathbb{R},\mathbb{C})$. We must show that the expression

$$h^{-1}\,\|E(t+h)(c)-C(h)\circ E(t)(c)\|$$
$$= h^{-1}\left[\int_{-\infty}^{+\infty}|u(x,t+h)-C(h)[u(x,t)]|^2dx\right]^{1/2}$$

converges uniformly to zero for $t \in [0,T]$ as $h \to 0$, for all $c \in D_C$. To this end, we use differential equation (5.7) to rewrite the expression

$$h^{-1}\{u(x,t+h)-C(h)[u(x,t)]\} = h^{-1}[u(x,t+h)-u(x,t)] -$$
$$(\Delta x)^{-2}H(\Delta x)[u(x,t)] = u_t(x,t) + \frac{h}{2}u_{tt}(x,t+\nu h) -$$
$$(\Delta x)^{-2}H(\Delta x)[u(x,t)] = \frac{h}{2}u_{tt}(x,t+\nu h)+a(x)u_{xx}(x,t) +$$
$$a'(x)u_x(x,t)-(\Delta x)^{-2}H(\Delta x)[u(x,t)], \quad \nu \in [0,1].$$

The term

$$f(s) = H(s\cdot\Delta x)[u(x,t)]$$

has a series development in s:

$$f(s) = f(0)+sf'(0) + \frac{s^2}{2}f''(0) + \frac{s^3}{6}f'''(0) + \frac{s^4}{24}f^{(4)}(\theta s),$$
$$\theta \in [0,1]$$

$$f(1) = H(\Delta x)[u(x,t)] = f(0) + f'(0) + \frac{1}{2}f''(0) + \frac{1}{6}f'''(0)$$
$$+ \frac{1}{24}f^{(4)}(\theta).$$

We introduce the following abbreviations:

u	for	$u(x,t)$
u_+	for	$u(x+s\Delta x,t)$
u_-	for	$u(x-s\Delta x,t)$
a	for	$a(x)$

$$a_+ \quad \text{for} \quad a(x+s\Delta x/2)$$

$$a_- \quad \text{for} \quad a(x-s\Delta x/2).$$

Corresponding abbreviations hold for the x-derivatives. Using $'$ for $\partial/\partial x$, it follows that

$$f(s) = a_-u_- - (a_+ + a_-)u + a_+u_+$$

$$f'(s) = \Delta x[-\tfrac{1}{2}a'_-u_- - a_-u'_- - \tfrac{1}{2}(a'_+ - a'_-)u + \tfrac{1}{2}a'_+u_+ + a_+u'_+]$$

$$f''(s) = (\Delta x)^2[\tfrac{1}{4}a''_-u_- + a'_-u'_- + a_-u''_- - \tfrac{1}{4}(a''_+ + a''_-)u + \tfrac{1}{4}a''_+u_+ + a'_+u'_+ + a_+u''_+]$$

$$f'''(s) = (\Delta x)^3[-\tfrac{1}{8}a'''_-u_- - \tfrac{3}{4}a''_-u'_- - \tfrac{3}{2}a'_-u''_- - a_-u'''_- - \tfrac{1}{8}(a'''_+ - a'''_-)u$$

$$+ \tfrac{1}{8}a'''_+u_+ + \tfrac{3}{4}a''_+u'_+ + \tfrac{3}{2}a'_+u''_+ + a_+u'''_+]$$

$$f(0) = f'(0) = f'''(0) = 0$$

$$\tfrac{1}{2}f''(0) = (\Delta x)^2(a'u' + au'').$$

Since $a'(x)$ is a function with compact support, we can use the integral form of Example 1.10 (cf. also §9) to describe the asymptotic behavior of the solution $u(x,t)$ of initial value problem (5.7). Thus there exists a square integrable function $L \in C^0(\mathbb{R}, \mathbb{R})$ with

$$\left|\frac{\partial^{j+k}u(x,t)}{\partial x^j \partial t^k}\right| \leq L(x), \qquad 0 \leq j+k \leq 4, \quad x \in \mathbb{R}, \quad t \in [0,T].$$

Combining all this with the appropriately chosen constant M yields the estimates

$$h^{-1}\, \| E(t+h)(c) - C(h)\circ E(t)(c) \|$$

$$\leq \frac{h}{2}\|L\| + \|au''+a'u' - (\Delta x)^{-2}f(1)\|$$

$$\leq \frac{h}{2}\|L\| + \|(\Delta x)^{-2}\,\tfrac{1}{24}f^{(4)}(\theta)\| \leq [\tfrac{h}{2} + M(\Delta x)^2]\,\|L\|.$$

Since $\lambda = h/(\Delta x)^2$, it follows that the method has order of consistency $O(h)$.

Proof of (2): Let $f \in C^0(\mathbb{R},\mathbb{C})$ be square integrable. Then we can write

$$C(h)(f) = f(x) + \lambda\{a(x+\Delta x/2)[f(x+\Delta x)-f(x)] - a(x-\Delta x/2)[f(x)-f(x-\Delta x)]\}.$$

It follows that

$$\|C(h)(f)\|^2 = \|f\|^2 + \lambda\int_{-\infty}^{+\infty} f(x)\overline{\{\dots\}}dx + \lambda\int_{-\infty}^{+\infty} \overline{f(x)}\{\dots\}dx + \lambda^2\int_{-\infty}^{+\infty}|\{\dots\}|^2 dx.$$

By Lemma 6.1(1),

$$\lambda\int_{-\infty}^{+\infty} f(x)\overline{\{\dots\}}dx = \lambda\int_{-\infty}^{+\infty} \overline{f(x)}\{\dots\}dx = -\lambda\int_{-\infty}^{+\infty} a(x)|f(x+\Delta x/2)-f(x-\Delta x/2)|^2 dx.$$

Lemma 6.1(2) enables the estimate

$$\lambda^2\int_{-\infty}^{+\infty}|\{\dots\}|^2 dx \le 4\lambda^2\int_{-\infty}^{+\infty} a(x)^2|f(x+\Delta x/2)-f(x-\Delta x/2)|^2 dx.$$

Altogether, we have

$$\|C(h)(f)\|^2 \le \|f\|^2 - 2\lambda\int_{-\infty}^{+\infty} a(x)[1-2\lambda a(x)]|f(x+\Delta x/2) - f(x-\Delta x/2)|^2 dx.$$

It follows from the stability condition that $2\lambda a(x) \le 1$. Therefore the integral is not negative, and hence

$$\|C(h)(f)\| \le \|f\|.$$

This, together with the submultiplicativity of the operator norm, implies that

$$\|C(h)^n(f)\| = \|C(h)\circ C(h)^{n-1}(f)\|$$

$$\leq \|C(h)^{n-1}(f)\| \leq \dots \leq \|f\| .$$

Since $C^o(\mathbb{R},\mathbb{C})$ is dense in B, the conclusion follows. □

We next investigate method (6.3) for the totally implicit case. The main advantage lies in the fact that λ is no longer subject to any conditions. As a result, h and Δx can be fixed independently of each other.

Theorem 6.6: *Totally implicit difference method for a parabolic differential equation.* The family $M_D = \{C(h) \mid h \in (0,h_o]\}$ where

$$C(h) = [I-\lambda H(\Delta x)]^{-1}$$

is, for all $\lambda \in \mathbb{R}_+$, a consistent and stable difference method for Problem (5.7) of consistency order $O(h)$. In particular, for all $\lambda \in \mathbb{R}_+$, $h \in (0,h_o]$, and $n \in \mathbb{N}$ with $nh \leq T$:

$$\|C(h)^n\|_2 \leq 1.$$

Proof: The invertibility of $I - \lambda H(\Delta x)$ follows from the stability condition $\|C(h)\| \leq 1$ by way of Theorem 4.21. That M_D is a consistent difference method of order $O(h)$ can be shown in a manner similar to that of Theorem 6.5(1). To establish stability, choose an arbitrary square integrable function $f \in C^o(\mathbb{R},\mathbb{C})$ and let $g = C(h)(f)$. Then we can write

$$f(x) = g(x)-\lambda\{a(x+\Delta x/2)[g(x+\Delta x)-g(x)]$$

$$- a(x-\Delta x/2)[g(x)-g(x-\Delta x)]\}.$$

It follows that

$$\|f\|^2 = \|g\|^2 - \lambda\int_{-\infty}^{+\infty} \overline{g(x)}\{\ldots\}dx - \lambda\int_{-\infty}^{+\infty} g(x)\overline{\{\ldots\}}dx + \lambda^2\int_{-\infty}^{+\infty}|\{\ldots\}|^2dx.$$

By Lemma 6.1(1) we have

$$-\lambda\int_{-\infty}^{+\infty} \overline{g(x)}\{\ldots\}dx = -\lambda\int_{-\infty}^{+\infty} g(x)\overline{\{\ldots\}}dx = \lambda\int_{-\infty}^{+\infty} a(x)|g(x+\Delta x/2)-g(x-\Delta x/2)|^2dx.$$

It follows from this that

$$\|f\|^2 = \|g\|^2 + 2\lambda\int_{-\infty}^{+\infty} a(x)|g(x+\Delta x/2) - g(x-\Delta x/2)|^2dx + \lambda^2\int_{-\infty}^{+\infty}|\{\ldots\}|^2dx.$$

Since $a(x) > 0$ and is bounded, the two integrals are non-negative. Hence

$$\|f\| \geq \|g\| = \|C(h)(f)\|$$

and the conclusion follows from this because of the submultiplicativity of the operator norm. □

For the parabolic differential equation (5.7), every value $u(\hat{x},\hat{t})$ depends on *all* initial values $\phi(x)$, $x \in \mathbb{R}$ (cf. Example 1.10). The domain of dependency of the point $(\hat{x},\hat{t})$ thus consists of the entire real line. To discuss the domain of dependency of the corresponding difference method, we divide the interval $[0,T]$ into n pieces of length $h = T/n$. Then

$$\Delta x = (T/n\lambda)^{1/2}.$$

To compute the approximation $w(0,h)$ for $u(0,h)$ with the explicit difference method ($\alpha = 1$), we only need initial values from the interval $[-\Delta x,\Delta x]$, but for $w(0,T)$ we need initial values from the domain of dependency

$$[-n\Delta x,\ n\Delta x] = [-(nT/\lambda)^{1/2},\ (nT/\lambda)^{1/2}]$$

which depends on n. A passage to the limit $h \to 0$ (equivalent to $n \to \infty$) deforms the dependency domain of the explicit difference method into that of the differential equation, i.e. $(-\infty,\infty)$. For positive step sizes, there always is dependence only on the values in a finite interval. The situation is different for implicit methods $(0 \leq \alpha < 1)$. There, after one step, $w(0,h)$ already depends on the initial values from all of the real line.

We next present the simplest difference methods for the initial value problem in Example 5.8. Recall the hyperbolic differential equation (5.9)

$$u_t(x,t) = a(x)u_x(x,t), \qquad a \in C^\infty(\mathbb{R}, \mathbb{R}),\ |a(x)| \leq K.$$

The "naive" discretization

$$h^{-1}[u(x,t+h)-u(x,t)] \approx \frac{1}{2}a(x)(\Delta x)^{-1}[u(x+\Delta x,t)-u(x-\Delta x,t)] \quad (6.7)$$

leads to a difference method which is unstable for all $\lambda = h/\Delta x > 0$. Therefore, we must look for other discretizations which do possess the desired stability properties. Here we will consider the methods of *Friedrichs* and *Courant-Isaacson-Rees*.

The *Friedrichs* method begins with the discretization

$$h^{-1}\{u(x,t+h) - \frac{1}{2}[u(x+\Delta x,t) + u(x-\Delta x,t)]\}$$

$$\approx \frac{1}{2}a(x)(\Delta x)^{-1}[u(x+\Delta x,t) - u(x-\Delta x,t)].$$

This leads to the difference method

$$C(h) = \frac{1-\lambda a(x)}{2} T_{\Delta x}^{-1} + \frac{1+\lambda a(x)}{2} T_{\Delta x}, \quad \lambda = \frac{h}{\Delta x}.$$

Theorem 6.8: Let the function $a(x)$ satisfy the inequality $|a'(x)| \leq \tilde{K}$, $x \in \mathbb{R}$. Then the Friedrichs method for

$$0 < \lambda \leq 1/K \quad \text{(stability condition)}$$

is a consistent and stable difference method for problem $P(L^2(\mathbb{R},\mathbb{C}),T,A)$ of Example 5.8 of consistency order $O(h)$. In particular, for all $h \in (0,h_0]$ and $n \in \mathbb{N}$ with $nh \leq T$,

$$||C(h)^n|| \leq (1+\tilde{K}h)^{n/2} \leq \exp(\tfrac{1}{2}\tilde{K}T).$$

Proof: To show consistency, we choose $D_C = C_0^\infty(\mathbb{R},\mathbb{C})$. We set

$$f(s) = \tfrac{1}{2}[u(x+s\Delta x,t) + u(x-s\Delta x,t)]$$
$$+ \tfrac{\lambda}{2}a(x)[u(x+s\Delta x,t) - u(x-s\Delta x,t)].$$

We obviously have

$$f(1) = C(h)[u(x,t)] = f(0)+f'(0)+\tfrac{1}{2}f''(\theta), \quad \theta \in [0,1].$$

Using the same abbreviations as in the proof of Theorem 6.5, we have

$$f'(s) = \tfrac{1}{2}\Delta x(u'_+-u'_-) + \tfrac{\lambda}{2}\Delta x\cdot a\cdot(u'_++u'_-)$$

$$f''(s) = \tfrac{1}{2}(\Delta x)^2(u''_+ + u''_-) + \tfrac{\lambda}{2}(\Delta x)^2\cdot a\cdot(u''_+-u''_-)$$

$$f(0) = u$$

$$f'(0) = h\cdot a\cdot u'.$$

Since the initial values are functions with compact support, the solutions $u(x,t)$ of the differential equation will be the same. Thus there exists a square integrable function $L \in C^0(\mathbb{R}, \mathbb{R})$ with

$$\left|\frac{\partial^{j+k}u(x,t)}{\partial x^j \partial t^k}\right| \le L(x), \qquad 0 \le j+k \le 2, \quad x \in \mathbb{R}, \quad t \in [0,T].$$

Combining all this gives the estimates

$$h^{-1}|u(x,t+h) - C(h)[u(x,t)]|$$
$$\le h^{-1}|u(x,t)+hu_t(x,t)+\tfrac{1}{2}h^2u_{tt}(x,t+\nu h)-f(0)-f'(0)-\tfrac{1}{2}f''(\theta)|$$
$$\le \tfrac{1}{2}(h+\lambda^{-1}\Delta x+K\Delta x)|L(x)|, \qquad \nu \in [0,1].$$

It follows from this that the difference method is of consistency order $O(h)$.

Next we show stability. Choose $f \in B$.

$$C(h)(f) = \tfrac{1}{2}[f(x+\Delta x)+f(x-\Delta x)] + \tfrac{\lambda}{2}a(x)[f(x+\Delta x)-f(x-\Delta x)].$$

It follows that

$$\|C(h)(f)\|^2 = \frac{1}{4}\int_{-\infty}^{+\infty}|f(x+\Delta x)+f(x-\Delta x)|^2dx$$
$$+ \frac{\lambda}{4}\int_{-\infty}^{+\infty} a(x)[f(x+\Delta x)+f(x-\Delta x)][\overline{f(x+\Delta x)}-\overline{f(x-\Delta x)}]dx$$
$$+ \frac{\lambda}{4}\int_{-\infty}^{+\infty} a(x)[\overline{f(x+\Delta x)}+\overline{f(x-\Delta x)}][f(x+\Delta x)-f(x-\Delta x)]dx$$
$$+ \frac{\lambda^2}{4}\int_{-\infty}^{+\infty} a(x)^2|f(x+\Delta x)-f(x-\Delta x)|^2dx.$$

Since $|a(x)| \le K$ and $\lambda \le 1/K$, we have $\lambda^2a(x)^2 \le 1$. Combine the first and last summands using the identity

$$|\alpha + \beta|^2 + |\alpha - \beta|^2 = 2(|\alpha|^2 + |\beta|^2).$$

Expanding the parentheses in the second and third summands leads to a cancellation of half of the products. Altogether we obtain

$$\|C(h)(f)\|^2 \le \frac{1}{2}\int_{-\infty}^{+\infty}(|f(x+\Delta x)|^2 + |f(x-\Delta x)|^2)dx$$

$$+\frac{\lambda}{2}\int_{-\infty}^{+\infty} a(x)(|f(x+\Delta x)|^2 - |f(x-\Delta x)|^2)dx$$

$$= \|f\|^2 + \frac{\lambda}{2}\int_{-\infty}^{+\infty} a(x+\Delta x)|f(x+\Delta x)|^2 dx$$

$$-\frac{\lambda}{2}\int_{-\infty}^{+\infty} a(x-\Delta x)|f(x-\Delta x)|^2 dx$$

$$+\frac{\lambda}{2}\int_{-\infty}^{+\infty}[a(x)-a(x+\Delta x)]|f(x+\Delta x)|^2 dx$$

$$-\frac{\lambda}{2}\int_{-\infty}^{+\infty}[a(x)-a(x-\Delta x)]|f(x-\Delta x)|^2 dx.$$

The second and third summands cancel each other; the fourth and fifth we combine by a shift in the domain of integration:

$$\|C(h)(f)\|^2 \le \|f\|^2 + \frac{\lambda}{2}\int_{-\infty}^{+\infty}[a(x-\Delta x)-a(x+\Delta x)]|f(x)|^2 dx.$$

Using

$$|a(x-\Delta x)-a(x+\Delta x)| \le 2\Delta x|a'(\theta)| \le 2\Delta x\tilde{K}$$

we finally obtain

$$\|C(h)(f)\|^2 \le \|f\|^2\,(1+\tilde{K}\lambda\Delta x) = \|f\|^2(1+\tilde{K}h)$$

and

$$\|C(h)\| \le (1+\tilde{K}h)^{1/2} \le \exp(\tfrac{1}{2}\,\tilde{K}h)$$

$$\|C(h)^n\| \le (1+\tilde{K}h)^{n/2} \le \exp(\tfrac{1}{2}\,\tilde{K}nh) \le \exp(\tfrac{1}{2}\,\tilde{K}T). \quad \square$$

A different and long known difference method for the hyperbolic differential equation (5.9) is that of *Courant-Isaacson-Rees*. It begins with the discretization

$$h^{-1}[u(x,t+h)-u(x,t)]$$

$$\approx (\Delta x)^{-1}\{a^{+}(x)[u(x+\Delta x,t)-u(x,t)]-a^{-}(x)[u(x,t)-u(x-\Delta x,t)]\}$$

$$a^{+}(x) = \begin{cases} a(x) & \text{for } a(x) \geq 0 \\ 0 & \text{otherwise} \end{cases}$$

$$a^{-}(x) = \begin{cases} -a(x) & \text{for } a(x) < 0 \\ 0 & \text{otherwise.} \end{cases}$$

This leads to the difference method

$$C(h) = \lambda a^{-}(x)T_{\Delta x}^{-1}+\{1-\lambda[a^{+}(x)+a^{-}(x)]\}I+\lambda a^{+}(x)T_{\Delta x}, \quad \lambda = h/\Delta x.$$

Theorem 6.9: Let the function $a(x)$ satisfy a global Lipschitz condition with respect to the norm $\|\cdot\|_2$ and let $|a(x)| \leq K$. Then the Courant-Isaacson-Rees method, for

$$0 < \lambda \leq 1/K \qquad \text{(stability condition)}$$

is a consistent and stable difference method for problem $P(L^2(\mathbb{R},\mathbb{C}),T,A)$ of Example 5.8 of consistency order $O(h)$.

Proof: The proof of consistency we leave to the reader. The stability of the method follows immediately from the methods of Section 8, since the method is *positive definite* in the terminology of that chapter. □

The dependencies of the various methods are shown pictorially in Figure 6.10. The arc in the naive method indicates that the x and t derivatives are not related to each other.

"naive" method Friedrichs method Courant-Isaacson-Rees method

Figure 6.10

The instability of discretization (6.7) is relatively typical for naive discretizations of hyperbolic differential equations. Stability is often achieved by means of additional smoothing terms, which remind one of parabolic equations. We clarify this in the case at hand by listing the unstable method and the two stabilizations, one below the other:

(1) $u(x,t+h)-u(x,t) = \frac{1}{2}\lambda a(x)[u(x+\Delta x,t)-u(x-\Delta x,t)].$

(2) $u(x,t+h)-u(x,t) = \frac{1}{2}\lambda a(x)[u(x+\Delta x,t)-u(x-\Delta x,t)]$

$+ \frac{1}{2}[u(x+\Delta x,t)-2u(x,t)+u(x-\Delta x,t)].$

(3) $u(x,t+h)-u(x,t) = \frac{1}{2}\lambda a(x)[u(x+\Delta x,t)-u(x-\Delta x,t)]$

$+ \frac{1}{2}\lambda|a(x)|[u(x+\Delta x,t)-2u(x,t)+u(x-\Delta x,t)].$

The additional terms in (2) and (3) may be regarded as a discretization of $\varepsilon(x,h)u_{xx}(x,t)$. They are called *numerical viscosity*. They do not influence the consistency order because $\varepsilon(x,h) = O(h)$. For higher order methods, the determination of suitable viscosity terms is more difficult. On the one hand, they should exert sufficiently strong smoothing, while on the other, they should vanish with order of higher powers of h.

Let us again consider the *domains of dependence* and

determinancy for the differential equation and difference method at hand. For the sake of clarity, let $a(x) \equiv$ constant. The domain of determinancy for a segment on the x-axis is then a parallelogram whose right and left sides are formed by characteristics, since the solution of the differential equation is constant on the characteristics. The domain of dependence of a point, therefore, consists of only a point on the x-axis (cf. Example 1.5). For the discussion of the domains of dependence and determinancy of the difference method, we divide the interval $[0,T]$ into n pieces of length $h = T/n$. Then $\Delta x = T/(n\lambda)$. For

$$\lambda < 1/|a| \quad \text{and} \quad \lambda > 1/|a|$$

one needs initial values from the interval $[-\Delta x, \Delta x]$ to compute the value $w(0,h)$. The determination of $w(0,T)$ requires initial values from the interval

$$[-n\Delta x, n\Delta x] = [-T/\lambda, T/\lambda],$$

which is independent of n. Thus the domain of determinancy is a triangle and the domain of dependence is a nondegenerate interval, which contains the "dependency point" of the differential equation only for $\lambda < 1/|a|$ and not for $\lambda > 1/|a|$. For $\lambda = 1/|a|$, the domains of determinancy and dependence of the differential equation and the difference method are identical, since only a term in $T_{\Delta x}^{-1}$ or $T_{\Delta x}$ remains in the expressions

$$C(h) = \frac{1-\lambda a}{2} T_{\Delta x}^{-1} + \frac{1+\lambda a}{2} T_{\Delta x}$$

$$C(h) = \lambda a^{-} T_{\Delta x}^{-1} + (1-\lambda|a|)I + \lambda a^{+} T_{\Delta x}.$$

In this case, the difference method becomes a *characteristic method.*

This situation is exploited in the *Courant-Friedrichs-Lewy condition* (cf. Courant-Friedrichs-Lewy, 1928) for testing the stability of a difference method. This condition is one of necessity and reads as follows:

> A difference method is stable only if the domain of dependence of the differential equation is contained in the domain of dependence of the difference equation upon passage to the limit $h \to 0$.

The two methods discussed previously, for example, are stable for exactly those λ-values for which the Courant-Friedrichs-Lewy condition is satisfied. Such methods are frequently called *optimally stable*. However, there also exist methods in which the ratio $\lambda = h/\Delta x$ must be restricted much more strongly than the above condition would require.

7. Inhomogeneous initial value problems

So far we have only explained the meaning of consistency, stability, and convergence for difference methods for homogeneous problems

$$\begin{aligned} u'(t) &= A(u(t)), \qquad t \in [0,T] \\ u(0) &= c. \end{aligned}$$

Naturally we also want to use such methods in the case of an inhomogeneous problem

$$\begin{aligned} u'(t) &= A(u(t)) + q(t), \qquad t \in [0,T] \\ u(0) &= c. \end{aligned} \tag{7.1}$$

We will show that consistency and stability, in the sense of Definition 5.10, already suffice to guarantee the convergence of the methods, even in the inhomogeneous case (7.1). Thus it will turn out that a special proof of consistency and stability is not required for inhomogeneous problems.

For this section, let $P(B,T,A)$ be an arbitrary but fixed, properly posed problem (cf. Definition 5.3). The continuous mappings $q\colon [0,T] \to B$, together with the norm

$$\|q\| = \max_{t\in[0,T]} \|q(t)\|$$

form a Banach space $B_T = C^0([0,T],B)$.

Theorem 7.2: Let $c \in B$, $q \in B_T$, $\theta \in [0,1]$, and a consistent and stable difference method $M_D = \{C(h) \mid h \in (0,h_o]\}$ be given. Further let the sequences $h_j \in (0,h_o]$ and $n_j \in \mathbb{N}$, $j = 1(1)\infty$, be such that $n_jh_j \leq T$, $\lim_{j\to\infty} h_j = 0$, and $\lim_{j\to\infty} n_jh_j = \hat{t} \in [0,T]$. Then the solution $u^{(n_j)}$ of the difference equations

$$u^{(\nu)} = C(h_j)(u^{(\nu-1)}) + h_jq(\nu h_j-\theta h_j), \quad \nu = 1(1)n_j$$
$$u^{(0)} = c$$

converges to

$$u(\hat{t}) = E(\hat{t})(c) + \int_0^{\hat{t}} E(\hat{t}-s)(q(s))ds.$$

u is called the *generalized solution* of problem (7.1).

Proof: We restrict ourselves to the cases $\theta = 0$ and $\hat{t} \leq n_jh_j < \hat{t}+h_j$, $j = 1(1)\infty$, and leave the others to the reader. Further, we introduce some notational abbreviations for the purposes of the proof:

$$C = C(h_j), \quad t_\nu = \nu h_j, \quad q_\nu = q(\nu h_j), \quad n = n_j.$$

We will now show the following:

(1) $E(t-s)[q(s)]$ is continuous and bounded on $\{(t,s) \mid t \in [0,T],\ s \in [0,t]\}$.

(2) $\lim_{j\to\infty} \int_0^{\hat t} E(nh_j-s)[q(s)]ds = \int_0^{\hat t} E(\hat t-s)[q(s)]ds.$

(3) For every $\varepsilon > 0$ there exists a $j_0 \in \mathbb{N}$ such that $\|E(t_{n-\nu})(q_\nu) - C^{n-\nu}(q_\nu)\| < \varepsilon \quad j \geq j_0,\ \nu \leq n.$

Proof of (1): Let $\|E(t)\| \leq L$, $t \in [0,T]$. For fixed t and s we consider differences of the form

$$D = E(t-s)[q(s)] - E(\tilde t-\tilde s)[q(\tilde s)]$$
$$= E(t-s)[q(s)] - E(\tilde t-\tilde s)[q(s)] - E(\tilde t-\tilde s)[q(\tilde s)-q(s)].$$

Either

$$E(t-s) = E(\tilde t-\tilde s)\circ E(t-s-\tilde t+\tilde s)$$

or

$$E(\tilde t-\tilde s) = E(t-s)\circ E(\tilde t-\tilde s-t+s).$$

In either case,

$$\|D\| \leq L\,\|E(|t-s-\tilde t+\tilde s|)[q(s)] - q(s)\| + L\,\|q(\tilde s)-q(s)\|.$$

By Theorem 5.5(3), $E(|t-s-\tilde t+\tilde s|)[q(s)]$ is a continuous function of $\tilde s-\tilde t$. Since $q(\tilde s)$ also is continuous, the right side of the inequality converges to zero as $(\tilde t,\tilde s) \to (t,s)$. $E(t-s)[q(s)]$ is also continuous in t and s simultaneously. Since the set $\{(t,s) \mid t \in [0,T],\ s \in [0,t]\}$ is compact, $\|E(t-s)(q(s))\|$ assumes its maximum there.

Proof of (2): By Theorem 4.19(1) we have

$$\int_0^{\hat t} E(nh_j-s)[q(s)]ds = E(nh_j-\hat t)\left[\int_0^{\hat t} E(\hat t-s)[q(s)]ds\right].$$

Since every generalized solution of the homogeneous problem is continuous by Theorem 5.5(3), the conclusion follows at once.

Proof of (3): Let $||E(t)|| \le L$ and $||C(h_j)^n|| \le L$, $t \in [0,T]$, $j = 1(1)\infty$. By the uniform continuity of q, there exists a $\delta > 0$ such that

$$||q(t)-q(\tilde{t})|| < \frac{\varepsilon}{4L}; \quad t,\tilde{t} \in [0,T], \ |t-\tilde{t}| < \delta.$$

Furthermore, there are finitely many μ such that $0 \le \mu\delta \le T$. The finitely many homogeneous initial value problems with initial values $q(\mu\delta)$ can be solved with M_D. Therefore there exists a $j_o \in \mathbb{N}$ such that for $j \ge j_o$ and $\mu\delta \le \hat{t}$

$$||E(\hat{t}-\mu\delta)[q(\mu\delta)] - C^{n-\nu}[q(\mu\delta)]|| < \frac{\varepsilon}{4}.$$

Here the choice of ν depends on μ, so that

$$\nu h_j \le \mu\delta < (\nu+1)h_j.$$

The functions $E(s)[q(\mu\delta)]$ are uniformly continuous in s. Therefore there exists a $\tilde{\delta} > 0$ such that for all t, $\tilde{t}$ with $|t-\tilde{t}| < \tilde{\delta}$

$$||E(t)[q(\mu\delta)] - E(\tilde{t})[q(\mu\delta)]|| < \frac{\varepsilon}{4}.$$

In particular, by choosing a larger j_o if necessary, one can obtain

$$||E(t)[q(\mu\delta)] - E(\hat{t}-\mu\delta)[q(\mu\delta)]|| < \frac{\varepsilon}{4}$$

$$t \in [\hat{t}-\mu\delta-2h_j, \ \hat{t}-\mu\delta+2h_j], \quad j \ge j_o.$$

Since

$$t_{n-\nu} = nh_j-\nu h_j = \hat{t}-\mu\delta+(nh_j-\hat{t}) + (\mu\delta-\nu h_j)$$

we always have

$$||E(t_{n-\nu})[q(\mu\delta)] - E(\hat{t}-\mu\delta)[q(\mu\delta)]|| < \frac{\varepsilon}{4}.$$

Combining all these inequalities, we get

$$\begin{aligned}\|E(t_{n-\nu})(q_\nu) - C^{n-\nu}(q_\nu)\| &\le \|E(t_{n-\nu})(q_\nu) - E(t_{n-\nu})[q(\mu\delta)]\| \\ &+ \|E(t_{n-\nu})[q(\mu\delta)] - E(\hat{t}-\mu\delta)[q(\mu\delta)]\| \\ &+ \|E(\hat{t}-\mu\delta)[q(\mu\delta)] - C^{n-\nu}[q(\mu\delta)]\| \\ &+ \|C^{n-\nu}[q(\mu\delta)] - C^{n-\nu}(q_\nu)\| \\ &\le L\frac{\varepsilon}{4L} + \frac{\varepsilon}{4} + \frac{\varepsilon}{4} + L\frac{\varepsilon}{4L} = \varepsilon.\end{aligned}$$

This completes the proof of (1), (2), and (3).

The solution of the difference equation is

$$u^{(n)} = C^n(c) + h_j \sum_{\nu=1}^{n} C^{n-\nu}(q_\nu).$$

It follows from Theorem 5.11 that

$$\lim_{j\to\infty} C^n(c) = E(t)(c).$$

Because of (2), it suffices to show

$$\lim_{j\to\infty} h_j \sum_{\nu=1}^{n} C^{n-\nu}(q_\nu) = \lim_{j\to\infty} \int_0^t E(nh_j-s)[q(s)]ds,$$

and for that, we use the estimate

$$\begin{aligned}&\|h_j \sum_{\nu=1}^{n} C^{n-\nu}(q_\nu) - \int_0^t E(nh_j-s)[q(s)]ds\| \\ &\qquad \le \|h_j \sum_{\nu=1}^{n} C^{n-\nu}(q_\nu) - h_j \sum_{\nu=1}^{n} E(t_{n-\nu})(q_\nu)\| \\ &\qquad + \|h_j \sum_{\nu=1}^{n} E(t_{n-\nu})(q_\nu) - \int_0^{nh_j} E(nh_j-s)[q(s)]ds\| \\ &\qquad + \|\int_0^{nh_j} E(nh_j-s)[q(s)]ds - \int_0^t E(nh_j-s)[q(s)]ds\| .\end{aligned}$$

The three differences on the right side of the inequality converge separately to zero as $j \to \infty$. For the first difference, this follows from (3); for the second, because it is the difference between a Riemann sum and the corresponding integral (cf. Theorem 4.19(2)). □

The generalized solutions of (7.1) are not necessarily differentiable, and thus are not solutions of (7.1) in each and every case. The solutions obtained are differentiable only if $c \in D_E$ and q are sufficiently "smooth". Exactly what that means will now be made precise.

Definition 7.3: We define

$$D_{\tilde{A}} = \{c \in B \mid u(t) = E(t)(c) \text{ is differentiable for } t = 0\},$$

$$\tilde{A} : D_{\tilde{A}} \to B \text{ given by } \tilde{A}(c) = u'(0). \quad \square$$

Remark: For $c \in D_{\tilde{A}}$, u is differentiable on all of $[0,T]$. For if $t_1 \geq t_2$, then

$$\frac{1}{h}\,[u(t_1)-u(t_2)] = E(t_2)\{\tfrac{1}{h}[E(t_1-t_2)(c)-c]\}. \quad \square$$

There is a simple relationship between A and $\tilde{A}$. Every solution u of $P(B,T,A)$ is also a solution of $P(B,T,\tilde{A})$, i.e.

$$u'(t) = A(u(t)) = \tilde{A}(u(t)).$$

In passing from $P(B,T,A)$ to $P(B,T,\tilde{A})$, we cannot lose any solutions, though we may potentially gain some. The space on which the operators $E_0(t)$ are defined may be enlarged under some circumstances. The operators $E(t)$, however, remain unchanged. Also nothing is changed insofar as the stability, consistency, and convergence properties of the difference methods are concerned. It can be shown that $\tilde{A}$ is a closed mapping, i.e., that the graph of $\tilde{A}$ in $B \times B$ is closed. This implies that $A = \tilde{A}$ whenever A is closed to begin with. Since we shall not use this fact, we won't comment on the proof [but see Richtmyer-Morton (1967), 3.6 and Yosida (1968), Ch. IX]. In our examples in Section 5, A is always closed.

Theorem 7.4: Let $\tilde{q} \in B_T$ and

$$q(t) = \int_0^\infty \phi(r)E(r)[\tilde{q}(t)]dr$$

where $\phi \in C^\infty(\mathbb{R}, \mathbb{R})$ and $\text{Support}(\phi) \subset (0,T)$. Then

$$\{\int_0^t E(t-s)[q(s)]ds\}' = q(t) + \tilde{A}\{\int_0^t E(t-s)[q(s)]ds\}.$$

Remark: q is called a regularization of $\tilde{q}$. It can be shown that with the proper choice of ϕ, q and $\tilde{q}$ differ arbitrarily little. For $c \in D_E$,

$$u(t) = E(t)(c) + \int_0^t E(t-s)[q(s)]ds$$

is obviously a solution of

$$u'(t) = \tilde{A}(u(t)) + q(t), \quad t \in [0,T]$$
$$u(0) = c. \qquad \square$$

Proof of Theorem 7.4: For $t \in [T,2T]$ define

$$E(t) = E(t/2)\circ E(t/2).$$

There exists an $\varepsilon > 0$ such that $\text{Support}(\phi) \subset [2\varepsilon,T-2\varepsilon]$. Let

$$f(t) = \int_0^t E(t-s)[q(s)]ds = \int_0^t \int_\varepsilon^{T-\varepsilon} \phi(r)E(t+r-s)[\tilde{q}(s)]drds.$$

For $|h| < \varepsilon$ and $t+h \geq 0$ we obtain

$$f(t+h) = I_1(h) + I_2(h)$$

$$I_1(h) = \int_0^t \int_\varepsilon^{T-\varepsilon} \phi(r)E(t+h+r-s)[\tilde{q}(s)]drds$$

$$I_2(h) = \int_t^{t+h} \int_\varepsilon^{T-\varepsilon} \phi(r)E(t+h+r-s)[\tilde{q}(s)]drds.$$

We make the substitution $\tilde{r} = r+h$ and exchange the order of integration (cf. Theorem 4.19(5)), so that

$$I_1(h) = \int_{\varepsilon+h}^{T+h-\varepsilon} \phi(\tilde{r}-h)\int_0^t E(t+\tilde{r}-s)[\tilde{q}(s)]dsd\tilde{r}$$

$$I_1(h) = \int_0^T \phi(\tilde{r}-h)\int_0^t E(t+\tilde{r}-s)[\tilde{q}(s)]dsd\tilde{r}.$$

I_1 has the derivative

$$I_1'(0) = -\int_0^T \phi'(\tilde{r})\int_0^t E(t+\tilde{r}-s)[\tilde{q}(s)]dsd\tilde{r}.$$

The second integral we split one more time, to get

$$I_2(h) = \int_t^{t+h} \int_{\varepsilon+h}^{T+h-\varepsilon} \phi(\tilde{r})E(t+\tilde{r}-s)[\tilde{q}(s)]d\tilde{r}ds$$

$$+ \int_t^{t+h}\int_{\varepsilon+h}^{T+h-\varepsilon} [\phi(\tilde{r}-h)-\phi(\tilde{r})]E(t+\tilde{r}-s)[\tilde{q}(s)]d\tilde{r}ds.$$

In the first summand we can again change the limits of the innermost integral to 0 and T. The integrand then no longer depends on h. The summand obviously can be differentiated with respect to h. Since $|\phi'(r)|$ is bounded, the second summand is of order $O(|h|^2)$, and hence differentiable for $h = 0$. It follows that

$$I_2'(0) = \int_0^T \phi(\tilde{r})E(\tilde{r})[\tilde{q}(t)]d\tilde{r} = q(t)$$

$$f'(t) = I_1'(0) + I_2'(0) = q(t) - \int_0^T \phi'(\tilde{r})\int_0^t E(t+\tilde{r}-s)[\tilde{q}(s)]dsd\tilde{r}.$$

Now let

$$\begin{aligned} g(h) &= E(h)\left(\int_0^t E(t-s)[q(s)]ds\right) \\ &= \int_0^t\int_\varepsilon^{T-\varepsilon} \phi(r)E(t+h+r-s)[\tilde{q}(s)]drds = I_1(h) \end{aligned}$$

$$g'(0) = I_1'(0).$$

Therefore

$$\int_0^t E(t-s)[q(s)]ds \in D_{\tilde{A}}$$

and

$$\tilde{A}\left(\int_0^t E(t-s)[q(s)]ds\right) = -\int_0^T \phi'(\tilde{r})\int_0^t E(t+\tilde{r}-s)[\tilde{q}(s)]dsd\tilde{r}. \qquad \square$$

8. Difference methods with positivity properties

The literature contains various (inequivalent!) definitions of difference methods of *positive type* (cf., e.g., Friedrichs 1954, Lax 1961, Collatz 1966, Törnig-Ziegler 1966). The differences arise because some consider methods in function spaces with the maximum norm, and others in function spaces with the L^2-norm. A number of classical methods fit into both categories. We will distinguish the two by referring to *positive* difference methods in the first case, and to *positive definite* difference methods in the second.

In the hyberbolic case, with a few unimportant exceptions, even if the initial value problem under consideration has a C^∞-solution, these methods all converge only to first order (cf. Lax 1961). However, they allow for very simple error estimates and they can be carried over to nonlinear problems with relative ease.

We consider positive difference methods primarily on the following vector spaces:

$$B_{1n} = \{f \in C^0(\mathbb{R},\mathbb{C}^n) \mid \lim_{x\to\infty} ||f(x)||_\infty = 0\}$$

$$B_{2n} = \{f \in C^0(\mathbb{R},\mathbb{C}^n) \mid f \ 2\pi\text{-periodic}\}$$

$$B_{3n} = \{f \in B_{2n} \mid f(x) = f(-x), \quad x \in \mathbb{R}\}$$

$$B_{4n} = \{f \in B_{2n} \mid f(x) = -f(-x), \quad x \in \mathbb{R}\}$$

$$B_{5n} = \{f \in C^0([-\pi/2,3\pi/2],\mathbb{C}^n) \mid f \text{ satisfies the equations in Def. 8.1}\}.$$

Definition 8.1: *Functional equations.*

$$\left.\begin{aligned}(\beta_o + \alpha_o x)f(-x) &= (\beta_0 - \alpha_0 x)f(x)\\(\beta_\pi + \alpha_\pi x)f(\pi+ x) &= (\beta_\pi - \alpha_\pi x)f(\pi- x).\end{aligned}\right\} \quad x \in (0,\pi/2)$$

Here we let α_0, β_0, α_π, β_π be real, nonnegative constants with $\alpha_0+\beta_0 > 0$ and $\alpha_\pi+\beta_\pi > 0$. □

B_{5n} naturally depends on the given constants. Since we shall think of these as fixed for the duration, we suppress them from the notation and use the abbreviation B_{5n}. Because of the various functional equations, the functions in B_{2n} through B_{5n} are determined by their values on a part of the interval of definition. Therefore we define the *primary intervals* to be

$$G_1 = \mathbb{R}, \quad G_2 = [-\pi,\pi], \quad G_3 = G_4 = G_5 = [0,\pi]$$

and the *intervals of definition* to be

$$D_1 = D_2 = D_3 = D_4 = \mathbb{R}, \quad D_5 = [-\pi/2, 3\pi/2].$$

For $\mu = 1(1)5$ we obviously have

$$\sup_{x\varepsilon D_\mu} \|f(x)\|_\infty = \max_{x\varepsilon G_\mu} \|f(x)\|_\infty, \qquad f \varepsilon B_{\mu n}.$$

These suprema are norms in $B_{\mu n}$, which we write $\|f\|_\infty$. The spaces $B_{\mu n}$ become Banach spaces by virtue of these norms and are subspaces of $C^0(D_\mu, ¢^n)$. They are useful for the study of functions on G_μ which satisfy certain boundary conditions. Thus if $f \varepsilon B_{\mu n} \cap C^1(D_\mu, ¢^n)$ we have the following relations, by cases:

$\mu = 2$: $f(-\pi) = f(\pi)$ and $f'(-\pi) = f'(\pi)$

$\mu = 3$: $f'(0) = f'(\pi) = 0$

$\mu = 4$: $f(0) = f(\pi) = 0$

$\mu = 5$: $\alpha_0 f(0) - \beta_0 f'(0) = \alpha_\pi f(\pi) + \beta_\pi f'(\pi) = 0.$

Obviously the boundary conditions for $\mu = 3$ and $\mu = 4$ are special cases of the boundary conditions for $\mu = 5$. The spaces B_{3n} and B_{4n} also become special cases of B_{5n} if we restrict the intervals of definition D_μ to $[-\pi/2, 3\pi/2]$.

Lemma 8.2: Let $g \varepsilon B_{5n} \cap C^1(D_5, ¢^n)$, $f \varepsilon C^3(D_5, ¢^n)$ and $g(x) = f(x)$ for all $x \varepsilon G_5$. Further let $\beta_0 \neq 0$ or $f''(0) = 0$ and simultaneously, $\beta_\pi \neq 0$ or $f''(\pi) = 0$. Then $g \varepsilon C^2(D_5, ¢^n)$ and

$$\sup_{x\varepsilon[-\Delta x, \pi+\Delta x]} \|f(x) - g(x)\|_\infty = O((\Delta x)^3), \qquad \Delta x \varepsilon (0, \pi/2).$$

Proof: $g \varepsilon B_{5n} \cap C^1(D_5, ¢^n)$ implies

$$\alpha_0 g(0) - \beta_0 g'(0) = \alpha_0 f(0) - \beta_0 f'(0) = 0$$

$$\alpha_\pi g(\pi) + \beta_\pi g'(\pi) = \alpha_0 f(\pi) + \beta_\pi f'(\pi) = 0.$$

Furthermore, by 8.1

$$g(x) = \begin{cases} f(x) & \text{for } x \varepsilon [0,\pi] \\ f(-x)(\beta_0+\alpha_0 x)/(\beta_0-\alpha_0 x) & \text{for } x \varepsilon [-\pi/2,0] \\ f(2\pi-x)[\beta_\pi-\alpha_\pi(x-\pi)]/[\beta_\pi+\alpha_\pi(x-\pi)] & \text{for } x \varepsilon [\pi,3\pi/2]. \end{cases}$$

First let $\beta_0 \neq 0$. For $x < 0$ we obtain

$$g'(x) = 2\alpha_0\beta_0 f(-x)/(\beta_0-\alpha_0 x)^2 - f'(-x)(\beta_0+\alpha_0 x)/(\beta_0-\alpha_0 x)$$

$$g''(x) = 4\alpha_0^2\beta_0 f(-x)/(\beta_0-\alpha_0 x)^3 - 4\alpha_0\beta_0 f'(-x)/(\beta_0-\alpha_0 x)^2$$
$$+ f''(-x)(\beta_0+\alpha_0 x)/(\beta_0-\alpha_0 x)$$

$$g''(-0) = 4\alpha_0^2 f(0)/\beta_0^2 - 4\alpha_0 f'(0)/\beta_0 + f''(0)$$
$$= 4\alpha_0[\alpha_0 f(0) - \beta_0 f'(0)]/\beta_0^2 + f''(0)$$
$$= f''(0) = g''(+0).$$

In the exceptional case $\beta_0 = 0$, we have for $x < 0$ that

$$g(x) = -f(-x)$$
$$g''(x) = -f''(-x)$$
$$g''(-0) = -f''(0) = 0 = f''(0) = g''(+0).$$

That $g''(\pi+0) = g''(\pi-0)$ is shown similarly. Thus $g \,\varepsilon\, C^2(D_5,\mathbb{C}^n)$. Since the restriction of g to $[-\pi/2,0]$ is three times continuously differentiable, the Taylor series for $x \,\varepsilon\, [-\Delta x,0]$ becomes

$$f(x) = f(0) + xf'(0) + \frac{x^2}{2}f''(0) + \frac{x^3}{6}f'''(\theta_1 x)$$
$$g(x) = g(0) + xg'(0) + \frac{x^2}{2}g''(0) + \frac{x^3}{6}g'''(\theta_2 x) \qquad \theta_1,\theta_2 \,\varepsilon\, (0,1).$$

Since $f(0) = g(0)$, $f'(0) = g'(0)$, and $f''(0) = g''(0)$, we obtain

$$\|f(x) - g(x)\|_\infty = \frac{|x|^3}{6} \|f'''(\theta_1 x)-g'''(\theta_2 x)\|_\infty \le M(\Delta x)^3.$$

M depends only on f, α_0, and β_0. There is an analogous estimate for $x \in (\pi,\pi+\Delta x]$. Combining the two, one obtains a bound which depends only on f, α_0, β_0, α_π, and β_π. □

Lemma 8.3: The following vector spaces are dense in the spaces $B_{\mu n}$.

(1) $C_0^\infty(\mathbb{R},\mathbb{C}^n)$ in B_{1n}.

(2) $C^\infty(\mathbb{R},\mathbb{C}^n) \cap B_{\mu n}$ in $B_{\mu n}$ for $\mu = 2,3,4$.

(3) $\{f \in B_{5n} | f \in C^1(D_5,\mathbb{C}^n)$ and $f_{/G_5} \in C^\infty(G_5,\mathbb{C}^n)\}$ in B_{5n}.

The proofs are left to the reader (cf. Theorem 4.8).

Definition 8.4: Let G, D be real intervals with $G \subset D$, and let $B \subset C^0(D,\mathbb{C}^n)$ be a Banach space with

$$\max_{x\in G} \|f(x)\|_\infty = \sup_{x\in D} \|f(x)\|_\infty, \qquad f \in B$$

and $P(B,T,A)$ a properly posed initial value problem with $M_D = \{C(h) | h \in (0,h_0]\}$ a given difference method. The method is called *positive* if it satisfies conditions (1) through (4) below. We assume suitable choices of

$k \in \mathbb{N}$; $\lambda,K,L \in \mathbb{R}_+$

$E,A_\nu,B_\nu \in C^0(G \times (0,h_0],MAT(n,n,\mathbb{R}))$, $\nu = -k(1)k$

$M \in C^0(D,MAT(n,n,\mathbb{R}))$.

(1) For all $h \in (0,h_0]$, $f \in B$, and $x \in G$, with $g = C(h)(f)$ and $\Delta x = h/\lambda$ or $\Delta x = \sqrt{h/\lambda}$, we have

$$E(x,h)g(x) = \sum_{\nu=-k}^{k} [A_\nu(x,h)f(x+\nu\Delta x)+B_\nu(x,h)g(x+\nu\Delta x)]$$

(2) $E(x,h) = \sum_{\nu=-k}^{k} [A_\nu(x,h) + B_\nu(x,h)]$

(3) $M(x)$ is always regular. $f(\cdot) \to M(\cdot)f(\cdot)$ is a mapping of B into itself. For every fixed $x \in G$, the similarity transformation $N \to M(x)^{-1}NM(x)$, $N \in MAT(n,n,\mathbb{R})$, carries all matrices $A_\nu(x,h)$ and $B_\nu(x,h)$ into *diagonal matrices* with *nonnegative* elements. The image of the matrix $\Sigma A_\nu(x,h)$ has diagonal elements which are greater than or equal to 1.

(4) The following inequalities hold:

$$\|A_\nu(x,h)\|_\infty \le K, \quad \|B_\nu(x,h)\|_\infty \le K, \quad x \in G, h \in (0,h_0]$$

$$\|M(x)\|_\infty \le K, \quad \|M(x)^{-1}\|_\infty \le K, \quad x \in D.$$

In the case $\Delta x = h/\lambda$, let $\|M(x) - M(y)\|_\infty \le L[k(k+1)]^{-1}|x-y|$; otherwise, let $M \equiv$ *constant*. □

Condition (2) is not as restrictive as it may appear at first glance, for the relation

$$\Sigma[A_\nu(x,h) + B_\nu(x,h)] = E(x,h) + O(h).$$

is valid for every at least first order consistent method. If $Q(h)$ is a suitably bounded operator, then for $C(h) - hQ(h)$ we have

$$\Sigma[A_\nu(x,h) + B_\nu(x,h)] = E(x,h).$$

Now by the Kreiss perturbation theorem (Theorem 5.13), the methods $C(h)$ and $C(h) - hQ(h)$ are either both stable or both unstable.

Condition (3) demands that all diagonal elements of

$$M(x)^{-1}[\Sigma A_\nu(x,h)]M(x)$$

be greater than or equal to 1. Since the equations in (1)

can be multiplied by any arbitrary positive constant without affecting any of the remaining properties, 1 may be replaced by any arbitrary positive constant, independent of x or h.

Theorem 8.5: A positive difference method $C(h)$ is stable. Furthermore, when M = constant,

$$\|C(h)^m\| \le \|M\|_\infty \|M^{-1}\|_\infty, \quad h \in (0,h_o],\ m \in \mathbb{N},\ mh \le T.$$

Proof: Let $h \in (0,h_o]$ be arbitrary but fixed, let $f \in B$ and $g = C(h)(f)$. Define

$$\tilde{f}(x) = M(x)^{-1}f(x) = (\tilde{f}_1(x), \ldots, \tilde{f}_n(x))$$
$$\tilde{g}(x) = M(x)^{-1}g(x) = (\tilde{g}_1(x), \ldots, \tilde{g}_n(x)).$$

$\tilde{f}$ and $\tilde{g}$ are related as follows:

$$\tilde{g} = [M(x)^{-1} \circ C(h) \circ M(x)](\tilde{f}) = \tilde{C}(h)(\tilde{f}).$$

Step 1: $\tilde{C}(h)$ is stable. For $x \in G$, it follows from (1) that

$$E(x,h)M(x)\tilde{g}(x) = \Sigma A_\nu(x,h)M(x)\tilde{f}(x+\nu\Delta x) + \Sigma B_\nu(x,h)M(x)\tilde{g}(x+\nu\Delta x) + R_1 + R_2$$

$$R_1 = \Sigma A_\nu(x,h)[M(x+\nu\Delta x)-M(x)]\tilde{f}(x+\nu\Delta x)$$

$$R_2 = \Sigma B_\nu(x,h)[M(x+\nu\Delta x)-M(x)]\tilde{g}(x+\nu\Delta x).$$

Multiply by $M(x)^{-1}$, write the equation in component form, and apply conditions (2) and (3) to obtain, for $j = 1(1)n$,

$$\varepsilon_j(x,h)\tilde{g}_j(x) = \Sigma[\alpha_{\nu j}(x,h)\tilde{f}_j(x+\nu\Delta x)+\beta_{\nu j}(x,h)\tilde{g}(x+\nu\Delta x)]+r_{1j}+r_{2j}$$

$$\mathrm{diag}(\alpha_{\nu j}(x,h)) = M(x)^{-1}A_\nu(x,h)M(x)$$

$$\mathrm{diag}(\beta_{\nu j}(x,h)) = M(x)^{-1}B_\nu(x,h)M(x)$$

$$\varepsilon_j(x,h) = \Sigma[\alpha_{\nu j}(x,h)+\beta_{\nu j}(x,h)].$$

r_{1j} and r_{2j} are the j-th components of $M(x)^{-1}R_1$ and $M(x)^{-1}R_2$. They are, of course, dependent on x and h. The numbers ε_j, $\alpha_{\nu j}$, and $\beta_{\nu j}$ are nonnegative, by (3). Now there exist x and j such that $\|\tilde{g}\| = |\tilde{g}_j(x)|$. Apply the triangle inequality and the estimates given by (4) to obtain

$$\varepsilon_j(x,h)\ \|\tilde{g}\| \leq \|\tilde{f}\|\ \Sigma\alpha_{\nu j}(x,h)$$
$$+ \|\tilde{g}\|\ \Sigma\ \beta_{\nu j}(x,h) + K^2L\lambda^{-1}h(\|\tilde{f}\| + \|\tilde{g}\|).$$

The last summand does not appear when $M = \textit{constant}$. Letting

$$\tilde{K} = \begin{cases} K^2L\lambda^{-1} & \text{for } M \not\equiv \text{const.} \\ 0 & \text{otherwise} \end{cases}$$

we have

$$[\Sigma\ \alpha_{\nu j}(x,h) - \tilde{K}h]\ \|\tilde{g}\| \leq [\Sigma\alpha_{\nu j}(x,h) + \tilde{K}h]\ \|\tilde{f}\|\ .$$

By (3), $\Sigma\alpha_{\nu j}(x,h) \geq 1$. Since the function $(z-\tilde{K}h)/(z+\tilde{K}h)$ is monotone increasing for $z \geq 1$, we have

$$(1-\tilde{K}h)\ \|\tilde{g}\| \leq (1+\tilde{K}h)\ \|\tilde{f}\|.$$

Without loss of generality, we assume that $1-\tilde{K}h_0 > 0$. Then there is a $\tilde{\tilde{K}} \geq 0$ such that

$$\frac{1+\tilde{K}h}{1-\tilde{K}h} \leq \exp(\tilde{\tilde{K}}h), \qquad h \in (0,h_0].$$

It follows that

$$\|\tilde{C}(h)\| \leq \exp(\tilde{\tilde{K}}h)$$
$$\|\tilde{C}(h)^m\| \leq \|\tilde{C}(h)\|^m \leq \exp(\tilde{\tilde{K}}\ m\ h).$$

Thus, for all m such that $mh \leq T$, we have

$$\|\tilde{C}(h)^m\| \leq \exp(\tilde{\tilde{K}}T).$$

Hence $\tilde{C}(h)$ is stable. Should $\tilde{K} = \tilde{\tilde{K}} = 0$, $\|\tilde{C}(h)^m\| \leq 1$.

Step 2: C(h) is stable.

$$\|C(h)^m\| = \|M(x)\circ[M(x)^{-1}\circ C(h)\circ M(x)]^m\circ M(x)^{-1}\|$$
$$= \|M(x)\circ\tilde{C}(h)^m\circ M(x)^{-1}\| \leq \|M(x)\|\,\|\tilde{C}(h)^m\|\,\|M(x)^{-1}\|$$
$$\leq K^2\,\|\tilde{C}(h)^m\|\,.$$

Hence C(h) is stable also. □

Example 8.6: *Positive difference methods for a parabolic initial boundary value problem with mixed boundary conditions.* Let $a \in C^\infty(\mathbb{R}, \mathbb{R}_+)$ be 2π-periodic with $a(-x) = a(x)$, $x \in \mathbb{R}$, and let α_o, β_o, α_π, and β_π be chosen as in Definition 8.1. Consider the problem

$$\begin{aligned} u_t(x,t) &= [a(x)u_x(x,t)]_x - q(x,t), && x \in (0,\pi),\ t \in (0,T) \\ u(x,0) &= \phi(x), && x \in [0,\pi] \\ \alpha_o u(0,t) - \beta_o u_x(0,t) &= 0 && \\ \alpha_\pi u(\pi,t) + \beta_\pi u_x(\pi,t) &= 0 && t \in [0,T]. \end{aligned}$$

The restriction to homogeneous boundary conditions is not a true restriction since one can always use the substitution

$$v(x,t) = u(x,t) + \delta_o + \delta_1 x + \delta_2 x^2$$

to change the inhomogeneous boundary conditions

$$\begin{aligned} \alpha_o u(0,t) - \beta_o u_x(0,t) &= \gamma_o \\ \alpha_\pi u(\pi,t) + \beta_\pi u_x(\pi,t) &= \gamma_\pi \end{aligned}$$

into the homogeneous boundary conditions

$$\begin{aligned} \alpha_o v(0,t) - \beta_o v_x(0,t) &= 0 \\ \alpha_\pi v(\pi,t) + \beta_\pi v_x(\pi,t) &= 0 \end{aligned}$$

One obtains δ_o, δ_1, and δ_2 as solutions of the system of equations

$$\begin{pmatrix} -\alpha_0 & \beta_0 & 0 \\ -\alpha_\pi & -\alpha_\pi \pi - \beta_\pi & -\alpha_\pi \pi^2 - 2\beta_\pi \pi \end{pmatrix} \begin{pmatrix} \delta_0 \\ \delta_1 \\ \delta_2 \end{pmatrix} = \begin{pmatrix} \gamma_0 \\ \gamma_\pi \end{pmatrix}.$$

Since the system is underdetermined, we add the conditions

$$\delta_2 = 0 \quad \text{if} \quad \alpha_0 + \alpha_\pi > 0$$
$$\delta_0 = 0 \quad \text{if} \quad \alpha_0 = \alpha_\pi = 0.$$

The inhomogeneous term $q(x,t)$ naturally must be transformed accordingly.

The Banach space $B = B_{51}$ suits the above problem. We choose

$$D_A = \{f \in B \mid f_{/G_5} \in C^2(G_5,\mathbb{C})\}$$

and define $A : D_A \to B$ by the assignment

$$f(x) \to [a(x)f'(x)]', \qquad x \in G_5.$$

If we choose $\phi \in D_E$, where

$$D_E = \{f \in B \mid f \in C^1(D_5,\mathbb{C}),\ f_{/G_5} \in C^\infty(G_5,\mathbb{C})\},$$

the initial value problem for $q = 0$ is uniquely solvable in the classical sense and the solution u belongs to $C^\infty(G_5 \times [0,T],\mathbb{C})$. It follows from Section 7 that the restriction to the homogeneous case $q \equiv 0$ is not a true restriction.

With the help of Theorem 8.5 it is now very easy to establish stability for the difference method (6.3). We need only check for which λ the method is *positive*. To do this, we rewrite (6.3) appropriately:

$$g(x)+(1-\alpha)\lambda[a(x+\Delta x/2)+a(x-\Delta x/2)]g(x)$$
$$= \alpha\lambda a(x-\Delta x/2)f(x-\Delta x)+\{1-\alpha\lambda[a(x+\Delta x/2)+a(x-\Delta x/2)]\}f(x) +$$

$$\alpha\lambda a(x+\Delta x/2)f(x+\Delta x)+(1-\alpha)\lambda a(x-\Delta x/2)g(x-\Delta x) + (1-\alpha)\lambda a(x+\Delta x/2)g(x+\Delta x),$$

$\lambda = h/(\Delta x)^2$.

Conditions (1) and (2) are satisfied for all λ independently of α. To establish (3), we choose $M \equiv 1$. The only condition that depends on λ and α turns out to be

$$1 - \alpha\lambda[a(x+\Delta x/2) + a(x-\Delta x/2)] \geq 0.$$

This is satisfied for $\alpha = 0$ (totally implicit case, method of Laasonen) for all $\lambda \in \mathbb{R}_+$. For $0 < \alpha \leq 1$, the *positivity condition*

$$0 < \lambda \cdot \max_{x\in[0,\pi]} a(x) \leq 1/2\alpha$$

results. For $\alpha = 0$ and $\alpha = 1$, this corresponds to the conclusions of Theorem 6.5 and Theorem 6.6. Note, however, that in Section 6 we used the L^2-norm $\|\cdot\|_2$. Condition (4) is satisfied because $a(x)$ is a bounded function. Consistency is established with the help of Lemma 8.2. In the most general case, the truncation error of the discretization is $O(h) + O(\Delta x)$. The details we leave to the reader. □

Example 8.7: *A positive difference method of consistency order* $O(h^2)$ *for a parabolic initial value problem.* Let $a \equiv$ constant > 0. Consider the problem

$$\begin{aligned} u_t(x,t) &= au_{xx}(x,t) - q(x,t) \\ u(x,0) &= \phi(x) \end{aligned} \qquad x \in \mathbb{R}, \quad t \in (0,T).$$

It can be formulated as a properly posed initial value problem in the space $B = B_{1n}$. The following discretization is the foundation of a difference method of consistency order $O(h^2)$:

$$\frac{1}{12}\,h^{-1}[u(x+\Delta x,t+h) - u(x+\Delta x,t)] +$$
$$\frac{5}{6}\,h^{-1}[u(x,t+h) - u(x,t)] +$$
$$\frac{1}{12}\,h^{-1}[u(x-\Delta x,t+h) - u(x-\Delta x,t)] =$$
$$\frac{1}{2}\,(\Delta x)^{-2}a[u(x+\Delta x,t+h) - 2u(x,t+h) + u(x-\Delta x,t+h)] +$$
$$\frac{1}{2}\,(\Delta x)^{-2}a[u(x+\Delta x,t) - 2u(x,t) + u(x-\Delta x,t)] -$$
$$[\tfrac{1}{12}\,q(x+\Delta x,t+h/2) + \tfrac{5}{6}q(x,t+h/2) + \tfrac{1}{12}\,q(x-\Delta x,t+h/2)]+R(h,\Delta x).$$

We first show that $R(h,\Delta x) = O(h^2) + O((\Delta x)^4)$. Using the abbreviation $\tilde{t} = t+h/2$ and the error estimates

$$h^{-1}[u(y,t+h)-u(y,t)] = u_t(y,\tilde{t}) + O(h^2)$$
$$(\Delta x)^{-2}[u(x+\Delta x,s)-2u(x,s)+u(x-\Delta x,s)] = u_{xx}(x,s)$$
$$+ \frac{1}{12}\,(\Delta x)^2 u_{xxxx}(x,s) + O((\Delta x)^4)$$

and the differential equation, it follows from the above discretization that

$$\frac{1}{12}\,u_t(x+\Delta x,\tilde{t}) + \frac{5}{6}u_t(x,\tilde{t}) + \frac{1}{12}u_t(x-\Delta x,\tilde{t}) + O(h^2)$$
$$= \frac{1}{12}au_{xx}(x+\Delta x,\tilde{t}) + \frac{5}{6}au_{xx}(x,\tilde{t}) + \frac{1}{12}au_{xx}(x-\Delta x,\tilde{t}) + O(h^2)$$
$$= \frac{1}{2}a\ \{u_{xx}(x,t+h)+u_{xx}(x,t) + \frac{1}{12}(\Delta x)^2[u_{xxxx}(x,t+h)+u_{xxxx}(x,t)]\}$$
$$+ O((\Delta x)^4).$$

Expanding further,

$$\frac{1}{12}\,a[u_{xx}(x+\Delta x,\tilde{t}) + u_{xx}(x-\Delta x,\tilde{t})]$$
$$= \frac{1}{6}au_{xx}(x,\tilde{t}) + \frac{1}{12}(\Delta x)^2 au_{xxxx}(x,\tilde{t}) + O((\Delta x)^4),$$
$$\frac{1}{2}a[u_{xx}(x,t+h)+u_{xx}(x,t)] = au_{xx}(x,\tilde{t}) + O(h^2),$$
$$\frac{1}{24}(\Delta x)^2 a[u_{xxxx}(x,t+h)+u_{xxxx}(x,t)] = \frac{1}{12}(\Delta x)^2 au_{xxxx}(x,\tilde{t})$$
$$+ O(h^2)O((\Delta x)^2).$$

The conclusion now follows by simple substitution.

The difference method for the homogeneous differential equation, when expressed in the notation of Definition 8.4, reads

$$\begin{aligned}(\tfrac{5}{6}+\lambda a)g(x) = (\tfrac{1}{12} + \tfrac{1}{2}\lambda a)f(x-\Delta x) + (\tfrac{5}{6}-\lambda a)f(x) \\ + (\tfrac{1}{12} + \lambda\tfrac{1}{2}\, a)f(x+\Delta x) + (-\tfrac{1}{12} + \tfrac{1}{2}\lambda a)g(x-\Delta x) \\ + (-\tfrac{1}{12} + \tfrac{1}{2}\lambda a)g(x+\Delta x),\end{aligned}$$

where $\lambda = h/(\Delta x)^2$. Conditions (1), (2), and (4) are satisfied for all λ. To establish (3), we choose $M \equiv 1$. We obtain the *positivity condition*

$$\tfrac{1}{6} \le \lambda a \le \tfrac{5}{6}.$$

Thus the difference method is stable for these λ, and since $\lambda = h/(\Delta x)^2$, it has consistency order $O(h^2)$.

In the inhomogeneous case ($q \not\equiv 0$) we should, according to Section 7, add the term

$$-h_j q(x,\nu h_j - \theta h_j), \qquad \theta \in [0,1].$$

But this would reduce the consistency order of the method to $O(h)$, even for $\theta = 1/2$. Therefore, we add a term

$$\begin{aligned}-h_j[\tfrac{1}{12}q(x+\Delta x,(\nu-1/2)h_j) + \tfrac{5}{6}q(x,(\nu-1/2)h_j) \\ + \tfrac{1}{12}\, q(x-\Delta x,(\nu-1/2)h_j)].\end{aligned}$$

corresponding to our discretization. The dependencies of the difference method are depicted in Figure 8.8. □

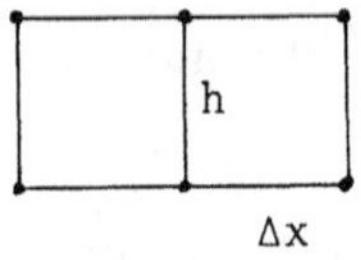

Figure 8.8

Example 8.9: *A positive difference method for a hyperbolic initial value problem.* Let $A \in C^\infty(\mathbb{R}, \mathrm{MAT}(n,n,\mathbb{R}))$ be real diagonalizable with bounded norm. Let $M(x)$ be the matrix which diagonalizes $A(x)$ and let it, together with its inverse, have bounded norm, and let $\|M(x)-M(y)\|_\infty \le L|x-y|$, $x,y \in \mathbb{R}$. It follows that $\rho(A(x))$ is also bounded. Consider the problem

$$\begin{aligned} u_t(x,t) &= A(x)u_x(x,t) + q(x,t) \\ u(x,0) &= \phi(x) \end{aligned} \qquad x \in \mathbb{R}, \quad t \in (0,T).$$

It can be formulated in the usual way as a properly posed initial value problem in the space $B = B_{1n}$.

The *Friedrichs* method (cf. Theorem 6.8) now becomes

$$C(h) = \tfrac{1}{2}\{[I-\lambda A(x)]T_{\Delta x}^{-1} + [I+\lambda A(x)]T_{\Delta x}\}, \quad \lambda = h/\Delta x.$$

We transform it into the notation of Definition 8.4:

$$g(x) = \tfrac{1}{2}[I-\lambda A(x)]f(x-\Delta x) + \tfrac{1}{2}[I+\lambda A(x)]f(x+\Delta x).$$

Conditions (1), (2), and (4) are satisfied for all $\lambda \in \mathbb{R}_+$. Condition (3) results in the *positivity condition*

$$0 < \lambda \sup_{x \in \mathbb{R}} \rho(A(x)) \le 1.$$

The *Courant-Isaacson-Rees* method (cf. Theorem 6.9) becomes

$$g(x) = \lambda A^-(x)f(x-\Delta x) + \{I-\lambda[A^+(x)+A^-(x)]\}f(x) + \lambda A^+(x)f(x+\Delta x), \quad \lambda = h/\Delta x.$$

Here

$$\begin{aligned} A(x) &= A^+(x) - A^-(x) \\ A^+(x) &= M(x)D^+(x)M(x)^{-1} \\ A^-(x) &= M(x)D^-(x)M(x)^{-1} \\ D^+(x) &= \mathrm{diag}(\max\{\lambda_i(x),0\}) \\ D^-(x) &= \mathrm{diag}(-\min\{\lambda_i(x),0\}) \end{aligned}$$

where $\lambda_i(x)$ are the eigenvalues of $A(x)$ with $\lambda_1(x) \leq \lambda_2(x) \leq \dots \leq \lambda_n(x)$. Conditions (1), (2), and (4) again are satisfied for all $\lambda \in \mathbb{R}_+$. To establish (3), we have to show that the diagonal matrices

$$\begin{aligned} M(x)^{-1}A^-(x)M(x) &= D^-(x) \\ M(x)^{-1}A^+(x)M(x) &= D^+(x) \\ M(x)^{-1}\{I-\lambda[A^+(x)+A^-(x)]\}M(x) &= \mathrm{diag}(1-\lambda|\lambda_i(x)|) \end{aligned}$$

have only nonnegative elements. For $D^-(x)$ and $D^+(x)$ this holds by definition. For the third matrix, we must have

$$0 < \lambda \sup_{x \in \mathbb{R}} \rho(A(x)) \leq 1 \qquad \textit{(positivity condition).} \qquad \square$$

Example 8.10: *Initial boundary value problem for the wave equation with variable coefficients.* Let $a \in C^\infty(\mathbb{R},\mathbb{R}_+)$ be 2π-periodic with $a(-x) = a(x)$, $x \in \mathbb{R}$. Consider the problem

$$\begin{aligned} u_{tt}(x,t) &= a(x)[a(x)u_x(x,t)]_x - q(x,t), && x \in (0,\pi),\ t \in (0,T) \\ u(0,t) &= u(\pi,t) = 0, && t \in [0,T] \\ u(x,0) &= \phi(x),\ u_t(x,0) = \psi(x), && x \in [0,\pi]. \end{aligned}$$

The equation can be rewritten as a first order system. The substitution

$$v_1(x,t) = a(x)u_x(x,t)$$
$$v_2(x,t) = u_t(x,t)$$

yields

$$\frac{\partial v_1}{\partial t}(x,t) = a(x)\frac{\partial v_2}{\partial x}(x,t)$$
$$\frac{\partial v_2}{\partial t}(x,t) = a(x)\frac{\partial v_1}{\partial x}(x,t) - q(x,t) \qquad x \in (0,\pi), \quad t \in (0,T)$$
$$v_2(0,t) = v_2(\pi,t) = 0, \qquad t \in [0,T]$$
$$v_1(x,0) = a(x)\phi'(x), \; v_2(x,0)=\psi(x), \; x \in [0,\pi].$$

In contrast to Problem 1.8, the boundary values are specified for only one component, namely v_2. The other component is free. However, in the homogeneous case $q(x,t) \equiv 0$, the differential equations imply

$$\frac{\partial v_1}{\partial x}(0,t) = \frac{\partial v_1}{\partial x}(\pi,t) = 0, \quad t \in [0,T].$$

Since these conditions result from the differential equations, they are distinct from independent boundary conditions. For $\phi,\psi \in C^\infty([0,\pi],\mathbb{R})$ with $\psi(0) = \psi(\pi) = 0$, the problem is uniquely solvable. A suitable Banach space is given by $B = B_{31} \times B_{41}$. However, we must then require

$$q(0,t) = q(\pi,t) = 0, \quad t \in [0,T].$$

For simplicity, we switch to vector notation. We only examine the homogeneous problem

$$v_t(x,t) = A(x)v_x(x,t), \qquad x \in (0,\pi), \quad t \in (0,T)$$
$$v(x,t) = (v_1(x,t),v_2(x,t)), \quad A(x) = \begin{pmatrix} 0 & a(x) \\ a(x) & 0 \end{pmatrix}.$$

The differential equation can be discretized as follows ($\alpha \in [0,1]$):

$$h^{-1}[v(x,t+h)-v(x,t)] \approx \tfrac{1}{2}\alpha A(x)(\Delta x)^{-1}[v(x+\Delta x,t)-v(x-\Delta x,t)]$$
$$+ \tfrac{1}{2}(1-\alpha)A(x)(\Delta x)^{-1}[v(x+\Delta x,t+h)-v(x-\Delta x,t+h)].$$

A positive method results from the addition of a suitable numerical viscosity (cf. §6) to the right side. One additional term, for example, is

$$\frac{r_1}{2h}[v(x+\Delta x,t) - 2v(x,t) + v(x-\Delta x,t)]$$
$$+ \frac{r_2}{2h}[v(x+\Delta x,t+h) - 2v(x,t+h) + v(x-\Delta x,t+h)]$$

where

$$\lambda = h/\Delta x,\quad s = \max_{x \in \mathbb{R}} \rho(A(x)),\quad r_1 = \alpha\lambda s,\quad r_2 = (1-\alpha)\lambda s.$$

This viscosity may be regarded as a discretization of $sh\lambda^{-1}v_{xx}$. We have:

$$2(1+r_2)v(x,t+h) \approx [r_1 I+\alpha\lambda A(x)]v(x+\Delta x,t)+(2-2r_1)v(x,t)$$
$$+ [r_1 I-\alpha\lambda A(x)]v(x-\Delta x,t)+[r_2 I+(1-\alpha)\lambda A(x)]v(x+\Delta x,t+h)$$
$$+ [r_2 I-(1-\alpha)\lambda A(x)]v(x-\Delta x,t+h).$$

This approximation can be used immediately to define a difference method. For $\alpha = 1$ it is explicit, and otherwise implicit. It is easily shown that for $r_1 \le 1$, i.e., for $\alpha\lambda s \le 1$, this results in a positive method. If $\max \rho(A(x))$ is not known exactly, one can choose a larger s. The condition $\alpha\lambda s \le 1$ is more restrictive then, for λ must be chosen to be smaller.

In an actual execution, one would restrict oneself to the lattice points $x = \nu\pi/N$, $\nu = 0(1)N$. The implicit methods then lead to a linear system of equations in $2N$ unknowns, namely the components of $v(x,t+h)$. The first components of $v(0,t+h)$ and $v(\pi,t+h)$ are always zero because of the boundary conditions. The matrix of the system is a band matrix. By exploiting the special structure of $A(x)$, the system of equations can be reduced to smaller systems with triangular matrices. Thus relatively little effort is needed to eliminate the unknowns. Nevertheless, it is more advantageous as a rule to use the explicit scheme. The additional effort required by the implicit method is typically not worthwhile for a hyperbolic differential equation. For parabolic equations the situation is completely different. The solutions of parabolic equations often increase in smoothness with increasing t. Then one wants to use very large values of h in the difference method, perhaps 100 or 1,000 times as large as initially. But that is only possible with methods which are stable for every $\lambda = h/(\Delta x)^2$, and that only occurs with implicit methods. For hyperbolic differential equations, one should not expect the solutions to increase in smoothness. The size of the derivative remains about the same. The coordinates t and x are equally important. If the truncation error is not to become too large, the ratio $\lambda = h/\Delta x$ must be kept more or less constant. In our case, a commendable choice is

$$\alpha = 1, \quad r_1 \approx 1, \quad \lambda \approx 1/s, \quad h \approx \frac{\Delta x}{s}. \qquad \square$$

For positive difference methods, stability is measured with respect to the norm $\|\cdot\|_\infty$. However, there exist difference methods which are not stable with respect to $\|\cdot\|_\infty$, but are stable with respect to $\|\cdot\|_2$. Since a direct stability proof is usually quite tedious in each particular case, one would like to have handy stability criteria to apply in this case, also. This leads to the definition of a *positive definite* method.

Definition 8.11: Let $B = L^2(\mathbb{R},\mathbb{C}^n)$, let $P(B,T,A)$ be a properly posed initial value problem, and let $M_D = \{C(h) | h \in (0,h_0]\}$ be a corresponding difference method. It is called *positive definite* if the following conditions (1) through (4) are satisfied.

(1) For all $h \in (0,h_0]$, $f \in B$, and $x \in \mathbb{R}$, where $g = C(h)(f)$ and $\Delta x = h/\lambda$, $\lambda \in \mathbb{R}_+$, it is true that

$$g(x) = \sum_{\nu=-k}^{k} A_\nu(x) f(x+\nu\Delta x).$$

Here $A_\nu \in C^0(\mathbb{R},\mathrm{MAT}(n,n,\mathbb{R}))$, $\nu = -k(1)k$.

(2) $$I = \sum_{\nu=-k}^{k} A_\nu(x), \qquad x \in \mathbb{R}.$$

(3) All matrices $A_\nu(x)$, $x \in \mathbb{R}$, $\nu = -k(1)k$ are symmetric and positive semidefinite.

(4) All matrices $A_\nu(x)$ satisfy a Lipschitz condition with respect to the norm $\|\cdot\|_2$:

$$\|A_\nu(x) - A_\nu(y)\| \leq L|x-y|, \quad x,y \in \mathbb{R}, \quad \nu = -k(1)k. \qquad \square$$

For positive definite methods, in contrast to positive methods, the matrices $A_\nu(x)$ are not allowed to depend on h. It can be shown that as a consequence of this, condition (2) is satisfied by all consistent methods. In practice, the conditions for a positive definite method are not as demanding, since the simultaneous diagonalizability of all the matrices is not required, unlike the case of positive methods. The following theorem is the analog of Theorem 8.5.

Theorem 8.12: *Friedrichs.* A positive definite difference method is stable. Furthermore,

$$\|C(h)\| = 1 + O(h), \qquad h \in (0,h_o].$$

The proof requires the following lemma.

Lemma 8.13: Let $H \in \mathrm{MAT}(n,n,\mathbb{C})$ be Hermitian and positive semidefinite; then

$$|z^H Hw| \le \frac{1}{2}(z^H Hz + w^H Hw), \qquad z,w \in \mathbb{C}^n$$

where $z^H = \bar{z}^T$.

Proof: Let $\{\phi_1,\ldots,\phi_n\}$ be an orthonormal basis of $\mathbb{C}^n$ with

$$H\phi_i = \lambda_i\phi_i, \qquad i = 1(1)n.$$

Let

$$z = \sum_{i=1}^{n} z_i\phi_i, \qquad w = \sum_{i=1}^{n} w_i\phi_i.$$

Since $\lambda_i \ge 0$ and

$$|\bar{\xi}\eta| \le \tfrac{1}{2}(|\xi|^2 + |\eta|^2), \qquad \xi,\eta \in \mathbb{C}$$

we have the following estimate

$$|z^H Hw| = \Big|\sum_{i=1}^{n} \lambda_i\bar{z}_i w_i\Big| \le \sum_{i=1}^{n} \lambda_i|\bar{z}_i w_i|$$

$$\le \frac{1}{2}\sum_{i=1}^{n} \lambda_i(|z_i|^2 + |w_i|^2) = \tfrac{1}{2}(z^H Hz + w^H Hw). \qquad \square$$

Proof of Theorem 8.12: By 8.11(3) and Lemma 8.13 we may estimate

$$\|g\|^2 = \int_{-\infty}^{+\infty} g(x)^H g(x)dx = \int_{-\infty}^{+\infty} [\sum_{\nu=-k}^{k} g(x)^H A_\nu(x) f(x+\nu\Delta x)]dx$$

$$\leq \frac{1}{2}\int_{-\infty}^{+\infty} [\sum_{\nu=-k}^{k} g(x)^H A_\nu(x) g(x)]dx$$

$$+ \frac{1}{2}\int_{-\infty}^{+\infty} [\sum_{\nu=-k}^{k} f(x+\nu\Delta x)^H A_\nu(x) f(x+\nu\Delta x)]dx.$$

By 8.11(2), the first summand is equal to $\frac{1}{2}\|g\|^2$. For the second summand, we have the further estimates

$$\int_{-\infty}^{+\infty} [\sum_{\nu=-k}^{k} f(x+\nu\Delta x)^H A_\nu(x) f(x+\nu\Delta x)]dx = \sum_{\nu=-k}^{k}\int_{-\infty}^{+\infty} f(x)^H A_\nu(x-\nu\Delta x) f(x)dx$$

$$= \|f\|^2 + \sum_{\nu=-k}^{k}\int_{-\infty}^{+\infty} f(x)^H [A_\nu(x-\nu\Delta x) - A_\nu(x)] f(x)dx$$

$$\leq \|f\|^2(1 + L\sum_{\nu=-k}^{k} |\nu|\Delta x).$$

Noting $\lambda = h/\Delta x$ and letting $K = \lambda^{-1}Lk(k+1)$, we get the estimate

$$\|g\|^2 \leq [1 + L\Delta x\ k(k+1)]\ \|f\|^2 = (1+Kh)\ \|f\|^2.$$

It follows from this that

$$\|C(h)\| \leq (1+Kh)^{1/2} \leq \exp(\tfrac{1}{2}Kh) = 1 + O(h);$$

$$\|C(h)^m\| \leq \exp(\tfrac{1}{2}mKh), \quad m \in \mathbb{N}, \quad mh \leq T. \quad \square$$

Example 8.14: *Positive definite difference methods for a hyperbolic initial value problem.* We again consider the hyperbolic initial value problem of Example 8.9

$$\begin{aligned} u_t(x,t) &= A(x)u_x(x,t) + q(x,t) \\ u(x,0) &= \phi(x) \end{aligned} \qquad x \in \mathbb{R}, \quad t \in (0,T).$$

Here let $A \in C^\infty(\mathbb{R}, \mathrm{MAT}(n,n,\mathbb{R}))$ be symmetric with bounded norm and satisfy a Lipschitz condition with respect to $\|\cdot\|_2$. It follows that $\rho(A(x))$ is bounded. The problem is properly posed in the Banach space $B = L^2(\mathbb{R},\mathbb{C}^n)$. The *Friedrichs* method

$$C(h) = \tfrac{1}{2}\{[I-\lambda A(x)]T_{\Delta x}^{-1} + [I+\lambda A(x)]T_{\Delta x}\}, \quad \lambda = h/\Delta x$$

is positive definite if the matrices $I-\lambda A(x)$ and $I+\lambda A(x)$ are positive semidefinite (condition (3)). This again leads to the condition

$$0 < \lambda \sup_{x\in\mathbb{R}} \rho(A(x)) \leq 1. \tag{8.15}$$

The other conditions in Definition 8.11 are satisfied for all $\lambda \in \mathbb{R}_+$. In the *Courant-Isaacson-Rees* method,

$$C(h) = \lambda A^-(x)T_{\Delta x}^{-1} + \{I-\lambda[A^+(x)+A^-(x)]\}I + \lambda A^+(x)T_{\Delta x},$$
$$\lambda = h/\Delta x$$

it must be shown that the matrices

$$A^-(x) = M(x)D^-(x)M(x)^{-1}$$
$$A^+(x) = M(x)D^+(x)M(x)^{-1}$$
$$I-\lambda[A^+(x)+A^-(x)] = I - \lambda\,\mathrm{diag}(|\lambda_i(x)|)$$

are symmetric and positive semidefinite and satisfy a Lipschitz condition with respect to $\|\cdot\|_2$. For this we make the assumption that $M(x)$ is always orthogonal. Then the coefficient matrices are obviously symmetric. By 8.15, they are also positive semidefinite. The Lipschitz conditions are immediate only if $M(x) \equiv$ constant. They are in fact satisfied under substantially more general conditions, but we shall not enter into a discussion of that here. □

In the previous examples, one could add terms of the form

$$b(x)u(x,t), \quad b \in C^0(\mathbb{R},\mathbb{R}) \text{ bounded}$$

or

$$B(x)u(x,t), \quad B \in C^{0}(\mathbb{R}, \mathrm{MAT}(n,n,\mathbb{R})) \quad \text{bounded in norm}$$

without creating any substantive changes in the situation. An additional term $hb(x)I$ or $hB(x)$ then appears in the difference operators, and the stability of the new difference methods then follows from Theorem 5.13 (*Kreiss*). For parabolic differential equations, the addition of a term $\tilde{a}(x)u_x(x,t)$ suggests itself. This term would be discretized as

$$\tilde{a}(x)[u(x+\Delta x,t) - u(x,t)]/\Delta x.$$

All in all, this leads to an operator which differs from the original operator by $O(\Delta x) = O(h^{1/2})$. Perturbations of this type are not covered by the Kreiss theorem (but see Richtmyer-Morton 1967, Section 5.3).

9. Fourier transforms of difference methods

The discretization of pure initial boundary value problems for parabolic or hyperbolic differential equations with constant coefficients leads to particularly simple difference methods. When the underlying Banach space is $L^2(\mathbb{R},\mathbb{C}^n)$, $L^2((0,2\pi),\mathbb{C}^n)$, or $L^2((0,\pi),\mathbb{C}^n)$, the requisite stability tests can be simplified considerably by the introduction of *Fourier transforms*. We begin with a review of basic facts about *Fourier series* and *Fourier integrals*. Proofs can be found in Yosida (1966), Ch. VI, and elsewhere.

In order to be able to apply the translation operator to functions from the spaces $L^2((0,2\pi),\mathbb{C}^n)$ or $L^2((0,\pi),\mathbb{C}^n)$, we extend the functions to all of $\mathbb{R}$ by making the following definition for all $x \in \mathbb{R}$:

$$f(x+2\pi) = f(x) \qquad \text{for} \quad f \in L^2((0,2\pi),\mathbb{C}^n)$$

$$\left.\begin{array}{l} f(x+2\pi) = f(x) \\ f(x) = -f(-x) \end{array}\right\} \quad \text{for} \quad f \in L^2((0,\pi),\mathbb{C}^n).$$

As a result of these definitions, the space $L^2(0,\pi),\mathbb{C}^n)$ becomes a closed subspace of $L^2((0,2\pi),\mathbb{C}^n)$. Departing from ordinary practice, we will call elements of $L^2((0,2\pi),\mathbb{C}^n)$ and $L^2((0,\pi),\mathbb{C}^n)$ ν-times continuously differentiable ($\nu = 0(1)\infty$), if such is the case for their extensions. For each ν-times continuously differentiable function f we then have

$$f^{(\mu)}(0) = f^{(\mu)}(2\pi), \qquad \mu = 0(1)\nu \quad \text{if } f\in L^2((0,2\pi),\mathbb{C}^n)$$

$$f^{(2\mu)}(0) = f^{(2\mu)}(\pi) = 0, \ \mu = 0(1)\tfrac{\nu}{2} \quad \text{if } f\in L^2((0,\pi),\mathbb{C}^n).$$

The two spaces $L^2((0,2\pi),\mathbb{C}^n)$ and $L^2((0,\pi),\mathbb{C}^n)$, like the spaces B_{2n} and B_{4n} in Section 8, thus have boundary conditions built in. The difference between the present and previous Banach spaces is in the norm.

Theorem 9.1: (1) The mapping $\mathscr{F}_{2\pi,n}: L^2((0,2\pi),\mathbb{C}^n) \to \ell^2(\mathbb{C}^n)$ defined by $f \to \{a(\nu)\}_{\nu\in\mathbb{Z}}$, where

$$a(\nu) = (2\pi)^{-1/2} \int_0^{2\pi} f(x)\exp(-i\nu x)\,dx \tag{9.2}$$

is an isometric isomorphism of $L^2((0,2\pi),\mathbb{C}^n)$ onto $\ell^2(\mathbb{C}^n)$; i.e., $\mathscr{F}_{2\pi,n}$ is linear, injective, surjective, and satisfies the condition

$$\|\mathscr{F}_{2\pi,n}(f)\| = \|f\|, \qquad f \in L^2((0,2\pi),\mathbb{C}^n).$$

(2) Let $a \in \ell^2(\mathbb{C}^n)$ and

$$f_\mu(x) = (2\pi)^{-1/2} \sum_{\nu=-\mu}^{\mu} a(\nu)\exp(i\nu x), \qquad \mu \in \mathbb{N}, \quad x \in (0,2\pi).$$

Then the sequence $\{f_\mu\}_{\mu\in\mathbb{N}}$ is a Cauchy sequence in $L^2((0,2\pi),\mathbb{C}^n)$ and converges to an element $f \in L^2((0,2\pi),\mathbb{C}^n)$. The assignment $a \to f$ defines a mapping of $\ell^2(\mathbb{C}^n)$ to $L^2((0,2\pi),\mathbb{C}^n)$ which is the inverse of $\mathcal{F}_{2\pi,n}$.

If f is continuous and of bounded variation, then in addition to Theorem 9.1 it can be shown that the infinite series

$$(2\pi)^{-1/2} \sum_{\nu=-\infty}^{\infty} a(\nu)\exp(i\nu x) \tag{9.3}$$

formed from the coefficients $a(\nu)$, given by (9.2), converges uniformly to f. In many contexts, however, convergence in the mean, i.e., in the sense of the norm of $L^2((0,2\pi),\mathbb{C}^n)$ is sufficient. In any of these cases, the above infinite series is called the *Fourier series* of f. From a physical point of view, the representation says that f may be represented as a superposition of (complex) harmonic oscillations (these are the oscillations with frequencies $\nu = 0, \pm 1, \pm 2, \ldots$). $\mathcal{F}_{2\pi,n}$ maps the space $L^2((0,\pi),\mathbb{C}^n)$, regarded as a subspace of $L^2((0,2\pi),\mathbb{C}^n)$, onto the set

$$\{a \in \ell^2(\mathbb{C}^n) \mid a(\nu) = -a(-\nu)\}.$$

The expansion in (2) is then a pure sine expansion:

$$f_\mu(x) = (2\pi)^{-1/2} \sum_{\nu=-\mu}^{\mu} a(\nu)\exp(i\nu x)$$

$$= 2i(2\pi)^{-1/2} \sum_{\nu=1}^{\mu} a(\nu)\sin(\nu x).$$

In the case of nonperiodic functions, Fourier integrals replace Fourier series.

Theorem 9.4: Let $f \in L^2(\mathbb{R},\mathbb{C}^n)$ and

$$a_\mu(y) = (2\pi)^{-1/2}\int_{-\mu}^{\mu} f(x)\exp(-iyx)dx$$
$$\mu \in \mathbb{N}, \quad y \in \mathbb{R}.$$
$$b_\mu(y) = (2\pi)^{-1/2}\int_{-\mu}^{\mu} f(x)\exp(iyx)dx$$

Then the sequences $\{a_\mu\}_{\mu\in\mathbb{N}}$ and $\{b_\mu\}_{\mu\in\mathbb{N}}$ are Cauchy sequences in $L^2(\mathbb{R},\mathbb{C}^n)$ and converge to elements $a,b \in L^2(\mathbb{R},\mathbb{C}^n)$. The mappings defined by the assignments $f \to a$ and $f \to b$ are isometric automorphisms of $L^2(\mathbb{R},C^n)$; i.e., the mappings are linear, injective, surjective, and satisfy the condition $\|a\| = \|f\| = \|b\|$. The second mapping is the inverse of the first. We denote the first by $\mathcal{F}_n$ and the second by $\mathcal{F}_n^{-1}$.

In order to simplify our notation, we will be somewhat imprecise in the sequel and write

$$\mathcal{F}_n(f)(y) = (2\pi)^{-1/2}\int_{-\infty}^{+\infty} f(x)\exp(-iyx)dx = a(y) \tag{9.5}$$

$$\mathcal{F}_n^{-1}(a)(x) = (2\pi)^{-1/2}\int_{-\infty}^{+\infty} a(y)\exp(ixy)dy = f(x) \tag{9.6}$$

This ignores the fact that the integrals converge only *in the mean*, in general. Pointwise convergence of the integrals only occurs in special cases, e.g., when f or a has compact support. Representation (9.6) is called the *Fourier integral* of f. From a physical point of view, it says that $f(x)$ cannot be built out of harmonic oscillations alone, but that (complex) oscillations of all frequencies y arise. Therefore, the infinite series (9.3) has to be replaced by an integral, where the "infinitesimal" factor $a(y)dy$ corresponds to the previous coefficient $a(\nu)$. The following lemmas describe important computational rules for Fourier series and Fourier integrals.

Lemma 9.7: For $\Delta x \in \mathbb{R}_+$, define $\varepsilon_{\Delta x}\colon \mathbb{R} \to \mathbb{C}$ by $\varepsilon_{\Delta x}(x) = \exp(ix\Delta x)$. Then

(1) $\mathscr{F}_{2\pi,n}[T_{\Delta x}(f)](\nu) = \varepsilon_{\Delta x}(\nu)\, \mathscr{F}_{2\pi,n}(f)(\nu),$

$$f \in L^2((0,2\pi),\mathbb{C}^n), \quad \nu \in \mathbb{Z}.$$

(2) $\mathscr{F}_n[T_{\Delta x}(f)] = \varepsilon_{\Delta x}(\cdot)\, \mathscr{F}_n(f), \qquad f \in L^2(\mathbb{R},\mathbb{C}^n).$

Proof: Conclusion (1) follows from the relation

$$(2\pi)^{-1/2} \int_0^{2\pi} T_{\Delta x}(f)(x)\exp(-i\nu x)\,dx$$

$$= (2\pi)^{-1/2} \int_0^{2\pi} f(x)\exp(-i\nu x)\exp(i\nu\Delta x)\,dx.$$

To prove (2), let $\Phi \in C^0(\mathbb{R},\mathbb{C}^n)$ be a function with compact support, i.e., $\phi(x) = 0$ for $x \in \mathbb{R} - [-\mu,\mu]$ for suitable $\mu \in \mathbb{N}$. Then we can rewrite

$$\int_{-\mu-\Delta x}^{\mu+\Delta x} \Phi(x+\Delta x)\exp(-iyx)\,dx = \int_{-\mu}^{\mu} \Phi(x)\exp[-iy(x-\Delta x)]\,dx$$

$$= \varepsilon_{\Delta x}(y)\int_{-\mu}^{\mu} \Phi(x)\exp(-iyx)\,dx.$$

The conclusion follows from this via Theorem 4.10(1), since the mappings $\mathscr{F}_n$, $T_{\Delta x}$, and $f \to \varepsilon_{\Delta x}(\cdot)f(\cdot)$ are bounded. □

Lemma 9.8: Let $\eta(x) = ix$. If the function $f \in C^\infty(\mathbb{R},\mathbb{C}^n)$ satisfies the growth condition

$$\sup_{x \in \mathbb{R}} \|P(x)f^{(j)}(x)\| < \infty.$$

for all $j \in \mathbb{N}$ and all polynomials P, then for all $q \in \mathbb{N}$

$$\mathscr{F}_n(f^{(q)}) = \eta(\cdot)^q\, \mathscr{F}_n(f).$$

Since the second lemma will not be used in the investigation of difference methods, we dispense with the proof and with the potential substantial weakenings of the hypotheses. Instead, we shall apply it to a simple example which is designed to show how valuable Fourier transforms can be, even for solving differential equations with constant coefficients.

Example 9.9: *Parabolic differential equation in the sense of Petrovski.* Let $q \in \mathbb{N}$ and $a \in \mathbb{C}$ and consider the differential equation

$$u_t(x,t) = a(\frac{\partial}{\partial x})^q u(x,t). \qquad (9.10)$$

The solution $u(x,t)$ and $u_t(x,t)$ are assumed to satisfy the growth condition of Lemma 9.8 for each fixed t. We apply Fourier transforms to the variable x, letting t play the role of a parameter:

$$v(y,t) = \mathscr{F}_1[u(\cdot,t)](y)$$

$$v_t(y,t) = \mathscr{F}_1[u_t(\cdot,t)](y).$$

Then it follows from Lemma 9.8 that

$$v_t(y,t) = \mathscr{F}_1[a(\frac{\partial}{\partial x})^q u(x,t)] = a\eta(y)^q \mathscr{F}_1[u(\cdot,t)](y)$$

$$= a\eta(y)^q v(y,t).$$

Thus v satisfies an ordinary differential equation with respect to t, and can be represented in the form

$$v(y,t) = \exp[a(iy)^q t]v(y,0)$$

Transforming back, we get

$$u(x,t) = \mathscr{F}_1^{-1}\{\exp[a(iy)^q t] \mathscr{F}_1[u(\cdot,0)](y)\}(x). \qquad (9.11)$$

Thus we have obtained the first integral representation of the solution. The *Petrovski* condition $\mathrm{Re}[a(iy)^q] \leq 0$, which for real a is equivalent to

$$q \text{ odd or } a(-1)^{q/2} \leq 0 \qquad \text{(cf. Example 1.13)},$$

quarantees that $v(y,t)$ does not grow faster than $v(y,0)$. For a pure imaginary a and q even, the Petrovski condition is always satisfied.

Next we want to simplify the representation (9.11) for $a \in \mathbb{R}$ and the two cases $q = 1$ and $q = 2$.

<u>q = 1</u>: $$u(x,t) = \frac{1}{2\pi}\int_{-\infty}^{+\infty} \exp(iayt)\Big[\int_{-\infty}^{+\infty} u(\tilde{x},0)\exp(-i\tilde{x}y)d\tilde{x}\Big]\cdot \exp(ixy)dy.$$

With the change $x \to x-at$, it follows that

$$u(x-at,t) = \frac{1}{2\pi}\int_{-\infty}^{+\infty}\Big[\int_{-\infty}^{+\infty} u(\tilde{x},0)\exp(-i\tilde{x}y)d\tilde{x}\Big]\exp(ixy)dy$$

and therefore that $u(x-at,t) = u(x,0)$. Thus we obtain the solution representation

$$u(x,t) = u(x+at,0).$$

<u>q = 2</u>: $$\begin{aligned} u(x,t) &= \frac{1}{2\pi}\int_{-\infty}^{+\infty}\exp(-ay^2t)\Big[\int_{-\infty}^{+\infty} u(\tilde{x},0)\exp(-i\tilde{x}y)d\tilde{x}\Big]\exp(ixy)dy \\ &= \frac{1}{2\pi}\int_{-\infty}^{+\infty}\int_{-\infty}^{+\infty} u(\tilde{x},0)\exp(-ay^2t+ixy-i\tilde{x}y)d\tilde{x}\,dy. \end{aligned}$$

Because of the rapid decay of $u(\tilde{x},0)$ for $t > 0$, the order of integration may be changed, yielding

$$u(x,t) = \frac{1}{2\pi}\int_{-\infty}^{+\infty} u(\tilde{x},0)\Big[\int_{-\infty}^{+\infty} \exp(-ay^2t+ixy-i\tilde{x}y)dy\Big]d\tilde{x}.$$

Since

$$-ay^2t + iy(x-\tilde{x}) = -at\Big(y-i\,\frac{x-\tilde{x}}{2at}\Big)^2 - \frac{(x-\tilde{x})^2}{4at}$$

the inner integral may be simplified further:

$$\int_{-\infty}^{+\infty}\exp[-ay^2t+iy(x-\tilde{x})]dy = \exp\Big[-\frac{(x-\tilde{x})^2}{4at}\Big]\int_{-\infty}^{+\infty}\exp(-atw^2)dw.$$

The right-hand integral is actually an integral along a line parallel to the x-axis, but according to the Cauchy integral theorem this is the same as the integral along the x-axis. Thus we get

$$\int_{-\infty}^{+\infty} \exp[-ay^2t+iy(x-\tilde{x})]dy = (at)^{-1/2}\exp[-\frac{(x-\tilde{x})^2}{4at}]\int_{-\infty}^{+\infty} \exp(-z^2)dz.$$

It is well known (cf. Abramowitz-Stegun 1965) that

$$\int_{-\infty}^{+\infty} \exp(-z^2)dz = \sqrt{\pi} .$$

Therefore we obtain the following representation of the solution (cf. also Example 1.10):

$$u(x,t) = \frac{1}{\sqrt{4\pi at}}\int_{-\infty}^{+\infty} u(\tilde{x},0)\exp[-\frac{(\tilde{x}-x)^2}{4at}]d\tilde{x}, \qquad t > 0.$$

For $a = ib$, where $b \in \mathbb{R}$, and $q = 2$, Equation (9.10) corresponds to a Schrödinger equation (cf. Example 1.13) or to a differential equation from elasticity theory, describing the vibration of a thin board (cf. Example 9.29). In this case, it is frequently formulated as a fourth order differential equation. The substitution $u = u_1 + iu_2$ leads to the system

$$\begin{aligned} \partial u_1/\partial t &= -b\partial^2 u_2/\partial x^2 \\ \partial u_2/\partial t &= b\partial^2 u_1/\partial x^2. \end{aligned}$$

From this it follows that

$$\partial^2 u_2/\partial t^2 = -b^2\partial^4 u_2/\partial x^4.$$

We want to simplify representation (9.11) a bit more. As above, for $q = 2$ and $t > 0$ one obtains

$$u(x,t) = \frac{1}{2\pi}\int_{-\infty}^{+\infty} u(\tilde{x},0)\{\int_{-\infty}^{+\infty} \exp[-iby^2t+i(x-\tilde{x})y]dy\}d\tilde{x}.$$

Since

$$-iby^2t + i(x-\tilde{x})y = -ibt[y^2 - \frac{x-\tilde{x}}{bt}y] =$$

$$= -ibt[(y - \frac{x-\tilde{x}}{2bt})^2 - \frac{(x-\tilde{x})^2}{4b^2t^2}] = -ibt(y - \frac{x-\tilde{x}}{2bt})^2 + i\frac{(x-\tilde{x})^2}{4bt}$$

the inner integral can be simplified further:

$$\int_{-\infty}^{+\infty} \exp[-iby^2t+i(x-\tilde{x})y]dy = \exp[i\frac{(x-\tilde{x})^2}{4bt}] \int_{-\infty}^{+\infty} \exp(-ibtw^2)dw$$

$$= 2(|b|t)^{-1/2}\exp[i\frac{(x-\tilde{x})^2}{4bt}]\int_0^{+\infty} \exp[-i\ \mathrm{sgn}(b)z^2]\,dz.$$

It is shown in Abramowitz-Stegun 1965 that

$$\int_0^{+\infty} \exp(\pm iz^2)dz = \frac{1}{2}\sqrt{\pi}\exp(\pm i\pi/4).$$

It follows that

$$\int_{-\infty}^{+\infty} \exp[-iby^2t + i(x-\tilde{x})y]dy$$

$$= \sqrt{\pi/(|b|t)}\exp[-i\ \mathrm{sgn}(b)\pi/4]\exp[i\ \frac{(x-\tilde{x})^2}{4bt}].$$

Altogether, for $t > 0$ we obtain the following representation of the solution:

$$u(x,t) = (4\pi|b|t)^{-1/2}\exp[-i\ \mathrm{sgn}(b)\pi/4]\int_{-\infty}^{+\infty} u(\tilde{x},0)\exp[i\frac{(x-\tilde{x})^2}{4bt}]d\tilde{x}.$$

Instead, one could also write

$$u(x,t) = (4\pi|bt|)^{-1/2}\exp[-i\ \mathrm{sgn}(bt)\pi/4]\int_{-\infty}^{+\infty} u(\tilde{x},0)\exp[i\frac{(x-\tilde{x})^2}{4bt}]d\tilde{x}.$$

This formula also holds for $t < 0$. The sign of b can be chosen arbitrarily. The solution obviously has an infinite domain of dependence. □

We are now ready to investigate the stability properties of difference methods with the aid of Fourier transforms. From now on, we assume the following:

J is one of the intervals $\mathbb{R}$, $(0,2\pi)$, or $(0,\pi)$

$B = L^2(J,\mathbb{C}^n)$

$P(B,T,A)$ is a properly posed initial value problem.

Definition 9.12: Let $M_D = \{C(h) \mid h \in (0,h_o]\}$, where

$$C(h) = \left[\sum_{\nu=-k}^{k} B_\nu(x,h) T_{\Delta x}^{\nu}\right]^{-1} \circ \left[\sum_{\nu=-k}^{k} A_\nu(x,h) T_{\Delta x}^{\nu}\right]$$

be a difference method. Here $A_\nu(x,h)$, $B_\nu(x,h) \in \mathrm{MAT}(n,n,\mathbb{R})$ and $\Delta x = h/\lambda$ or $\Delta x = \sqrt{h/\lambda}$, $\lambda \in \mathbb{R}_+$.

(1) The rule

$$G(h,y,x) = \left[\sum_{\nu=-k}^{k} \exp(i\nu y\Delta x) B_\nu(x,h)\right]^{-1} \left[\sum_{\nu=-k}^{k} \exp(i\nu y\Delta x) A_\nu(x,h)\right]$$

defines a mapping of $(0,h_o] \times \mathbb{R} \times J$ to $\mathrm{MAT}(n,n,\mathbb{C})$, called an *amplification matrix* for M_D. It is $(2\pi/\Delta x)$-periodic in y.

(2) M_D is called a difference method with *coordinate-free* coefficients if A_ν and B_ν do not depend on x. In that case we abbreviate $G(h,y,x)$ to $G(h,y)$. □

The following theorem gives necessary and sufficient conditions for the stability of a difference method with coordinate-free coefficients. It is proven with the help of a Fourier transformation.

Theorem 9.13: A difference method with coordinate-free coefficients is stable if and only if there exists a constant $K \in \mathbb{R}_+$ such that the spectral norm, $\|\cdot\|_2$, of the amplification matrix satisfies

$$\|G(h,y)^\mu\|_2 \le K, \quad h \in (0,h_o],\ \mu \in \mathbb{N},\ \mu h \le T,\ y \in \mathbb{F}.$$

Here

$$\mathbb{F} = \begin{cases} \mathbb{R} & \text{for } J = \mathbb{R} \\ \mathbb{Z} & \text{for } J = (0,2\pi) \\ \mathbb{N} & \text{for } J = (0,\pi). \end{cases}$$

Proof: *Case 1,* $\mathbb{F} = \mathbb{R}$: We show first that the condition

$$\|G(h,y)^{\mu}\|_2 \leq K$$

is sufficient for stability. So for $f \in B$, define $g_\mu = C(h)^{\mu}(f)$. With the aid of Lemma 9.7(2) it follows that

$$\mathscr{F}_n(g_\mu) = G(h,\cdot)^{\mu}\,\mathscr{F}_n(f).$$

By Theorem 9.4, $\mathscr{F}_n$ is isometric, so that we have

$$\|g_\mu\| = \|\mathscr{F}_n(g_\mu)\| = \|(G(h,\cdot)^{\mu}\,\mathscr{F}_n(f)\|$$

$$\leq \max_{y\in\mathbb{R}} \|G(h,y)^{\mu}\|_2 \cdot \|\mathscr{F}_n(f)\| \leq K\,\|f\|\,.$$

The proof in the reverse direction will be carried out indirectly. Thus we assume that for arbitrary $K \in \mathbb{R}_+$ there exist a $w \in \mathbb{R}$, an $\tilde{h} \in (0,h_o]$, and an $\ell \in \mathbb{N}$ with $\ell h \leq T$, such that the inequality

$$\|G(\tilde{h},w)^{\ell}\|_2 > K$$

is satisfied. We set

$$S(y) = G(\tilde{h},y)^{\ell}$$

$$\lambda = \|S(w)\|_2.$$

Then there exists a $v \in \mathbb{C}^n$ such that

$$v^H S(w)^H S(w) v = \lambda^2 v^H v\,,$$

and a continuous, square integrable function $\hat{f}\colon \mathbb{R} \to C^n$ with $\hat{f}(w) = v$. It follows that

$$\hat{f}(w)^H S(w)^H S(w)\hat{f}(w) > K^2\hat{f}(w)^H\hat{f}(w).$$

Since $\hat{f}$ is continuous, there exists a nondegenerate interval $[y_1,y_2]$ such that

$$\hat{f}(y)^H S(y)^H S(y)\hat{f}(y) > K^2\hat{f}(y)^H\hat{f}(y), \quad y \in [y_1,y_2]. \quad (9.14)$$

We define

$$f(y) = \begin{cases} \hat{f}(y) & \text{for } y \in [y_1,y_2] \\ 0 & \text{for } y \in \mathbb{R} - [y_1,y_2]. \end{cases}$$

$$g = \mathcal{F}_n^{-1}(f).$$

By (9.14) we have $\|S(\cdot)f\|^2 > K^2\,\|f\|$, so upon applying Theorem 9.4 and Lemma 9.7(2), we obtain

$$\begin{aligned}\|S(\cdot)f\| &= \|G(\tilde{h},\cdot)^{\ell}\mathcal{F}_n(g)\| = \|\,\mathcal{F}_n[C(\tilde{h})^{\ell}(g)]\,\| \\ &= \|C(\tilde{h})^{\ell}(g)\| > K\,\|f\| = K\,\|g\|\,.\end{aligned}$$

Therefore, the difference method cannot be stable.

Case 2, $\mathbb{F} = \mathbb{Z}$ or $\mathbb{F} = \mathbb{N}$: The proof is analogous to Case 1. Instead of the Fourier integral $\mathcal{F}_n(f)$, we have the Fourier series $\mathcal{F}_{2\pi,n}(f)$. Instead of Theorem 9.4, we apply Theorem 9.1. Lemma 9.7(2) is replaced by Lemma 9.7(1). □

It follows from the preceding theorem that a difference method which is stable in the space $L^2(\mathbb{R},\mathbb{C}^n)$ is also stable in the spaces $L^2((0,2\pi),\mathbb{C}^n)$ and $L^2((0,\pi),\mathbb{C}^n)$. However, the converse of this statement need not be true, although the difference methods where it fails are all pathological ones. As a result, in practice one tests the boundedness of the norm of the powers of the amplification matrix for $y \in \mathbb{R}$, and not for $y \in \mathbb{Z}$ or $y \in \mathbb{N}$, even when the methods belong to the spaces $L^2((0,2\pi),\mathbb{C}^n)$ or $L^2((0,\pi),\mathbb{C}^n)$.

The necessary and sufficient condition for the stability of a difference method with coordinate-free coefficients, given in Theorem 9.13, is one which is not easily checked, in general. The following theorem provides a simple necessary condition.

Theorem 9.15: *Von Neumann condition.* Let M_D be a difference method with coordinate-free coefficients. The eigenvalues of the amplification matrix $G(h,y)$ for M_D are denoted by

$$\lambda_j(h,y), \quad j = 1(1)n, \quad h \in (0,h_0], \quad y \in \mathbb{F}.$$

If M_D is stable, there exists a constant $\tilde{K} > 0$ such that

$$|\lambda_j(h,y)| \leq 1+\tilde{K}h, \quad j = 1(1)n, \quad h \in (0,h_0], \quad y \in \mathbb{F}.$$

Proof: Let M_D be stable. By Theorem 9.13 it follows that

$$\|G(h,y)^\mu\|_2 \leq K, \quad y \in \mathbb{F},\ h \in (0,h_0], \quad \mu \in \mathbb{N},\ \mu h \leq T.$$

Since

$$\rho(G(h,y)^\mu) \leq \|G(h,y)^\mu\|_2$$

it follows that

$$|\lambda_j(h,y)|^\mu \leq K$$

and hence, for $\mu h \geq T/2$, that

$$|\lambda_j(h,y)| \leq K^{1/\mu} \leq K^{2h/T} = \exp[2h(\log K)/T] \leq 1+\tilde{K}h. \qquad \square$$

Theorem 9.16: The von Neumann condition

$$|\lambda_j(h,y)| \leq 1+\tilde{K}h, \quad j = 1(1)n, \quad h \in (0,h_0], \quad y \in \mathbb{F}$$

of Theorem 9.15 is sufficient for the stability of difference method M_D if one of the following conditions is satisfied:

(1) The amplification matrix $G(h,y)$ is always normal.

(2) There exists a similarity transformation, independent of h, which simultaneously transforms all the matrices $A_\nu(h)$ and $B_\nu(h)$ to diagonal form.

(3) $G(h,y) = \hat{G}(\omega)$ where $\omega = y\Delta x$ and $\Delta x = h/\lambda$ or $\Delta x = \sqrt{h/\lambda}$. Further, for each $\omega \in \mathbb{R}$ one of the following three cases holds:

(a) $\hat{G}(\omega)$ has n different eigenvalues.

(b) $\hat{G}^{(\mu)}(\omega) = \gamma_\mu I$ for $\mu = 0(1)k-1$, $\hat{G}^{(k)}(\omega)$ has n different eigenvalues.

(c) $\rho(\hat{G}(\omega)) < 1$.

Three lemmas precede the proof of the theorem.

Lemma 9.17: Let $A \in MAT(n,n,\mathbb{C})$ be a matrix with $\rho(A) < 1$. Then there exists a matrix norm $\|\cdot\|_s$ with $\|A\|_s < 1$ and $\|B\cdot C\|_s \leq \|B\|_s\cdot\|C\|_s$ for all $B,C \in MAT(n,n,\mathbb{C})$.

For a proof, see Householder 1964.

Lemma 9.18: Let $G \in C^0(\mathbb{R},MAT(n,n,\mathbb{C}))$ and $\omega_0 \in \mathbb{R}$. Also let $G(\omega_0)$ have n different eigenvalues. Then there exists an $\varepsilon > 0$ and maps S, $D \in C^0((\omega_0-\varepsilon,\omega_0+\varepsilon), MAT(n,n,\mathbb{C}))$, such that for all $\omega \in (\omega_0-\varepsilon,\omega_0+\varepsilon)$:

(1) $D(\omega)$ is a diagonal matrix.

(2) $S(\omega)$ is regular.

(3) $D(\omega) = S(\omega)^{-1}G(\omega)S(\omega)$.

The proof is left to the reader. It is possible to weaken the hypotheses of the lemma somewhat.

Lemma 9.19: Let $G \in C^k(\mathbb{R},MAT(n,n,\mathbb{C}))$ and $\omega_0 \in \mathbb{R}$. For $\mu = 0(1)k-1$, let $G^{(\mu)}(\omega_0) = \gamma_\mu I$ and let $G^{(k)}(\omega_0)$ have n

different eigenvalues. Then there exists an $\varepsilon > 0$ and mappings $S, D \in C^0((\omega_0-\varepsilon,\omega_0+\varepsilon), MAT(n,n,\mathbb{C}))$, so that for all $\omega \in (\omega_0-\varepsilon,\omega_0+\varepsilon)$:

(1) $D(\omega)$ is a diagonal matrix.

(2) $S(\omega)$ is regular

(3) $D(\omega) = S(\omega)^{-1}G(\omega)S(\omega)$.

Proof: By Taylor's theorem,

$$G(\omega) = I\sum_{\mu=0}^{k-1} \frac{1}{\mu!}\,\gamma_\mu(\omega-\omega_0)^\mu + \frac{1}{k!}(\omega-\omega_0)^k\tilde{G}(\omega).$$

$\tilde{G}$ is continuous. $\tilde{G}(\omega_0)$ has n different eigenvalues. The conclusion follows by applying Lemma 9.18 to $\tilde{G}$. □

Proof of Theorem 9.16(1): Since the spectral radius and the spectral norm are the same for normal matrices, the bound for the eigenvalues implies a bound

$$||G(h,y)^\mu||_2 \le (1+\tilde{K}h)^\mu \le \exp(\tilde{K}\mu h) \le \exp(\tilde{K}T)$$

for the norm of the powers of the amplification matrix.

Proof of 9.16(2): Let S be the matrix which simultaneously transforms all the matrices $A_\nu(h)$ and $B_\nu(h)$ to diagonal form:

$$\begin{aligned} S^{-1}A_\nu(h)S &= D_\nu(h) \\ S^{-1}B_\nu(h)S &= \tilde{D}_\nu(h). \end{aligned} \qquad \nu = -k(1)k$$

Then,

$$S^{-1}G(h,y)S = \left[\sum_{\nu=-k}^{k}\exp(i\nu y\Delta x)\tilde{D}_\nu(h)\right]^{-1}\left[\sum_{\nu=-k}^{k}\exp(i\nu y\Delta x)D_\nu(h)\right].$$

The transformed matrix $S^{-1}G(h,y)S$ is normal and has the same eigenvalues as $G(h,y)$. It follows that

$$\|[S^{-1}G(h,y)S]^{\mu}\|_2 \le \exp(\tilde{K}T)$$
$$\|G(h,y)^{\mu}\|_2 \le \|S\|_2\, \|S^{-1}\|_2 \exp(\tilde{K}T).$$

Proof of 9.16(3): $\hat{G}(\omega)$, as a rational function of $\exp(i\omega)$, is arbitrarily often differentiable and 2π-periodic. We first prove the following assertion. For every $\omega_0 \in \mathbb{R}$ there exists an $\varepsilon > 0$ and a $K > 0$ such that for all $\nu \in \mathbb{N}$ and all $\omega \in (\omega_0-\varepsilon,\omega_0+\varepsilon)$,

$$\|\hat{G}(\omega)^{\nu}\|_2 \le K.$$

The constants ε and K depend on ω_0 to begin with. Since finitely many open intervals $(\omega_0-\varepsilon,\omega_0+\varepsilon)$ will cover the interval $[0,2\pi]$, we can find a different K so that the inequality holds for all $\omega \in \mathbb{R}$. To establish conclusion (3), we have to distinguish three cases, depending on whether hypothesis (a), (b), or (c) applies to ω_0.

Case (a): A special case of (b) for $k = 0$.

Case (b): $\hat{G}$ and ω_0 satisfy the conditions of Lemma 9.19. The quantity ε of the lemma we denote by 2ε here. For $\omega \in (\omega_0-2\varepsilon,\omega_0+2\varepsilon)$, $\hat{G}^{\nu}$ then has the representation

$$\hat{G}(\omega)^{\nu} = S(\omega)D(\omega)^{\nu}S(\omega)^{-1}.$$

Let

$$K = \max_{\omega\in[\omega_0-\varepsilon,\omega_0+\varepsilon]} \|S(\omega)\|_2\, \|S(\omega)^{-1}\|_2.$$

The diagonal of $D(\omega)$ contains the eigenvalues of $G(h,y)$. They depend only on ω, and not explicitly on h. It then follows from the von Neumann condition that

$$\left.\begin{array}{l}\|D(\omega)\|_2 \le 1\\ \|G(\omega)^{\nu}\|_2 \le K\end{array}\right. \quad \omega \in (\omega_0-\varepsilon,\omega_0+\varepsilon).$$

Case (c): $\hat{G}(\omega_0)$ satisfies the conditions of Lemma 9.17. Let

$$\|\hat{G}(\omega_0)\|_s = L < 1.$$

One can choose an $\varepsilon > 0$ so small that for all $\omega \in (\omega_0-\varepsilon, \omega_0+\varepsilon)$,

$$\|\hat{G}(\omega)\|_s \le \tfrac{1}{2}(L+1) < 1$$

$$\|\hat{G}(\omega)^\nu\|_s \le \tfrac{1}{2}(L+1) < 1, \quad \nu \in \mathbb{N}.$$

Since the spectral norm can be estimated by $\|\cdot\|_s$, it follows that

$$\|\hat{G}(\omega)^\nu\|_2 \le K, \quad \nu \in \mathbb{N},\ \omega \in (\omega_0-\varepsilon,\omega_0+\varepsilon). \quad \square$$

Methods which satisfy the von Neumann condition are called *weakly stable*. Some of these methods are not stable (cf. Examples 9.28 and 9.30). One could also call them *weakly unstable*. As a rule they do converge for solutions of problems which are sufficiently often differentiable. That is not a counterexample to the Lax-Richtmyer equivalence theorem (Theorem 5.11) since there the question is one of convergence for generalized solutions. In many cases, methods which are weakly stable but not stable are even of practical significance.

We will now compute the amplification matrices for a number of concrete difference methods with coordinate-free coefficients. That is the easiest way to check the stability or instability of such methods. In practice, the results are the same for any of the three spaces $L^2(\mathbb{R},\mathbb{C}^n)$, $L^2((0,2\pi),\mathbb{C}^n)$, and $L^2((0,\pi),\mathbb{C}^n)$. Therefore we restrict our examples to the first of these spaces. All the methods which we discussed in Sections 6 and 8 will reappear here. However, note that methods with coordinate-free coefficients only obtain for differential equations with constant coefficients. In the previous chapters, we considered equations with non-constant coefficients.

Stability in the following is always meant in the sense of the norm of $L^2(\mathbb{R},\mathbb{C}^n)$. Positive methods, however, are stable in the sense of the maximum norm. This is a far-reaching conclusion. Thus it is of great interest to compare the stability conditions developed here with the positivity conditions in Section 8. Stability analysis with Fourier transforms gains its real significance from the fact that there exist a number of good, stable methods which are neither positive nor positive definite, no matter what the ratio of the step sizes, $\lambda = h/\Delta x$ or $\lambda = h/(\Delta x)^2$, may be. In particular, this is true for several interesting higher order methods.

Our present restriction to differential equations with constant coefficients should not be misunderstood. In this special case, the theory is particularly simple and easily applied. But there are many other differential equations for which it is possible to analyze the difference methods with the aid of amplification matrices. We will return to this subject briefly later on.

In all the examples, $\lambda = h/\Delta x$ or $\lambda = h/(\Delta x^2)$, depending on the order of the differential equation. We always set $\omega = y\Delta x$. Since y runs through all the real numbers, so does ω.

Example 9.20: (cf. Equation 6.3 and Example 8.6).

Differential equation:

$$u_t(x,t) = au_{xx}(x,t), \quad a > 0.$$

Method:

$$C(h) = [I-(1-\alpha)\lambda H(\Delta x)]^{-1}[I+\alpha\lambda H(\Delta x)]$$

where

$$H(\Delta x) = a[T_{\Delta x}-2I+T_{\Delta x}^{-1}], \quad \alpha \in [0,1].$$

Amplification matrix:

$$G(h,y) = [1+\alpha\lambda\hat{H}(h,y)]/[1-(1-\alpha)\lambda\hat{H}(h,y)]$$

where

$$\hat{H}(h,y) = -2a(1-\cos\omega).$$

Like all matrices in $MAT(1,1,\mathbb{C})$, $G(h,y)$ is normal. Hence the von Neumann condition is necessary and sufficient for stability. We have $2 \geq \varepsilon = 1-\cos\omega \geq 0$. The condition $|G(h,y)| \leq 1$ therefore implies

$$2a\alpha\lambda\varepsilon - 1 \leq 1 + 2a(1-\alpha)\lambda\varepsilon,$$

so the *stability condition* reads

$$2a\lambda(2\alpha-1) \leq 1.$$

More precisely, this says that

$$\alpha \leq 1/2 \quad \text{or} \quad 2a\lambda \leq 1/(2\alpha-1).$$

For $\alpha = 1$ (explicit case) and $\alpha = 0$ (totally implicit case), these are precisely the conditions of Theorems 6.5 and 6.6 and Example 8.6. For $\alpha \in (0,1)$, the positivity condition

$$2a\lambda \leq 1/\alpha$$

is substantially more restrictive. The popular *Crank-Nicolson* method ($\alpha = 1/2$), for example, is stable for all $\lambda > 0$, but is positive only for $\lambda \leq 1/a$. For those combinations of λ and α for which $C(h)$ is stable, we have

$$\|C(h)\|_2 \leq 1$$

and for those combinations for which $C(h)$ is positive, we have

$$\|C(h)\|_\infty \leq 1.$$

Stability is thus "uniform" in λ and α. It follows from this that one can let h and Δx go to zero independently of each other, even though α can depend on h. The method converges in the norm $\|\cdot\|_2$ if

$$2a(2\alpha-1)h/(\Delta x)^2 \leq 1,$$

is always satisfied. If

$$2a\alpha h/(\Delta x)^2 \leq 1$$

then we also have convergence in the norm $\|\cdot\|_\infty$. At the beginning of the computation, when the step size h should in any case be small, the preferred choice is $\alpha = 1$. With increasing t and increasing smoothness of the solution, one would like to switch to larger values of h. Then it is preferable to choose $\alpha \in [1/2,1)$.

In Example 8.7 we investigated an especially precise method. For $a\lambda \geq 1/6$, it may be regarded as a special case of the method just considered. One must then choose $\alpha = 1/2 + 1/12a\lambda$. Thus, for $a\lambda = 1/6$, the method is the same as the explicit method, and as λ increases, it approximates the Crank-Nicolson method. The stability condition $2a\lambda \leq 1/(2\alpha-1)$ is always satisfied. The positivity condition

$2a\lambda \leq 1/\alpha$, however, leads to the condition $a\lambda \leq 5/6$. □

Example 9.21:

Differential equation:

$$u_t(x,t) = au_{xx}(x,t) + bu_x(x,t) + cu(x,t),$$

$$a > 0 \quad \text{and} \quad b,c \in \mathbb{R}.$$

Method:

$$C(h) = I + a\lambda(T_{\Delta x}-2I+T_{\Delta x}^{-1}) + \tfrac{1}{2}b\sqrt{h\lambda}\,(T_{\Delta x}-T_{\Delta x}^{-1}) + chI.$$

Amplification matrix:

$$G(h,y) = 1-2a\lambda(1-\cos\omega) + ib\sqrt{h\lambda}\sin\omega + ch.$$

The von Neumann condition once again is necessary and sufficient. We have

$$|G(h,y)|^2 = [1-2a\lambda(1-\cos\omega)]^2 + O(h).$$

Case 1: $2a\lambda > 1$. For $\omega = \pi$, we have $\lim_{h\to 0}|G(h,y)|^2 > 1$. The method is unstable.

Case 2: $2a\lambda \leq 1$. It follows from $|G(h,y)|^2 \leq 1 + O(h)$ that $|G(h,y)| \leq 1 + O(h)$. The method is stable. The stability condition $2a\lambda \leq 1$ holds independently of b and c. For the parameter c, this follows from Theorem 5.13. Thus the only surprise is in the lack of influence of b. The perturbation $b\sqrt{h\lambda}\,(T_{\Delta x}-T_{\Delta x}^{-1})$ is not insignificant from the viewpoint of Theorem 5.13. Nevertheless, it has no influence on stability in this case. When stability obtains, it is a matter of definition that the powers $|G(h,y)|^{\nu}$ are uniformly bounded. The bound, however, depends on a, b, and c. It is 1 for $b = c = 0$. When the fractions $|b|/a$ or $|c|/a$ are very large, the bound is very large. Then it also depends on the limits of the time interval T.

Let us consider the special case $c = 0$ and $|b|/a \gg 1$. The method is then very similar to the Friedrichs method for the first order differential equation arising from the limiting case $a = 0$. The true viscosity here is potentially even smaller than the numerical viscosity in the Friedrichs method. That leads to practical instability. In this, as in many similarly situated cases, it pays to investigate $G(h,y)$ more closely. We have

$$\begin{aligned}|G(h,y)|^2 &= 1-4a\lambda(1-\cos\omega) + 4a^2\lambda^2(1-\cos\omega)^2+b^2h\lambda\sin^2\omega \\ &= 1-4a\lambda(1-2a\lambda)(1-\cos\omega)-\lambda(4a^2\lambda-b^2h)\sin^2\omega.\end{aligned}$$

For $2a\lambda \le 1$ and $b^2h \le 4a^2\lambda$, it follows that $|G(h,y)| \le 1$. Error amplification only begins on the other side of this bound. The inequality $b^2h \le 4a^2\lambda = 4a^2h/(\Delta x)^2$ is equivalent to

$$\Delta x \le 2a/|b|.$$

Combined with the stability condition, this becomes

$$h/\Delta x \le 1/|b|.$$

This is the stability condition of the Friedrichs method.

For $c > 0$ and $\omega = 0$, we get $G(h,0) = 1+ch$. There is no additional condition that will yield $|G(h,0)| \le 1$. In any case, we must have $ch \ll 1$. Effectively, the bound for $|G(h,y)|^\nu$ now depends on the upper limit of the time interval $[0,T]$. This result is not surprising, since the differential equation has solutions which grow like $\exp(ct)$. The situation is more favorable for $c < 0$ and $b = 0$. For $2a\lambda \le ch/2$, we again have $|G(h,y)| \le 1$. This sharpens the stability condition somewhat. It is preferable to change the method a little. If we evaluate the term $cu(x,t)$ at time $t+h$, we obtain the difference operator

$$\tilde{C}(h) = \frac{1}{1-ch}\,[I+a\lambda(T_{\Delta x}-2I+T_{\Delta x}^{-1}) + \tfrac{1}{2}b\sqrt{h\lambda}\,(T_{\Delta x}-T_{\Delta x}^{-1})]$$

with amplification matrix

$$\tilde{G}(h,y) = \frac{1}{1-ch}\,[1-2a\lambda(1-\cos\omega) + ib\sqrt{h\lambda}].$$

Its size decreases as $-c$ grows. Subject to the conditions

$$c \le 0, \quad 2a\lambda \le 1 + \tfrac{1}{2}|c|h, \quad \Delta x \le 2a/|b|$$

we always have $|\tilde{G}(h,y)| \le 1$. □

Example 9.22: (cf. Thomée 1972).

Differential equation:

$$u_t(x,t) = au_{xx}(x,t), \quad a > 0.$$

Method:

$$\begin{aligned} C(h) = {} & [(1-10a\lambda)T_{\Delta x}^2 + (26-20a\lambda)T_{\Delta x} + (66+60a\lambda)I \\ & + (26-20a\lambda)T_{\Delta x}^{-1} + (1-10a\lambda)T_{\Delta x}^{-2}]^{-1} \circ \\ & [(1+10a\lambda)T_{\Delta x}^2 + (26+20a\lambda)T_{\Delta x} + (66-60a\lambda)I \\ & + (26+20a\lambda)T_{\Delta x}^{-1} + (1+10a\lambda)T_{\Delta x}^{-2}]. \end{aligned}$$

Amplification matrix:

$$G(h,y) = \frac{(1+10a\lambda)\cos 2\omega + (26+20a\lambda)\cos\omega + (33-30a\lambda)}{(1-10a\lambda)\cos 2\omega + (26-20a\lambda)\cos\omega + (33+30a\lambda)}$$

Thomée obtained the method by means of a spline approximation. It converges like $O(h^4) + O((\Delta x)^4)$. Unfortunately, it is not positive for any $\lambda > 0$, since the conditions $66-60a\lambda \ge 0$ and $20a\lambda - 26 \ge 0$ are contradictory. Letting

$$\eta = (33+30a\lambda) + (26-20a\lambda)\cos \omega + (1-10a\lambda)\cos 2\omega > 0$$

we have on the one hand that

$$G(h,y) = 1-20a\lambda(3-2\cos \omega - \cos 2\omega)/\eta \leq 1$$

and on the other,

$$G(h,y) = -1+2(33+26\cos \omega + \cos 2\omega)/\eta \geq -1.$$

The method is stable, therefore, for all $\lambda > 0$. □

Example 9.23: (cf. Equation 6.7).

Differential equation:

$$u_t(x,t) = au_x(x,t), \quad a \in \mathbb{R}.$$

Method:

$$C(h) = I + \frac{1}{2} a\lambda(T_{\Delta x} - T_{\Delta x}^{-1}).$$

Amplification matrix:

$$G(h,y) = 1 + ia\lambda \sin \omega.$$

The method is unstable, as already asserted in Section 6, for the norm of the amplification matrix is $1 + a^2\lambda^2 \sin^2\omega$. □

Example 9.24: *Friedrichs method.*

Differential equation:

$$u_t(x,t) = Au_x(x,t)$$

where $A \in MAT(n,n,\mathbb{R})$ is real diagonalizable.

Method:

$$C(h) = \frac{1}{2}[(I+\lambda A)T_{\Delta x} + (I-\lambda A)T_{\Delta x}^{-1}].$$

Amplification matrix:

$$G(h,y) = I \cos \omega + i\lambda A \sin \omega.$$

Since the coefficients of $C(h)$ are simultaneously diagonalizable, the von Neumann condition is also sufficient for stability. For every eigenvalue μ of A there is a corresponding eigenvalue $\tilde{\mu}(\lambda,\omega)$ of $G(h,y)$, and vice-versa. We have

$$\tilde{\mu}(\lambda,\omega) = \cos\omega + i\lambda\mu\sin\omega,$$

$$|\tilde{\mu}(\lambda,\omega)|^2 = 1 + (\lambda^2\mu^2-1)\sin^2\omega.$$

Thus the method is stable for

$$\lambda\rho(A) \le 1$$

This condition corresponds to the Courant-Friedrichs-Lewy condition in Section 6. When the method is stable, it is also positive, and is positive definite if A is symmetric. □

Example 9.25: *Courant-Isaacson-Rees method.*

Differential equation:

$$u_t(x,t) = Au_x(x,t)$$

where $A \in \mathrm{MAT}(n,n,\mathbb{R})$ is real diagonalizable.

Method:

$$C(h) = \lambda A^+ T_{\Delta x} + [I-\lambda A^+ - \lambda A^-] + \lambda A^- T_{\Delta x}^{-1}.$$

Amplification matrix:

$$G(h,y) = \lambda A^+\exp(i\omega) + I-\lambda A^+ - \lambda A^- + \lambda A^-\exp(-i\omega).$$

In analogy with the previous example, we compute $\tilde{\mu}(\lambda,\omega)$. For $\mu \ge 0$, μ is an eigenvalue of A^+ and 0 is the eigenvalue of A^- for the same eigenvector. We obtain

$$\tilde{\mu}(\lambda,\omega) = \lambda\mu\exp(i\omega) + 1-\lambda\mu,$$

$$|\tilde{\mu}(\lambda,\omega)|^2 = 1 + 2\lambda\mu(\lambda\mu-1)(1-\cos\omega).$$

Similarly, for $\mu \leq 0$ we get

$$\tilde{\mu}(\lambda,\omega) = 1-\lambda|\mu|+\lambda|\mu|\exp(-i\omega),$$

$$|\tilde{\mu}(\lambda,\omega)|^2 = 1+2\lambda|\mu|(\lambda|\mu|-1)(1-\cos\omega).$$

Stability holds exactly for

$$\lambda\rho(A) \leq 1.$$

This again is the Courant-Friedrichs-Lewy condition from Section 6. Here again stability implies positive, and potentially, positive definite. □

Example 9.26: *Lax-Wendroff method.*

Differential equation:

$$u_t(x,t) = Au_x(x,t)$$

where $A \in MAT(n,n,\mathbb{R})$ is real diagonalizable.

Method:

$$C(h) = I + \tfrac{1}{2}\lambda A(T_{\Delta x}-T_{\Delta x}^{-1}) + \tfrac{1}{2}\lambda^2A^2(T_{\Delta x}-2I+T_{\Delta x}^{-1}).$$

Amplification matrix:

$$G(h,y) = I + i\lambda A\sin\omega - \lambda^2A^2(1-\cos\omega).$$

Because the method converges like $O(h^2)$ it is very popular in practice. But it is positive or positive definite only in unimportant exceptional cases. Assume, for example, that A is symmetric and that $\mu \neq 0$ is an eigenvalue of A. $C(h)$ is positive or positive definite only if the three matrices

$$I-\lambda^2A^2,\quad \lambda A+\lambda^2A^2 \quad\text{and}\quad -\lambda A+\lambda^2A^2$$

are positive semidefinite. This means

$$1-\lambda^2|\mu|^2 \geq 0 \quad\text{and}\quad -|\mu| + \lambda|\mu|^2 \geq 0.$$

All eigenvalues $\mu \neq 0$ of A must have the same absolute value, and $\lambda = 1/|\mu|$ is the only possible choice for the step size ratio. In this special case, the method can be regarded as a characteristic method. In Section 11 we will show that the Lax-Wendroff method can be derived from the Friedrichs method by an extrapolation.

The von Neumann condition leads to some necessary and sufficient conditions for stability. We have

$$\tilde{\mu}(\lambda,\omega) = 1+i\lambda\mu \sin \omega-\lambda^2\mu^2(1-\cos \omega),$$

$$|\tilde{\mu}(\lambda,\omega)|^2 = 1+\lambda^4\mu^4(1- \cos \omega)^2-2\lambda^2\mu^2(1-\cos \omega) + \lambda^2\mu^2\sin^2\omega.$$

We substitute $\tilde{\omega} = \omega/2$ and obtain

$$1-\cos \omega = 2 \sin^2\tilde{\omega},$$

$$\sin^2\omega = 4 \sin^2\tilde{\omega} - 4 \sin^4\tilde{\omega},$$

$$|\tilde{\mu}(\lambda,\omega)|^2 = 1-4\lambda^2\mu^2(1-\lambda^2\mu^2)\sin^4\tilde{\omega},$$

Stability is now decided by the sign of $1-\lambda^2\mu^2$. In agreement with the Courant-Friedrichs-Lewy condition, we obtain the stability condition

$$\lambda\rho(A) \leq 1. \qquad \square$$

Example 9.27: (cf. Example 8.10).

Differential equation:

$$u_t(x,t) = Au_x(x,t)$$

where $A \in \mathrm{MAT}(n,n,\mathbb{R})$ is real diagonalizable.

Method:

$$C(h) = \{-[r_2I+(1-\alpha)\lambda A]T_{\Delta x}+(2+2r_2)I-[r_2I-(1-\alpha)\lambda A]T_{\Delta x}^{-1}\}^{-1}$$

$$\circ\ [(r_1I+\alpha\lambda A)T_{\Delta x}+(2-2r_1)I+(r_1I-\alpha\lambda A)T_{\Delta x}^{-1}]$$

where $\alpha \in [0,1]$, $r_1 = \alpha\lambda\rho(A)$, and $r_2 = (1-\alpha)\lambda\rho(A)$.

Amplification matrix:

$$G(h,y) = [(1+r_2-r_2 \cos\omega)I-i(1-\alpha)\lambda A \sin\omega]^{-1}$$
$$\cdot[(1-r_1+r_1\cos\omega)I+i\alpha\lambda A \sin\omega].$$

In Example 8.10 we only considered the special case

$$A = \begin{pmatrix} 0 & a \\ a & 0 \end{pmatrix},$$

albeit for nonconstant coefficients. Implicit methods are of practical significance for initial boundary value problems, at best. Nevertheless, in theory they can also be applied to pure initial value problems. For $\alpha\lambda\rho(A) \leq 1$, the method is positive. In particular, this is so for $\alpha = 0$ and $\lambda > 0$ arbitrary. We now compute the eigenvalues $\tilde{\mu}(\lambda,\omega)$ of the amplification matrix $G(h,y)$:

$$\tilde{\mu}(\lambda,\omega) = \frac{1-r_1(1-\cos\omega) + i\alpha\lambda\mu \sin\omega}{1+r_2(1-\cos\omega) - i(1-\alpha)\lambda\mu \sin\omega}$$

$$|\tilde{\mu}(\lambda,\omega)|^2 = \frac{[1-r_1(1-\cos\omega)]^2+\alpha^2\lambda^2\mu^2\sin^2\omega}{[1+r_2(1-\cos\omega)]^2+(1-\alpha)^2\lambda^2\mu^2\sin^2\omega}.$$

We have stability so long as the numerator is not greater than the denominator, for all $|\mu| \leq \rho(A)$. The difference D of the numerator and denominator is

$$D = -2(r_1+r_2)(1-\cos\omega) + (r_1^2-r_2^2)(1-\cos\omega)^2$$
$$-\lambda^2\mu^2\sin^2\omega + 2\alpha\,\lambda^2\mu^2\sin^2\omega.$$

Since $r_1+r_2 = \lambda\rho(A)$ and $r_1^2-r_2^2 = (2\alpha-1)\lambda^2\rho(A)^2$, we get

$$D = -2\lambda\rho(A)(1-\cos\omega) + (2\alpha-1)\lambda^2\rho(A)^2(1-\cos\omega)^2$$
$$+ (2\alpha-1)\lambda^2\mu^2\sin^2\omega.$$

For $\alpha \leq 1/2$, $D \leq 0$ and $|\tilde{\mu}(\lambda,\omega)| \leq 1$. Thus the method is stable for all $\lambda \geq 0$. It remains to investigate the case $\alpha > 1/2$. We have

$$D \leq \lambda\rho(A)[-2(1-\cos\omega)+(2\alpha-1)\lambda\rho(A)(1-\cos\omega)^2 + (2\alpha-1)\lambda\rho(A)\sin^2\omega].$$

To simplify matters, we again substitute $\tilde{\omega} = \omega/2$. We get the inequality

$$D \leq 4\lambda\rho(A)[-\sin^2\tilde{\omega} + (2\alpha-1)\lambda\rho(A)\sin^4\tilde{\omega} + (2\alpha-1)\lambda\rho(A)\sin^2\tilde{\omega} - (2\alpha-1)\lambda\rho(A)\sin^4\tilde{\omega}.$$

Thus $(2\alpha-1)\lambda\rho(A) \leq 1$ is sufficient for stability of the method $(D \leq 0)$. To obtain a necessary condition, we substitute the values $\omega = \pi$ and $\mu^2 = \rho(A)^2$ in the equation for D. We obtain

$$D = 4\lambda\rho(A)[-1 + (2\alpha-1)\lambda\rho(A)].$$

Thus, the given condition is also necessary. The stability condition

$$(2\alpha-1)\lambda\rho(A) \leq 1$$

in part for $\alpha \in (0,1)$ may deviate substantially from the positivity condition

$$\alpha\lambda\rho(A) \leq 1. \qquad \square$$

<u>Example 9.28</u>: (cf. Example 1.9).

Differential equation:

$$u_t(x,t) = Au_x(x,t)$$

where

$$A = a\begin{pmatrix} 0 & 1 \\ 1 & 0 \end{pmatrix}, \qquad a > 0.$$

Method:

$$C(h) = \left[I - \tfrac{1}{2}a\lambda B_1(T_{\Delta x} - T_{\Delta x}^{-1})\right]^{-1} \circ \left[I + \tfrac{1}{2}a\lambda B_2(T_{\Delta x} - T_{\Delta x}^{-1})\right]$$

where

$$B_1 = \begin{pmatrix} 0 & 0 \\ 1 & 0 \end{pmatrix}, \quad B_2 = \begin{pmatrix} 0 & 1 \\ 0 & 0 \end{pmatrix}.$$

Amplification matrix:

$$G(h,y) = [I - ia\lambda B_1 \sin\omega]^{-1}[I + ia\lambda B_2 \sin\omega].$$

In this case it is not very suggestive to represent the difference method with the translation operator, so we shall switch to componentwise notation. Let

$$u(x,t) = \begin{pmatrix} v(x,t) \\ w(x,t) \end{pmatrix}.$$

The method now reads

$$v(x,t+h) = v(x,t) + \tfrac{1}{2}a\lambda[w(x+\Delta x,t) - w(x-\Delta x,t)]$$

$$w(x,t+h) - \tfrac{1}{2}a\lambda[v(x+\Delta x,t+h) - v(x-\Delta x,t+h)] = w(x,t).$$

In the first equation, the space derivative is formed at time t, and in the second, at time $t+h$. In practice, one would first compute v on the new layer, and then w. Then one can use the new v-values in the computation of w. The method thus is practically explicit. It is not positive for any $\lambda > 0$. Since $B_1^2 = 0$,

$$\begin{aligned} G(h,y) &= [I + ia\lambda B_1 \sin\omega][I + ia\lambda B_2 \sin\omega] \\ &= I + ia\lambda(B_1+B_2)\sin\omega - a^2\lambda^2 B_1 B_2 \sin^2\omega \\ &= \begin{pmatrix} 1 & ia\lambda \sin\omega \\ ia\lambda \sin\omega & 1 - a^2\lambda^2 \sin^2\omega \end{pmatrix}. \end{aligned}$$

B_1 and B_2 obviously are not exchangable. Thus the coefficients of the method cannot be diagonalized simultaneously. In addition, we will show that $G(h,y)$ is not normal for all ω

and that double eigenvalues occur. The von Neumann condition therefore is only a necessary condition.

Let $\eta = a^2\lambda^2\sin^2\omega$. The eigenvalues of $G(h,y)$ satisfy the equation

$$\tilde{\mu}^2 - (2-\eta)\tilde{\mu}+1 = 0.$$

The solutions are

$$\tilde{\mu}_{1,2} = 1 - \tfrac{1}{2}\eta \pm (\tfrac{1}{4}\eta^2-\eta)^{1/2}.$$

Case 1: $a\lambda > 2$. If $\omega = \pi/2$, then $\eta > 4$. Both eigenvalues are real. $|1 - \frac{1}{2}\eta - (\frac{1}{4}\eta^2-\eta)^{1/2}|$ is greater than 1. The method is unstable.

Case 2: $a\lambda < 2$. If $\omega \neq \nu\pi$ where $\nu \in \mathbb{Z}$, the eigenvalues are different and of absolute value 1. For $\omega = \nu\pi$, $G(h,y) = I$. The derivative of $G(h,y)$ with respect to ω has distinct eigenvalues at these points, namely

$$\hat{\mu}_{1,2} = \pm ia\lambda.$$

By Theorem 9.16(3), the method is stable.

Case 3: $a\lambda = 2$. All eigenvalues of $G(h,y)$ have absolute value 1. The method is weakly stable. Suppose it were also stable. Then every perturbation of the method in the sense of Theorem 5.13 would also be stable. We replace matrix B_2 with $B_2(1+h)$ and obtain a method with amplification matrix

$$\begin{pmatrix} 1 & ia\lambda(1+h)\sin\omega \\ ia\lambda\sin\omega & 1-a^2\lambda^2(1+h)\sin^2\omega \end{pmatrix}.$$

In the special case $\omega = \pi/2$, we get, for $a\lambda = 2$,

$$\begin{pmatrix} 1 & 2i+2ih \\ 2i & 1-4-4h \end{pmatrix}.$$

The eigenvalues of this matrix include

$$\tilde{\mu} = -1 - 2h - 2(h+h^2)^{1/2}.$$

Obviously there is no positive constant K such that $|\tilde{\mu}| \leq 1+Kh$. The perturbed method is not stable. Thus we have shown: (1) For $a\lambda = 2$, one obtains a weakly stable method which is not stable, and (2) there is no theorem analogous to 5.13 for weakly stable methods. Thus, the stability condition for our method is $a\lambda < 2$. The Courant-Friedrichs-Lewy condition yields $a\lambda \leq 2$. The difference is without practical significance.

With respect to computational effort, accuracy, and stability conditions, this method is better than the three explicit methods for hyperbolic systems given in Examples 9.24, 9.25, and 9.27. A comparison with the Lax-Wendroff method is not possible, since the semi-implicit method considered here converges only like $O(h)$. Unfortunately the method is tailored specifically for the wave equation with coefficient matrix

$$A = a\begin{pmatrix} 0 & 1 \\ 1 & 0 \end{pmatrix}.$$

This becomes even clearer upon combining two time steps. Then

$$y(x,t+h) - 2y(x,t) + y(x,t-h)$$
$$= \frac{1}{4}a^2\lambda^2[y(x+2\Delta x,t) - 2y(x,t) + y(x-2\Delta x,t)]$$

where

$$v(x,t) = [y(x,t)-y(x,t-h)]/h$$

and

$$w(x,t) = \frac{a}{2}[y(x+\Delta x,t)-y(x-\Delta x,t)]/\Delta x.$$

Here one can compute forwards as well as backwards, i.e., λ can be replaced by $-\lambda$ and h by $-h$. So far such a time reversal has resulted in a stable method only with Massau's method. In all the other examples, numerical viscosity requires a fixed sign on h. This viscosity term is missing here. Reversibility also requires all the eigenvalues of the amplification matrix to have absolute value 1. Experience shows that methods of this type are no longer useful when the differential equations contain any nonlinearities whatsoever. Exceptions once again are the characteristic methods. □

Example 9.29: (cf. Examples 1.13 and 9.9).

Differential equation:

$$u_t(x,t) = iau_{xx}(x,t), \quad a \in \mathbb{R} - \{0\}.$$

Method:

$$C(h) = [I-(1-\alpha)\lambda H(\Delta x)]^{-1} \circ [I+\alpha\lambda H(\Delta x)]$$

where

$$H(\Delta x) = ia[T_{\Delta x} - 2I + T_{\Delta x}^{-1}], \quad \alpha \in [0,1].$$

Amplification matrix:

$$G(h,y) = [1+\alpha\lambda\hat{H}(h,y)]/[1-(1-\alpha)\lambda\hat{H}(h,y)]$$

where

$$\hat{H}(h,y) = 2ia(\cos\omega - 1).$$

Formally, the method is the same as the method for parabolic equations. Since

$$|G(h,y)|^2 = \frac{1+4\alpha^2a^2\lambda^2(1-\cos\omega)^2}{1+4(1-\alpha)^2a^2\lambda^2(1-\cos\omega)^2}$$

we obtain, independently of λ, the stability condition

$(1-\alpha)^2 \geq \alpha^2$ or $2\alpha \leq 1$. All stable methods of this type are implicit. As for parabolic equations, the truncation error is $O(h) + O((\Delta x)^2)$ for $\alpha < 1/2$, and $O(h^2) + O((\Delta x)^2)$ for $\alpha = 1/2$. Naturally we prefer to compute with real numbers. Therefore we set $v(x,t) = \mathrm{Re}(u(x,t))$ and $w(x,t) = \mathrm{Im}(u(x,t))$. This leads to the differential equations

$$\begin{aligned} v_t(x,t) &= -aw_{xx}(x,t) \\ w_t(x,t) &= av_{xx}(x,t) \end{aligned}$$

and the methods

$$\begin{aligned} &v(x,t+h)+(1-\alpha)a\lambda[w(x+\Delta x,t+h)-2w(x,t+h)+w(x-\Delta x,t+h)] \\ &\quad = v(x,t)-\alpha a\lambda[w(x+\Delta x,t)-2w(x,t)+w(x-\Delta x,t)], \\ &w(x,t+h)-(1-\alpha)a\lambda[v(x+\Delta x,t+h)-2v(x,t+h)+v(x-\Delta x,t+h)] \\ &\quad = w(x,t)+\alpha a\lambda[v(x+\Delta x,t)-2v(x,t)+v(x-\Delta x,t)]. \qquad \square \end{aligned}$$

Example 9.30: (cf. Examples 1.13 and 9.9).

Differential equation:

$$u_t(x,t) = Au_{xx}(x,t)$$

where

$$A = a\begin{pmatrix} 0 & -1 \\ 1 & 0 \end{pmatrix}, \quad a \in \mathbb{R} - \{0\}.$$

Method:

$$C(h) = \left[I-a\lambda B_1(T_{\Delta x}-2I+T_{\Delta x}^{-1})\right]^{-1} \circ \left[I-a\lambda B_2(T_{\Delta x}-2I+T_{\Delta x}^{-1})\right]$$

where

$$B_1 = \begin{pmatrix} 0 & 0 \\ 1 & 0 \end{pmatrix}, \quad B_2 = \begin{pmatrix} 0 & -1 \\ 0 & 0 \end{pmatrix}.$$

Amplification matrix:

$$G(h,y) = [I+2a\lambda B_1(1-\cos\omega)]^{-1}[I+2a\lambda B_2(1-\cos\omega)].$$

The differential equation is equivalent to the equation in

the previous example. The method under discussion is more easily explained in real terms. We again have a semi-implicit method with great similarity to Example 9.28. Rewriting the difference equations componentwise, we have

$$v(x,t+h) = v(x,t)-a\lambda[w(x+\Delta x,t)-2w(x,t)+w(x-\Delta x,t)]$$

$$w(x,t+h)-a\lambda[v(x+\Delta x,t+h)-2v(x,t+h)+v(x-\Delta x,t+h)] = w(x,t).$$

The computational effort is as for an explicit method. Since $B_1^2 = 0$, we can change the amplification matrix as follows:

$$G(h,y) = [I-2a\lambda B_1(1-\cos\omega)][I+2a\lambda B_2(1-\cos\omega)].$$

Making the substitution $\tilde{\omega} = \omega/2$, we get

$$G(h,y) = \begin{pmatrix} 1 & 4a\lambda\sin^2\tilde{\omega} \\ -4a\lambda\sin^2\tilde{\omega} & 1-16a^2\lambda^2\sin^4\tilde{\omega} \end{pmatrix}$$

Let $\eta = 16a^2\lambda^2\sin^4\tilde{\omega}$. $G(h,y)$ then has the eigenvalues

$$\tilde{\mu}_{1,2} = 1 - \tfrac{1}{2}\eta \pm (\tfrac{1}{4}\eta^2-\eta)^{1/2}.$$

The remaining analysis is entirely analogous to Example 9.28.

Case 1: $a^2\lambda^2 > 1/4$. If $\tilde{\omega} = \pi/2$, $|1-\tfrac{1}{2}\eta - (\tfrac{1}{4}\eta^2-\eta)^{1/2}| > 1$. The method is unstable.

Case 2: $a^2\lambda^2 < 1/4$. All eigenvalues have absolute value 1. For $\tilde{\omega} \neq \nu\pi$ $(\nu \in \mathbb{Z})$ they are distinct. At the exceptional points $\tilde{\omega} = \nu\pi$ we have $\hat{G}(\omega) = I$ and $\hat{G}'(\omega) = 0$. The second derivative with respect to ω is

$$\begin{pmatrix} 0 & 2a\lambda \\ -2a\lambda & 0 \end{pmatrix}.$$

This matrix has two distinct eigenvalues. By Theorem 9.16(3), the method is stable.

Case 3: $a^2\lambda^2 = 1/4$. The method is weakly stable. One again shows that it is unstable with the help of a perturbation of order $O(h)$.

The stability condition is $|a|\lambda < 1/2$. Since the differential equation does not have a finite domain of dependency (cf. the closed solution in 9.9), no comparison with the Courant-Friedrichs-Lewy criterion is possible. The method is not positive for any $\lambda > 0$. Richtmyer-Morton prefer the implicit method of the previous example, since the present stability condition, $\lambda = h/(\Delta x)^2 \leq 1/(2|a|)$, is very strong.

When the coefficients of a difference method do depend on the coordinates, one cannot automatically apply a Fourier transform to the method. Although the amplification matrix appears formally the same as for coordinate-free coefficients, it is not the Fourier transform of the method. In these cases, the amplification matrix can only be used to investigate *local stability* of the method. Then the variable coefficients of the difference method are "frozen" with respect to x, and the stability properties of the resulting method with coordinate-free coefficients become the subject of investigation.

The following theorem 9.31 shows that under certain additional conditions, local stability is necessary for stability. For simplicity, we restrict ourselves to explicit methods and $B = L^2(\mathbb{R},\mathbb{C}^n)$.

There also exist a number of sufficient stability criteria which depend on properties of the amplification matrix. We refer to the work of Lax-Wendroff (1962), Kreiss (1964), Widlund (1965), and Lax-Nirenberg (1966). The proofs are all very complicated. We therefore will restrict ourselves to a result on hyperbolic systems (Theorem 9.34) due to Lax and Nirenberg.

Theorem 9.31: Let $A_\nu \in C^0(\mathbb{R} \times (0,h_0], \mathrm{MAT}(n,n,\mathbb{R}))$, $\nu = -k(1)k$ and $M_D = \{C(h) \mid h \in (0,h_0]\}$ be an explicit difference method for a properly posed problem $P(B,T,A)$, where

$$C(h) = \sum_{\nu=-k}^{k} A_\nu(x,h) T_{\Delta x}^{\nu}.$$

Further assume that as $h \to 0$, each $A_\nu(x,h)$ converges uniformly on every compact subset of $\mathbb{R}$ to a mapping $A_\nu(x,0)$ of bounded norm. Then the stability of method M_D implies the stability of the method $\tilde{M}_D = \{\tilde{C}(h) \mid h \in (0,h_0]\}$, where

$$\tilde{C}(h) = \sum_{\nu=-k}^{k} A_\nu(\tilde{x},0) T_{\Delta x}^{\nu}$$

for every (fixed) point $\tilde{x}$ of $\mathbb{R}$.

Proof: The proof will be indirect and so we assume that there exists an $\tilde{x} \in \mathbb{R}$ for which the method $\tilde{M}_D$ is not stable. By Theorem 9.13, for each constant $K \in \mathbb{R}_+$ there exists a $y \in \mathbb{R}$, an $h \in (0,h_0]$, an $N \in \mathbb{N}$ with $Nh \leq T$, and a vector $V \in \mathbb{C}^n$, such that

$$\|\Gamma(\tilde{x})^N V\|_2 > K\,\|V\|_2, \tag{9.32}$$

where

$$\Gamma(x) = \sum_{\nu=-k}^{k} \exp(i\nu\xi) A_\nu(x,0), \qquad \xi = y\Delta x.$$

Since $A_\nu(x,0)$ is continuous, there is a $\delta \in \mathbb{R}_+$ such that inequality (9.32) also holds for all $x \in S_\delta = (\tilde{x}-\delta, \tilde{x}+\delta)$. We now fix ξ and pass to the limit $h \to 0$ (and hence $\Delta x \to 0$). Inequality (9.32) remains valid throughout for all $x \in S_\delta$.

Now let $\rho\colon \mathbb{R} \to \mathbb{R}$ be an infinitely often differentiable function with

$$\rho(x) = 0 \quad \text{for} \quad x \in \mathbb{R} - S_\delta$$
$$\rho(x) \not\equiv 0 \quad \text{in} \quad S_\delta.$$

Set

$$v(x) = V\rho(x)\exp(iyx).$$

Then

$$C(h)(v)(x) = \sum_{\nu=-k}^{k} A_\nu(x,h)V\rho(x+\nu\Delta x)\exp[iy(x+\nu\Delta x)]$$
$$= \Gamma(x)v(x) + \varepsilon_1(x,h).$$

Here $\varepsilon_1(x,h)$ is a function for which there is a $\tilde{\delta} \in \mathbb{R}_+$ such that: (1) $\varepsilon_1(x,h) = 0$ for $x \in \mathbb{R}-(\tilde{x}-\delta-\tilde{\delta},\tilde{x}+\delta+\tilde{\delta})$ and h sufficiently small, and (2) $\varepsilon_1(x,h)$ converges uniformly to zero as $h \to 0$ for $x \in (\tilde{x}-\delta-\tilde{\delta},\tilde{x}+\delta+\tilde{\delta})$. Applying $C(h)$ repeatedly, we obtain

$$C(h)^N(v)(x) = \Gamma(x)^N v(x) + \varepsilon_N(x,h).$$

Here $\varepsilon_N(x,h)$ has the same properties as $\varepsilon_1(x,h)$. Choose a sufficiently small h, and then it follows from (9.32) that

$$||C(h)^N|| \geq K.$$

This contradicts the stability of M_D. □

<u>Example 9.33</u>: Application of Theorem 9.31.

Differential equation (cf. Example 5.6):

$$u_t(x,t) = [a(x)u_x(x,t)]_x$$

where

$$a \in C^\infty(\mathbb{R}, \mathbb{R}), \quad a' \in C_0^\infty(\mathbb{R}, \mathbb{R}) \quad \text{and} \quad a(x) > 0, \quad x \in \mathbb{R}.$$

Method (cf. (6.3) for $\alpha = 1$):

$$C(h) = \lambda a(x-\Delta x/2)T_{\Delta x}^{-1} + [1-\lambda a(x+\Delta x/2)-\lambda a(x-\Delta x/2)]I$$
$$+ \lambda a(x+\Delta x/2)T_{\Delta x}, \qquad \lambda = h/(\Delta x)^2.$$

By Theorem 6.5, the condition

$$0 < \lambda \max_{x \in \mathbb{R}} a(x) \leq 1/2$$

is sufficient for the stability of the above method. It follows from Theorem 9.31 that this condition is also necessary. Thus, for fixed $\tilde{x} \in \mathbb{R}$, consider the method

$$\tilde{C}(h) = \lambda a(\tilde{x})T_{\Delta x}^{-1} + [1-2\lambda a(\tilde{x})]I + \lambda a(\tilde{x})T_{\Delta x}.$$

The corresponding amplification matrix is

$$\tilde{G}(h,y) = 1 + 2\lambda a(\tilde{x})[\cos\omega - 1].$$

Since the above stability condition is necessary and sufficient for

$$|\tilde{G}(h,y)| \leq 1, \quad y \in \mathbb{R}, \quad h \in (0,h_0],$$

the conclusion follows from Theorem 9.13. □

Theorem 9.34: *Lax-Nirenberg.* Let $M_D = \{C(h)|h > 0\}$ be a difference method for a properly posed problem $P(L^2(\mathbb{R}, \mathbb{R}^n), T, A)$, where

$$C(h) = \sum_{\mu=-k}^{k} B_\mu(x)T_{\Delta x}^{\mu}$$

and $\Delta x = h/\lambda$, with $\lambda > 0$ fixed. Let the following conditions be satisfied:

(1) $B_\mu \in C^2(\mathbb{R}, MAT(n,n,\mathbb{R}))$, $\mu = -k(1)k$.

(2) All elements of the matrices $B_\mu^{(\nu)}(x)$, $\nu = 0(1)2$, $\mu = -k(1)k$, $x \in \mathbb{R}$ are uniformly bounded.

(3) $\|G(h,y,x)\|_2 \leq 1$ for $h > 0$, $y \in \mathbb{R}$, $x \in \mathbb{R}$ where $G(h,y,x)$ is the amplification matrix for $C(h)$.

Then M_D is stable.

Remark: Although we have a real Banach space, we form the amplification matrix exactly as in the case of $L^2(\mathbb{R},\mathbb{C}^n)$, namely

$$G(h,y,x) = \sum_{\mu=-k}^{k} B_\mu(x)\exp(i\mu y\Delta x).$$

For fixed $x_0 \in \mathbb{R}$, it follows from condition (3) that

$$\|\sum_{\mu=-k}^{k} B_\mu(x_0)T_{\Delta x}^\mu\|_2 \leq 1.$$

For the proof, we embed $L^2(\mathbb{R}, \mathbb{R}^n)$ in $L^2(\mathbb{R},\mathbb{C}^n)$ in the canonical way. The conclusion then follows from

$$\sum_{\mu=-k}^{k} B_\mu(x_0)T_{\Delta x}^\mu = \mathcal{F}_n^{-1}\circ G(h,y,x_0)\circ \mathcal{F}_n. \qquad \square$$

Before we prove Theorem 9.34, we need to establish several further observations and lemmas. We begin by introducing a scalar product for $u,v \in L^2(\mathbb{R}, \mathbb{R}^n)$:

$$\langle u,v\rangle = \int_{\mathbb{R}} u(x)^T v(x)dx = \langle v,u\rangle.$$

With respect to this scalar product, there exists an operator $C(h)^T$ which is adjoint to $C(h)$, namely

$$C(h)^T = \sum_{\mu=-k}^{k} T_{\Delta x}^{-\mu}\circ B_\mu(x)^T.$$

Using the symmetry of the matrices $B_\mu(x)$, we obtain

$$\begin{aligned}\langle C(h)(u),v\rangle &= \sum_{\mu=-k}^{k} \int_{\mathbb{R}} [B_\mu(x)u(x+\mu\Delta x)]^T v(x)dx \\ &= \sum_{\mu=-k}^{k} \int_{\mathbb{R}} u(x)^T [B_\mu(x-\mu\Delta x)^T v(x-\mu\Delta x)]dx \\ &= \langle u,C(h)^T(v)\rangle.\end{aligned}$$

In particular,

$$\langle C(h)(u), C(h)(v)\rangle = \langle u, C(h)^T \circ C(h)(v)\rangle.$$

In addition to the difference operators $C(h)$, we also have to consider the difference operators with "frozen" coefficients. For fixed $\alpha \in \mathbb{R}$ let

$$C_\alpha(h) = \sum_{\mu=-k}^{k} B_\mu(\alpha) T_{\Delta x}^{\mu}$$

$$C_\alpha(h)^T = \sum_{\mu=-k}^{k} B_\mu(\alpha)^T T_{\Delta x}^{-\mu}.$$

As remarked above, it follows from (3) that

$$\| C_\alpha(h)(u) \|_2 \leq \| u \|_2,$$

which is to say

$$\langle C_\alpha(h)(u), C_\alpha(h)(u)\rangle \leq \langle u,u\rangle$$

or

$$\langle u, [I - C_\alpha(h)^T \circ C_\alpha(h)](u)\rangle \geq 0.$$

Our goal is to establish a similar inequality for the operators $C(h)$, instead of $C_\alpha(h)$.

Lemma 9.35:

$$I - C(h)^T \circ C(h) = Q(h) + \Delta x\, R(h)$$

where $Q(h)$ and $R(h)$ are bounded linear mappings of $L^2(\mathbb{R},\mathbb{R}^n)$ into itself. $\|R(h)\|_2$ is bounded independently of h and Δx. The coefficients of $Q(h)$ are independent of h. We then have

$$Q(h) = \sum_{\mu=-2k}^{2k} D_\mu(x) T_{\Delta x}^{\mu}.$$

Proof: The product $C(h)^T \circ C(h)$ consists of summands of the form

$$T_{\Delta x}^{r}\circ B_{-r}(x)^T B_s(x)\circ T_{\Delta x}^{s} = B_{-r}(x)^T B_s(x)\circ T_{\Delta x}^{r+s}$$

$$+ [B_{-r}(x+r\Delta x)^T B_s(x+r\Delta x)-B_{-r}(x)^T B_s(x)]T_{\Delta x}^{r+s}.$$

$Q(h)$ contains the first summand on the right side. The term in square brackets is divisible by Δx. The quotient is bounded independently of Δx because $||B_\mu'(x)||$ and $||B_\mu(x)||$ are bounded. □

Analogously to $C_\alpha(h)$, we can define

$$Q_\alpha(h) = \sum_{\mu=-2k}^{2k} D_\mu(\alpha)T_{\Delta x}^{\mu}.$$

Obviously,

$$Q_\alpha(h) = I-C_\alpha(h)^T\circ C_\alpha(h)$$

and therefore,

$$\langle u,Q_\alpha(h)(u)\rangle \geq 0.$$

<u>Lemma 9.36</u>: Let

$$E_\mu = \tfrac{1}{2}[D_\mu(\alpha) + D_{-\mu}(\alpha)]$$

$$F_\mu = \tfrac{1}{2}[D_\mu(\alpha) - D_{-\mu}(\alpha)].$$

Then:

(1) $$Q_\alpha(h) = \frac{1}{2}\sum_{\mu=-2k}^{2k}[E_\mu(T_{\Delta x}^{\mu}+T_{\Delta x}^{-\mu}) + F_\mu(T_{\Delta x}^{\mu}-T_{\Delta x}^{-\mu})]$$

(2) $$E_\mu^T = E_\mu,\quad F_\mu^T = -F_\mu,\quad \mu = -2k(1)2k.$$

Proof: (1) is trivial. As for (2), note that $Q_\alpha(h)$ consists of summands of the form

$$B_{-r}(\alpha)^T B_s(\alpha)T_{\Delta x}^{r+s}+B_s(\alpha)^T B_{-r}(\alpha)T_{\Delta x}^{-r-s},\quad \mu = r+s > 0$$

and

$$B_s(\alpha)^T B_s(\alpha)T_{\Delta x}^{0}$$

The corresponding summands in $2E_\mu$ are

$$B_{-r}(\alpha)^T B_s(\alpha) + B_s(\alpha)^T B_{-r}(\alpha), \qquad \mu > 0$$

or

$$B_s(\alpha)^T B_s(\alpha), \qquad \mu = 0.$$

These matrices are symmetric. $2F_\mu$ contains the antisymmetric terms

$$B_{-r}(\alpha)^T B_s(\alpha) - B_s(\alpha)^T B_{-r}(\alpha). \qquad \square$$

One important tool for the proof of Theorem 9.34 is a special partition of unity.

Lemma 9.37: There exists a $\phi \in C^\infty(\mathbb{R}, \mathbb{R})$ with the following three properties.

(1) $\text{Support}(\phi) = [-2/3, 2/3]$.

(2) $0 \leq \phi(x) \leq 1, \quad x \in \mathbb{R}$.

(3) $\sum_{\mu=-\infty}^{+\infty} \phi(x-\mu)^2 = 1, \quad x \in \mathbb{R}$.

Proof: Choose $\tilde{\phi} \in C^\infty(\mathbb{R}, \mathbb{R})$ (cf. Lemma 4.9) such that

$$\begin{aligned}
\tilde{\phi}(x) &= 1 && \text{for } |x| \leq 1/3 \\
\tilde{\phi}(x) &= 0 && \text{for } |x| \geq 2/3 \\
0 < \tilde{\phi}(x) &< 1 && \text{for } x \in (1/3, 2/3) \\
\tilde{\phi}(x) &= 1-\tilde{\phi}(x+1) && \text{for } x \in (-2/3, -1/3).
\end{aligned}$$

It follows that

$$\sum_{\mu=-\infty}^{+\infty} \tilde{\phi}(x-\mu) = 1, \qquad x \in \mathbb{R}.$$

All the derivatives of $\tilde{\phi}$ vanish at the zeroes of $\tilde{\phi}$. Therefore, $\phi(x) = \tilde{\phi}(x)^{1/2}$ is arbitrarily often differentiable. For if $|x_0| \geq 2/3$, i.e., x_0 a zero of $\tilde{\phi}$, then in some neighborhood of x_0,

$$\tilde{\phi}(x) = (x-x_0)^{2s} \psi_s(x)$$

$$\phi(x) = |x-x_0|^s \sqrt{\psi_s(x)}.$$

Here $s \in \mathbb{N}$ is arbitrary and ψ_s is continuous. It is easily shown that $\sqrt{\psi_s(x)}$ is continuous and ϕ is at least $(s-1)$ times differentiable at x_0. □

Like ϕ, the function ϕ' has compact support. Therefore, $|\phi'(x)|$ attains its maximum, which we denote by L in the following. By the Mean Value Theorem,

$$|\phi(x) - \phi(\tilde{x})| \leq L|x-\tilde{x}|.$$

for all $x,\tilde{x} \in \mathbb{R}$.

Lemma 9.38: Let $\eta_\nu(x) = \phi(\gamma x-\nu)$, $\nu = -\infty(1)\infty$ where ϕ is as in the preceding lemma. Then for all $h = \lambda\Delta x > 0$ and all $\gamma \in \mathbb{R}_+$ where $6k\gamma\Delta x \leq 1$,

$$|\langle u,Q(h)(u)\rangle - \sum_{\nu=-\infty}^{\infty} \langle \eta_\nu(\cdot)u(\cdot),Q(h)[\eta_\nu(\cdot)u(\cdot)]\rangle|$$

$$\leq M_1\gamma^2L^2(\Delta x)^2 \|u\|_2^2, \quad u \in L^2(\mathbb{R}, \mathbb{R}^n).$$

The constant $M_1 > 0$ depends only on the method M_D, and not on u, γ, h, or Δx.

Proof: For $\mu = -2k(1)2k$ we have

$$D_\mu(x)u(x+\mu\Delta x) - \sum_{\nu=-\infty}^{\infty} \eta_\nu(x)D_\mu(x)\eta_\nu(x+\mu\Delta x)u(x+\mu\Delta x)$$

$$= [1 - \sum_{\nu=-\infty}^{\infty} \eta_\nu(x)\eta_\nu(x+\mu\Delta x)]D_\mu(x)u(x+\mu\Delta x).$$

Using 9.37(3), we replace the 1 in the square brackets by

$$\frac{1}{2}\sum_{\nu=-\infty}^{\infty} \eta_\nu(x)^2 + \frac{1}{2}\sum_{\nu=-\infty}^{\infty} \eta_\nu(x+\mu\Delta x)^2.$$

The term in the brackets is then a sum of squares,

$$1 - \sum_{\nu=-\infty}^{\infty} \eta_\nu(x)\eta_\nu(x+\mu\Delta x) = \frac{1}{2}\sum_{\nu=-\infty}^{\infty} [\eta_\nu(x)-\eta_\nu(x+\mu\Delta x)]^2.$$

$\eta_\nu(x) = 0$ for $|\gamma x-\nu| \geq 2/3$. Since $|\mu| \leq 2k$ and $\Delta x \leq (6\gamma k)^{-1}$, $\eta_\nu(x)-\eta_\nu(x+\mu\Delta x) = 0$ for $|\gamma x-\nu| \geq 1$. For fixed x, at most two summands in the sum are different from zero. The Lipschitz condition on ϕ then yields

$$\frac{1}{2}\sum_{\nu=-\infty}^{\infty} [\eta_\nu(x)-\eta_\nu(x+\mu\Delta x)]^2 \leq 4k^2\gamma^2L^2(\Delta x)^2.$$

It follows from the Schwartz inequality that

$$|\langle u(x),D_\mu(x)u(x+\mu\Delta x)\rangle - \sum_{\nu=-\infty}^{\infty}\langle\eta_\nu(x)u(x),D_\mu(x)\eta_\nu(x+\mu\Delta x)u(x+\mu\Delta x)\rangle| \leq k^2\gamma^2L^2(\Delta x)^2\tilde{M}\,\|u\|_2^2.$$

Summing all these inequalities for $\mu = -2k(1)2k$ establishes the conclusion of the lemma. □

Lemma 9.39: Let $\beta > 0$ and $\psi \in C^0(\mathbb{R}, \mathbb{R})$ be such that $|\psi(x)-\psi(\tilde{x})| \leq \tilde{L}|x-\tilde{x}|$ for all $x,\tilde{x} \in (-2\beta,2\beta)$. Then for all $u \in L^2(\mathbb{R}, \mathbb{R}^n)$ with $\text{Support}(u) \subset [-\beta,\beta]$ and for all h with $2k\Delta x < \beta$,

$$|\langle\psi u,Q_\alpha(h)(\psi u)\rangle - \langle\psi^2u,Q_\alpha(h)(u)\rangle| \leq \tilde{K}\,\tilde{L}^2\,(\Delta x)^2\,\|u\|_2^2.$$

$\tilde{K}$ depends only on the method M_D.

Proof: We have (cf. Lemma 9.36):

$$Q_\alpha(h) = \frac{1}{2}\sum_{\mu=-2k}^{2k} [E_\mu(T_{\Delta x}^{\mu} + T_{\Delta x}^{-\mu}) + F_\mu(T_{\Delta x}^{\mu} - T_{\Delta x}^{-\mu})].$$

We now establish the inequality of the lemma for each of the summands of $Q_\alpha(h)$. First let

$$H = E_\mu(T^{\mu}_{\Delta x} + T^{-\mu}_{\Delta x})$$

and later let

$$H = F_\mu(T^{\mu}_{\Delta x} - T^{-\mu}_{\Delta x}).$$

In the first case, we obtain

$$\begin{aligned} D &= \langle\psi u, H(\psi u)\rangle - \langle\psi^2 u, H(u)\rangle \\ &= \int_{\mathbb{R}} \psi(x)[\psi(x+\mu\Delta x)-\psi(x)]u(x)^T E_\mu u(x+\mu\Delta x)\,dx \\ &\quad + \int_{\mathbb{R}} \psi(x)[\psi(x-\mu\Delta x)-\psi(x)]u(x)^T E_\mu u(x-\mu\Delta x)\,dx. \end{aligned}$$

In the second integral we can make the substitution $\tilde{x} = x-\mu\Delta x$ and then call $\tilde{x}$ x. Then we have

$$\begin{aligned} D &= \int_{\mathbb{R}} \psi(x)[\psi(x+\mu\Delta x)-\psi(x)]u(x)^T E_\mu u(x+\mu\Delta x)\,dx \\ &\quad - \int_{\mathbb{R}} \psi(x+\mu\Delta x)[\psi(x+\mu\Delta x)-\psi(x)]u(x+\mu\Delta x)^T E_\mu u(x)\,dx. \end{aligned}$$

By the symmetry of E_μ,

$$u(x)^T E_\mu u(x+\mu\Delta x) = u(x+\mu\Delta x)^T E_\mu u(x)$$

and therefore,

$$D = -\int_{\mathbb{R}} [\psi(x+\mu\Delta x)-\psi(x)]^2 u(x)^T E_\mu u(x+\mu\Delta x)\,dx$$

$$|D| \le \tilde{L}^2 4k^2(\Delta x)^2\, \|E_\mu\|_2\, \|u\|_2^2.$$

In the second case, we have

$$\begin{aligned} D &= \int_{\mathbb{R}} \psi(x)[\psi(x+\mu\Delta x)-\psi(x)]u(x)^T F_\mu u(x+\mu\Delta x)\,dx \\ &\quad - \int_{\mathbb{R}} \psi(x)[\psi(x-\mu\Delta x)-\psi(x)]u(x)^T F_\mu u(x-\mu\Delta x)\,dx. \end{aligned}$$

Using the fact that

$$u(x)^T F_\mu u(x+\mu\Delta x) = -u(x+\mu\Delta x)^T F_\mu u(x)$$

and making the same substitution, we get

$$D = -\int_{\mathbb{R}} [\psi(x+\mu\Delta x)-\psi(x)]^2 u(x)^T F_\mu u(x+\mu\Delta x)dx$$

$$|D| \leq \tilde{L}^2 4k^2(\Delta x)^2 \,||F_\mu||_2\, ||u||_2^2. \qquad \square$$

Proof of Theorem 9.34: We set

$$\gamma = h^{-1/2} \quad \text{and} \quad \beta = 1/\gamma = h^{1/2}.$$

If $h \leq h_0 = (\lambda/6k)^2$, then

$$6k\gamma\Delta x = 6k\gamma h/\lambda \leq 1$$

and

$$2k\Delta x = 2kh/\lambda < \beta.$$

Thus the conditions of Lemmas 9.38 and 9.39 are satisfied. We can approximate the matrices $D_\mu(x)$ on the interval $[-3\beta,3\beta]$ with a linear combination of the matrices $D_\mu(-3\beta)$ and $D_\mu(3\beta)$, namely,

$$D_\mu(x) = \frac{1}{6\beta}[(3\beta-x)D_\mu(-3\beta)+(3\beta+x)D_\mu(3\beta)] + Z(x,\mu,\beta).$$

Since the second derivatives of the elements of D_μ are bounded independently of x, we have $||Z(x,\mu,\beta)||_2 \leq M_2\beta^2$. The constant M_2 depends only on the functions D_μ, i.e., on the method M_D. On the smaller interval $[-2\beta,2\beta]$, we have

$$\frac{1}{6\beta}(3\beta \pm x) \geq \frac{1}{6}.$$

The functions

$$\psi_1(x) = [(3\beta-x)/6\beta]^{1/2}$$

$$\psi_2(x) = [(3\beta+x)/6\beta]^{1/2}$$

are continuously differentiable on this interval. The absolute values of the first derivatives of ψ_1 and ψ_2 are bounded by $1/\beta$. For $x,\tilde{x} \in [-2\beta, 2\beta]$, $j = 1,2$, it follows that

$$|\psi_j(x) - \psi_j(\tilde{x})| \le \frac{1}{\beta}|x-\tilde{x}|. \tag{9.40}$$

The above interpolation formula for $D_\mu(x)$ can now be written as

$$D_\mu(x) = \psi_1(x)^2 D_\mu(-3\beta) + \psi_2(x)^2 D_\mu(3\beta) + Z(x,\mu,\beta),$$
$$\mu = -2k(1)2k, \quad x \in [-2\beta, 2\beta].$$

For fixed but arbitrary $u \in L^2(\mathbb{R}, \mathbb{R}^n)$ we define $u_\nu = \eta_\nu(\cdot)u(\cdot)$, $\nu = -\infty(1)\infty$ (cf. Lemmas 9.37 and 9.38). We have $\text{Support}(u_\nu) \subset [(\nu-1)\beta, (\nu+1)\beta]$, and hence by Lemma 9.39 together with equation (9.40),

$$\begin{aligned}
&|\langle u_o,Q(h)(u_o)\rangle-\langle\psi_1u_o,Q_{-3\beta}(h)(\psi_1u_0)\rangle-\langle\psi_2u_0,Q_{3\beta}(h)(\psi_2u_o)\rangle| \\
&\quad \le |\langle u_o,Q(h)(u_o)\rangle-\langle\psi_1^2u_0,Q_{-3\beta}(h)(u_0)\rangle-\langle\psi_2^2u_0,Q_{3\beta}(h)(u_0)\rangle| \\
&\quad + |\langle\psi_1^2u_0,Q_{-3\beta}(h)(u_0)\rangle-\langle\psi_1u_0,Q_{-3\beta}(h)(\psi_1u_0)\rangle| \\
&\quad + |\langle\psi_2^2u_0,Q_{3\beta}(h)(u_0)\rangle-\langle\psi_2u_0,Q_{3\beta}(h)(\psi_2u_0)\rangle| \\
&\quad \le [(4k+1)M_2\beta^2 + 2\tilde{K}\beta^{-2}(\Delta x)^2]\,\|u_0\|_2^2 \\
&\quad = (2kM_2 + 2\tilde{K}/\lambda^2)h\,\|u_0\|_2^2 = M_3h\,\|u_0\|_2^2.
\end{aligned}$$

The scalar products

$$\langle\psi_1u_0,Q_{-3\beta}(h)(\psi_1u_0)\rangle$$
$$\langle\psi_2u_0,Q_{3\beta}(h)(\psi_2u_0)\rangle$$

are both nonnegative, which proves that

$$\langle u_o,Q(h)(u_o)\rangle \ge -M_3h\,\|u_o\|_2^2.$$

Analogously, we have for all $\nu = -\infty(1)\infty$ that

$$\langle u_\nu, Q(h)(u_\nu)\rangle \geq -M_3 h \, \|u_\nu\|_2^2.$$

Lemma 9.38 then implies that

$$\langle u, Q(h)(u)\rangle \geq -M_1\gamma^2 L^2(\Delta x)^2 \, \|u\|_2^2 - M_3 h \sum_{\nu=-\infty}^{\infty} \|u_\nu\|_2^2.$$

Since $\gamma^2(\Delta x)^2 = h/\lambda^2$ and since by Lemma 9.37(3),

$$\sum_{\nu=-\infty}^{\infty} \|u_\nu\|_2^2 = \|u\|_2^2,$$

we have that

$$\langle u, Q(h)(u)\rangle \geq -M_4 h \, \|u\|_2^2$$

where

$$M_4 = M_1 L^2/\lambda^2 + M_3.$$

Applying Lemma 9.35 yields

$$\langle u, I - C(h)^T \circ C(h)(u)\rangle \geq -M_4 h \, \|u\|_2^2 - M_5 h \, \|u\|_2^2$$

$$\langle u,u\rangle - \langle C(h)(u), C(h)(u)\rangle \geq -M_6 h\langle u,u\rangle$$

$$\langle C(h)(u), C(h)(u)\rangle \leq (1+M_6 h) \langle u,u\rangle$$

$$\|C(h)\|_2 \leq (1+M_6 h)^{1/2} \leq 1+M_7 h. \qquad \square$$

We will not give an application of Theorem 9.34 at this time. In the next section, we will return to initial value problems in several variables, and there we will use the theorem to help show that the generalization of the Lax-Wendroff method to variable coefficients is stable. There does not seem to be any simpler means of establishing the stability of that method.

10. Initial value problems in several space variables

So far we have only investigated partial differential equations in one time variable, t and *one* space variable, x. For *pure* initial value problems, the main results of the previous chapter can be extended effortlessly to partial differential equations in $\mathbb{R}^{m+1}$ with one time variable t and m space variables $x_1, \ldots, x_m$. We have avoided the case $m > 1$ until now for didactic and notational reasons. We will explain the situation for $m > 1$ in this section with the aid of typical examples.

Initial boundary value problems, in contrast to pure initial value problems, are substantially more complicated when $m > 1$. The additional difficulties, which we cannot discuss here, arise because of the varying types of boundaries. The problems resemble those which arise in the study of boundary value problems (undertaken in Part II).

Throughout this chapter we will use the notation

$$x = (x_1,\ldots,x_m) \in \mathbb{R}^m, \qquad y = (y_1,\ldots,y_m) \in \mathbb{R}^m$$

$$dx = dx_1 \ldots dx_m, \qquad \langle x,y\rangle = \sum_{\mu=1}^{m} x_\mu y_\mu .$$

In addition, we introduce the multi-indices

$$s = (s_1,\ldots,s_m) \in \mathbb{Z}^m$$

The translation operator $T_{\Delta x}$ of $\mathbb{R}^1$ (cf. Definition 6.2) is replaced by m different operators $T_{k\mu}$ in $\mathbb{R}^m$:

$$e^{(\mu)} = (e_1^{(\mu)},\ldots,e_m^{(\mu)}), \qquad \mu = 1(1)m$$

$$e_\nu^{(\mu)} = \delta_{\mu\nu}, \qquad \mu,\nu = 1(1)m$$

$$T_{k\mu}(x) = x + ke^{(\mu)}, \quad x \in \mathbb{R}^m, \quad k \in \mathbb{R}, \quad \mu = 1(1)m.$$

For all $f \in L^2(\mathbb{R}^m,\mathbb{C}^n)$ let

$$T_{k\mu}(f)(x) = f(T_{k\mu}(x)), \quad x \in \mathbb{R}^m.$$

With this definition, the translation operators become bijective continuous linear mappings of $L^2(\mathbb{R}^m,\mathbb{C}^n)$ into itself. They commute with each other. For all $\nu \in \mathbb{Z}$ we have

$$T_{k\mu}^{\nu} = T_{\nu k,\mu} .$$

Let $B \in C^0(\mathbb{R}^m,\mathrm{MAT}(n,n,\mathbb{C}))$ have bounded spectral norm $||B(x)||_2$. The map

$$f(\cdot) \to B(\cdot)f(\cdot)$$

is a bounded linear operator in $L^2(\mathbb{R}^m,\mathbb{C}^n)$. The commutativity relations for $T_{k\mu}$ and B are

$$T_{k\mu}\circ B(x) = B(T_{k\mu}(x))T_{k\mu}$$

$$B(x)\circ T_{k\mu} = T_{k\mu}\circ B(T_{-k\mu}(x)).$$

In many cases, B will satisfy a Lipschitz condition

$$||B(x)-B(y)||_2 \leq L\,||x-y||_2.$$

Then,

$$||B(x)\circ T_{k\mu}-T_{k\mu}\circ B(x)||_2 \leq L|k|.$$

For $k \in \mathbb{R}$ and arbitrary multi-indices s, we define

$$T_k^s = \prod_{\mu=1}^{m} T_{k\mu}^{s_\mu} .$$

The difference method

$$M_D = \{C(h)\,|\,h \in (0,h_0]\}$$

can now be written in the form

$$C(h) = \Big(\sum_s B_s(x,h)T_k^s\Big)^{-1}\circ\Big(\sum_s A_s(x,h)T_k^s\Big).$$

All sums, here and henceforth, extend only over finitely many multi-indices. Also we assume that for all s, x, and h,

$$A_s(x,h), B_s(x,h) \in \mathrm{MAT}(n,n,\mathbb{R})$$

$$k = h/\lambda \quad \text{or} \quad k = \sqrt{h/\lambda} \quad \text{where} \quad \lambda \in \mathbb{R}_+.$$

Analogously to Definition 9.12, we can assign to each difference method an amplification matrix

$$G(h,y,x) = \Big(\sum_s \exp(ik\langle s,y\rangle)B_s(x,h)\Big)^{-1}\Big(\sum_s \exp(ik\langle x,y\rangle)A_s(x,h)\Big).$$

If the matrices $A_s(x,h)$ and $B_s(x,h)$ are all independent of x, we speak of a method with space-free coefficients. Then we abbreviate

$$A_s(x,h),\ B_s(x,h),\ G(h,y,x)$$

to

$$A_s(h),\ B_s(h),\ G(h,y).$$

The stability of a method with space-free coefficients can again be determined solely on the basis of the amplification matrix $G(h,y)$. Theorems 9.13 and 9.15 extend word for word to the Banach spaces $L^2(\mathbb{R}^m,\mathbb{C}^n)$ if $\mathbb{F}$ is replaced by $\mathbb{R}^m$. Theorems 9.16, 9.31, and 9.34 also carry over in essence. All the proofs are almost the same as for the case $m = 1$. Basically, the only additional item we need is the m-dimensional Fourier transform, which is defined for all $f \in L^2(\mathbb{R}^m,\mathbb{C}^n)$ by

$$a_\nu(y) = (2\pi)^{-m/2}\int_{\|x\|_2\le\nu} f(x)\exp(-i\langle x,y\rangle)dx$$

$$\mathscr{F}_n(f) = \lim_{\nu\to\infty} a_\nu.$$

The limit is taken with respect to the topology of $L^2(\mathbb{R}^m,\mathbb{C}^n)$. As in the case $m = 1$, we have:

(1) $\mathcal{F}_n$ is bijective.

(2) $\|\mathcal{F}_n\| = \|\mathcal{F}_n^{-1}\| = 1$.

(3) $\mathcal{F}_n(T_k^s(f))(\cdot) = \exp(ik\langle s,\cdot\rangle)\,\mathcal{F}_n(f)(\cdot)$.

For differential equations with constant coefficients, the best stability criteria are obtained from the amplification matrix. Even when the coefficients are not constant, this route is still available in certain cases, for example, with hyperbolic systems. Also, one can define positive and positive definite methods. They are always stable.

For positive definite methods, we need

(1) $C(h) = \sum_s A_s(x)T_k^s$ with $k = h/\lambda$.

(2) $I = \sum_s A_s(x)$.

(3) All matrices $A_s(x)$ are real, symmetric, and positive semidefinite.

(4) For all multi-indices s and all $x,y \in \mathbb{R}^m$ we have

$$\|A_s(x)-A_s(y)\|_2 \le L\|x-y\|_2.$$

where $L > 0$.

We consider positive methods only in the scalar case $m = 1$. If $m > 1$, they are of little significance for systems of differential equations. This is due to Condition (3) of Definition 8.4, which implies that the coefficients of the difference operators commute. For $m > 1$, the coefficients of most systems of differential equations do not commute, and hence neither do the coefficients of the difference operators.

The positive methods occur in the Banach space

$$B = \{f \in C^0(\mathbb{R}^m, \mathbb{C}) \mid \lim_{\|x\|_\infty \to \infty} |f(x)| = 0\}$$

Here the norm is the maximum norm. The operators T_k^s are also defined in this space. For positive methods, we need

(1) $e(x,h)C(h) = \sum_s a_s(x,h)T_k^s + \sum_s b_s(x,h)T_k^s \circ C(h)$

$k = h/\lambda$ or $k = \sqrt{h/\lambda}$ where $\lambda \in \mathbb{R}_+$.

(2) $e(x,h) = \sum_s [a_s(x,h) + b_s(x,h)]$, $x \in \mathbb{R}^m$, $h \in (0,h_0]$

(3) $a_s, b_s \in C^0(\mathbb{R}^m, \mathbb{R})$

$a_s(x,h) \geq 0$, $b_s(x,h) \geq 0$

$\sum_s a_s(x,h) \geq 1$.

For $m > 1$, the so-called product methods occupy a special place. They arise from the "multiplication" of methods for $m = 1$. Their stability follows directly from the stability of the factors. More precisely, we have the following.

Theorem 10.1: Let B be a Banach space and

$$M_{D,\mu} = \{C_\mu(h) \mid h \in (0,h_0]\}, \quad \mu = 1(1)m$$

a family of difference methods for m (possibly different) properly posed problems. The difference method

$$M_D = \{C(h) \mid h \in (0,h_0]\}$$

is defined by

$$C(h) = C_1(h)C_2(h) \ldots C_m(h).$$

M_D is stable if one of the following two conditions is

satisfied.

(1) For fixed $h \in (0,h_o]$, the operators $C_\mu(h)$, $\mu = 1(1)m$, commute.

(2) There exists a $K > 0$ such that

$$\|C_\mu(h)\| \leq 1+Kh, \quad \mu = 1(1)m, \quad h \in (0,h_o].$$

Proof: If (1) holds, we can write

$$\|C(h)^n\| \leq \prod_{\mu=1}^{m} \|C_\mu(h)^n\| .$$

Since each of the m factors is bounded, so is the product. If (2) holds, we have the inequalities

$$\|C(h)^n\| \leq (1+Kh)^{mn} \leq \exp(mKT). \quad \square$$

We now present a number of methods for the case $m > 1$. In all the examples, $\lambda = h/k$ or $\lambda = h/k^2$, depending on the order of the differential equation.

Example 10.2:

Differential equation:

$$u_t(x,t) = \sum_{\mu=1}^{m} \partial_\mu[a_\mu(x)\partial_\mu u(x,t)], \quad \partial_\mu = \partial/\partial x_\mu$$

where $a_\mu \in C^1(\mathbb{R}^m, \mathbb{R})$, $0 < \beta \leq a_\mu(x) \leq K$

and $|\partial_\nu a_\mu(x)| \leq \tilde{K}$, $\nu = 1(1)m$.

Method:

$$C(h) = [I-(1-\alpha)\lambda H]^{-1}\circ[I+\alpha\lambda H]$$

where

$$H = \sum_{\mu=1}^{m} [a_\mu(x + \tfrac{1}{2}ke_\mu)(T_{k\mu}-I) + a_\mu(x- \tfrac{1}{2}ke_\mu)(T_{k\mu}^{-1}-I)]$$

and $\alpha \in [0,1]$.

Amplification matrix:

$$G(h,y,x) = [1+\alpha\lambda\hat{H}]/[1-(1-\alpha)\lambda\hat{H}]$$

where

$$\hat{H} = \sum_{\mu=1}^{m} \{a_\mu(x+\tfrac{1}{2}ke_\mu)[\exp(iky_\mu)-1] + a_\mu(x-\tfrac{1}{2}ke_\mu)[\exp(-iky_\mu)-1]\}.$$

For

$$2mK\alpha\lambda \leq 1$$

the method is positive, and hence stable. Subject to the usual regularity conditions, the global error is at most

$$O(h) + O(k^2) \quad \text{for} \quad \alpha \neq 1/2$$

$$O(h^2) + O(k^2) \quad \text{for} \quad \alpha = 1/2 \quad \text{(Crank-Nicolson Method)}.$$

If all a_μ are constant, then

$$\hat{H} = -2\sum_{\mu=1}^{m} a_\mu(1-\cos ky_\mu) \geq -4mK.$$

Precisely when

$$2mK(2\alpha-1)\lambda \leq 1$$

we have

$$|G(h,y)| \leq 1$$

and hence stability.

Theoretically speaking, there is nothing new in the case $m > 1$ that wasn't contained in the case $m = 1$. Practically speaking, the implicit methods ($\alpha < 1$) with few restricting stability conditions are very time consuming for $m > 1$. A large system of linear equations has to be solved for each time interval. For $m = 1$, the matrix of the system is triangular, and five arithmetic operations per lattice point and time interval are required for the solution. Thus the total effort required by an implicit method is not very large in the case $m = 1$. For $m > 1$, the matrix of the system of equations is no longer triangular. Even with an optimal ordering of the lattice points, we get a band matrix where the width of the band grows as the number of lattice points increases. The solution of the system then requires considerable effort. □

Example 10.3:

Differential equation:

$$u_t(x_1,x_2,t) = a\partial_{11}^2 u(x_1,x_2,t) + 2b\partial_1\partial_2 u(x_1,x_2,t) + c\partial_{22}^2 u(x_1,x_2,t)$$

where $a > 0$, $c > 0$, and $ac > b^2$.

Method:

$$C(h) = [I-(1-\alpha)\lambda H]^{-1}\circ[I+\alpha\lambda H]$$

where $\alpha \in [0,1]$ and

$$H = a(T_{k1}+T_{k1}^{-1}-2I) + \tfrac{1}{2}b(T_{k1}-T_{k1}^{-1})(T_{k2}-T_{k2}^{-1}) + c(T_{k2}+T_{k2}^{-1}-2I).$$

Amplification matrix:

$$G(h,y) = [1+\alpha\lambda\hat{H}]/[1-(1-\alpha)\lambda\hat{H}]$$

where

$$\hat{H} = -2a[1-\cos(ky_1)] - 2b\sin(ky_1)\cdot\sin(ky_2) - 2c[1-\cos(ky_2)].$$

The differential equation differs from the one in the previous example in the term $2b\partial_1\partial_2 u(x_1,x_2,t)$. Since $a > 0$, $c > 0$, and $ac > b^2$, it is nevertheless parabolic. When $b \neq 0$, the method is never positive, regardless of α and λ. A stability criterion is obtainable only through the amplification matrix.

We set $\tilde{\omega}_1 = \frac{1}{2}ky_1$ and $\tilde{\omega}_2 = \frac{1}{2}ky_2$, and get

$$\hat{H} = -4a \sin^2\tilde{\omega}_1 - 8b \sin\tilde{\omega}_1 \sin\tilde{\omega}_2 \cos\tilde{\omega}_1 \cos\tilde{\omega}_2 - 4c \sin^2\tilde{\omega}_2$$
$$= -4(a+c) + 4a \cos^2\tilde{\omega}_1 - 8b \sin\tilde{\omega}_1 \sin\tilde{\omega}_2 \cos\tilde{\omega}_1 \cos\tilde{\omega}_2 + 4c \cos^2\tilde{\omega}_2.$$

Also, for $\tilde{\omega}_1 \neq \ell\pi/2$ and $\tilde{\omega}_2 \neq \ell\pi/2$ ($\ell \in \mathbb{Z}$), let

$$\varepsilon = \operatorname{sgn}(\sin\tilde{\omega}_1 \sin\tilde{\omega}_2)$$
$$\eta = \operatorname{sgn}(\cos\tilde{\omega}_1 \cos\tilde{\omega}_2).$$

Thus we obtain two representations for $\hat{H}$,

$$\hat{H} = -4(\sqrt{a} \sin\tilde{\omega}_1 - \varepsilon\sqrt{c} \sin\tilde{\omega}_2)^2$$
$$-8[\sqrt{ac}|\sin\tilde{\omega}_1 \sin\tilde{\omega}_2| + b \sin\tilde{\omega}_1 \sin\tilde{\omega}_2 \cos\tilde{\omega}_1 \cos\tilde{\omega}_2]$$

and

$$\hat{H} = -4(a+c) + 4(\sqrt{a} \cos\tilde{\omega}_1 - \eta\sqrt{c} \cos\tilde{\omega}_2)^2$$
$$+8[\sqrt{ac}\,|\cos\tilde{\omega}_1 \cos\tilde{\omega}_2| - b \sin\tilde{\omega}_1 \sin\tilde{\omega}_2 \cos\tilde{\omega}_1 \cos\tilde{\omega}_2].$$

Since $|b| < \sqrt{ac}$, the first term in the square brackets is always the dominant one. Hence,

$$-4(a+c) \leq \hat{H} \leq 0.$$

Equality can occur at both ends of the expression. For

$$2(a+c)(2\alpha-1)\lambda \leq 1$$

we have $|G(h,y)| \leq 1$. This stability condition is independent of b. On the other hand, if $b^2 > ac$, there always exist $\tilde{\omega}_1$ and $\tilde{\omega}_2$ such that $\hat{H} > 0$ and $|G(h,y)| > 1$. The method is unstable for all α and λ. □

Example 10.4: *ADI-method.*

Differential equation:

$$u_t(x_1,x_2,t) = a[\partial_{11}^2 u(x_1,x_2,t) + \partial_{22}^2 u(x_1,x_2,t)]$$
$$+ b_1\partial_1 u(x_1,x_2,t) + b_2\partial_2 u(x_1,x_2,t)$$

where $a > 0$ and $b_1, b_2 \in \mathbb{R}$.

Method:

$$C(h) = C_1(h/2) \circ C_2(h/2)$$

$$C_\rho(h/2) = [I - \tfrac{1}{2}a\lambda(T_{k\rho} - 2I + T_{k\rho}^{-1}) - \tfrac{1}{4}b_\rho k\lambda(T_{k\rho} - T_{k\rho}^{-1})]^{-1}$$

$$\circ\ [I + \tfrac{1}{2}a\lambda(T_{k\sigma} - 2I + T_{k\sigma}^{-1}) + \tfrac{1}{4}b_\sigma k\lambda(T_{k\sigma} - T_{k\sigma}^{-1})],$$

$$\rho = 1,2 \quad \text{and} \quad \sigma = 3-\rho.$$

Amplification matrix:

$$G(h,y) = G_1(h,y)G_2(h,y)$$

$$G_\rho(h,y) = \frac{1 - a\lambda(1-\cos\omega_\rho) + \frac{1}{2}ib_\rho\sqrt{h\lambda}\sin\omega_\rho}{1 + a\lambda(1-\cos\omega_\rho) - \frac{1}{2}ib_\rho\sqrt{h\lambda}\sin\omega_\rho}$$

and

$$\omega_1 = ky_1 \quad \text{and} \quad \omega_2 = ky_2.$$

The abbreviation *ADI* stands for *Alternating Direction Implicit method.* The first ADI method was described by Peaceman-Rachford (1955). The method is of great practical significance for the following reasons. Suppose that one of the fractions $|b_\rho|/a$ is very large. To avoid practical instabilities, one must then choose k very small. Otherwise, one immediately encounters difficulties such as those in Example 9.21. With an explicit method, the stability condition ($2ma\lambda \leq 1$) demands

$$ah \leq \frac{1}{4}k^2.$$

Hence h has to be chosen extremely small. Although implicit methods allow h to be chosen substantially larger, one nevertheless has to solve very large systems of equations because the lattice point separation k is small. Of course, this is also true for ADI methods. The difference with other implicit methods is in the structure of the system of equations. In each of the factors C_1 and C_2, the systems of equations decompose into independent subsystems for the lattice points $x_2 \equiv$ *constant* and $x_1 \equiv$ *constant*, and the matrices of the subsystems are triangular. Only five to eight arithmetic operations per lattice point and half time interval are required, and that is a justifiable effort.

Note that G_1 does not belong to C_1. The factors of the amplification matrix are exchanged in the representation of the method. Such a representation of the amplification matrix is only possible with constant coefficients.

In practice, one deals mostly with initial boundary value problems. Stability then depends also on the nature of the region. Thus we must caution that the following remark is directly applicable only to rectangular regions. Rather different results can occur when the region is not rectangular or the differential equations do not have constant coefficients.

We have

$$|G_\rho(h,y)|^2 = \frac{[1-a\lambda(1-\cos\omega_\rho)]^2 + \frac{1}{4}b_\rho^2\, h\lambda\, \sin^2\omega_\rho}{[1+a\lambda(1-\cos\omega_\rho)]^2 + \frac{1}{4}b_\rho^2\, h\lambda\, \sin^2\omega_\rho} \leq 1$$

and hence

$$|G(h,y)| \leq 1.$$

C is stable for all λ, although the factors C_1 and C_2 are unstable for large λ. In order to solve the triangular system of equations without pivoting, we need $\frac{1}{2}a\lambda \geq \frac{1}{4}|b_\rho|k\lambda$. This means

$$k \leq 2\min(a/|b_1|,\ a/|b_2|).$$

If additionally, $a\lambda \leq 1$, then C is also positive. In practice, it suffices to limit k; h can be chosen arbitrarily large. □

Example 10.5:

Differential equation:

$$u_t(x_1,x_2,t) = ia\partial_1\partial_2 u(x_1,x_2,t), \quad a \in \mathbb{R} - \{0\}.$$

Method:

$$C(h) = [I-(1-\alpha)\lambda H]^{-1}\circ[I+\alpha\lambda H]$$

where $\alpha \in [0,1]$ and

$$H = \frac{1}{4}\, ia(T_{k1}-T_{k1}^{-1})(T_{k2}-T_{k2}^{-1}).$$

Amplification matrix:

$$G(h,y) = \frac{1-ia\alpha\lambda \sin\omega_1 \sin\omega_2}{1+ia(1-\alpha)\lambda \sin\omega_1 \sin\omega_2}$$

where $\omega_1 = ky_1$ and $\omega_2 = ky_2$. The differential equation is sometimes called a pseudo-parabolic equation. It corresponds to the real system

$$\frac{\partial}{\partial t} u_1(x_1,x_2,t) = -a\partial_1\partial_2 u_2(x_1,x_2,t)$$

$$\frac{\partial}{\partial t} u_2(x_1,x_2,t) = a\partial_1\partial_2 u_1(x_1,x_2,t).$$

It follows that for $u \in C^4(\mathbb{R},\mathbb{C})$

$$\frac{\partial^2}{\partial t^2} u_1(x_1,x_2,t) = -a^2\partial_{11}^2\partial_{22}^2 u_1(x_1,x_2,t).$$

Solutions of the differential equation can be computed with Fourier transforms, analogously to Example 9.9. The method is formally the method of Example 10.3. It is not positive, but is stable for $\alpha \leq 1/2$. □

Example 10.6: *Product method* for symmetric hyperbolic systems.

Differential equation:

$$u_t(x_1,x_2,t) = A_1(x)\partial_1 u(x_1,x_2,t) + A_2(x)\partial_2 u(x_1,x_2,t).$$

where

$$A_\mu \in C^2(\mathbb{R}^2, \mathrm{MAT}(n,n,\mathbb{R}))$$

$$A_\mu(x) \text{ symmetric}, \quad \rho(A_\mu(x)) \text{ bounded}$$

$$\|A_\mu(x)-A_\mu(\tilde{x})\|_2 \leq L\,\|x-\tilde{x}\|_2, \quad x,\tilde{x} \in \mathbb{R}^2, \quad \mu = 1,2.$$

Method:

$$\begin{aligned} C(h) = \tfrac{1}{4}\{ & [I+\lambda A_1(x)][I+\lambda A_2(x)]T_{k1}\circ T_{k2} \\ & + [I+\lambda A_1(x)][I-\lambda A_2(x)]T_{k1}\circ T_{k2}^{-1} \\ & + [I-\lambda A_1(x)][I+\lambda A_2(x)]T_{k1}^{-1}\circ T_{k2} \\ & + [I-\lambda A_1(x)][I-\lambda A_2(x)]T_{k1}^{-1}\circ T_{k2}^{-1}\}. \end{aligned}$$

The method can also be derived from the Friedrichs method for $m = 1$. To see this, consider the two systems of differential equations

$$u_t(x_1,x_2,t) = A_1(x)\partial_1 u(x_1,x_2,t)$$

$$u_t(x_1,x_2,t) = A_2(x)\partial_2 u(x_1,x_2,t).$$

In the first system, there is no derivative with respect to x_2, and in the second, there is none with respect to x_1. Thus, each system can be solved with the Friedrichs method. The variables x_2 and x_1, respectively, only play the role of parameters. The methods are

$$C_\mu(h) = \tfrac{1}{2}\,[I+\lambda A_\mu(x)]T_{k\mu} + \tfrac{1}{2}\,[I-\lambda A_\mu(x)]T_{k\mu}^{-1}.$$

For $\lambda \sup \|A_\mu(x)\| \leq 1$ the methods are positive definite. By Theorem 8.12, there is then a $K > 0$ such that

$$\|C_\mu(h)\| \leq 1+Kh, \qquad \mu = 1,2, \qquad h \in (0,h_0].$$

By Theorem 10.1, the product is stable.

$$\begin{aligned}\tilde{C}(h) &= C_1(h)\circ C_2(h)\\ &= \tfrac{1}{4}\{[I+\lambda A_1(x)][I+\lambda A_2(x+ke_1)]T_{k1}\circ T_{k2}\\ &\quad + [I+\lambda A_1(x)][I-\lambda A_2(x+ke_1)]T_{k1}\circ T_{k2}^{-1}\\ &\quad + [I-\lambda A_1(x)][I+\lambda A_2(x-ke_1)]T_{k1}^{-1}\circ T_{k2}\\ &\quad + [I-\lambda A_1(x)][I-\lambda A_2(x-ke_1)]T_{k1}^{-1}\circ T_{k2}^{-1}\}.\end{aligned}$$

$\tilde{C}$ and C agree up to terms of order $O(h)$. Hence C is also stable for

$$\lambda \max_{\mu=1,2} \sup_{x\in\mathbb{R}^2} \rho(A_\mu(x)) \leq 1.$$

The consistency of a product method also follows immediately from the consistency of the factors. We would like to demonstrate this fact by means of this example. Let $u \in C_0^2(\mathbb{R}^2,\mathbb{C}^n)$. We have

$$\begin{aligned}h^{-1}[\tilde{C}(h)-I](u) &= h^{-1}[C_1(h)-I](u) + h^{-1}[C_2(h)-I](u)\\ &\quad + h\{h^{-2}[C_1(h)-I]\circ[C_2(h)-I](u)\}.\end{aligned}$$

Since the Friedrichs method is consistent, the summands on the right side are approximations for

$$A_1(x)\partial_1 u(x,t),\ A_2(x)\partial_2 u(x,t)$$

and

$$hA_1(x)\partial_1[A_2(x)\partial_2 u(x,t)].$$

Thus, up to $O(h)$, the left side is an approximation for

$$A_1(x)\partial_1 u(x,t) + A_2(x)\partial_2 u(x,t).$$

This establishes consistency for $\tilde{C}$. For simplicity, $\tilde{C}$ was replaced by C. This doesn't affect consistency, since

$$\frac{4}{\lambda}\,[\tilde{C}(h)-C(h)](u)$$

$$= [I+\lambda A_1(x)][A_2(x+ke_1)-A_2(x)]T_{k1}\circ[T_{k2}-T_{k2}^{-1}](u)$$

$$+ [I-\lambda A_1(x)][A_2(x-ke_1)-A_2(x)]T_{k1}^{-1}\circ[T_{k2}-T_{k2}^{-1}](u).$$

The difference is obviously of order $O(h^2)$. It follows that

$$h^{-1}[\tilde{C}(h)-I](u) - h^{-1}[C(h)-I](u) = O(h).$$

Usually C and $\tilde{C}$ are not positive definite, because the matrix $A_1(x)A_2(x)$ is not symmetric. This deficiency can be remedied. Let C^* be formed from C by exchanging A_1 and A_2 as well as T_{k1} and T_{k2}. Then the method $(C + C^*)/2$ is positive definite. All of the methods mentioned here are too complicated for practical considerations. Further details on product methods can be found in Janenko (1971). □

Example 10.7: *m-dimensional Friedrichs method.*

Differential equation:

$$u_t(x,t) = \sum_{\mu=1}^{m} A_\mu(x)\partial_\mu u(x,t)$$

where

$$A_\mu \in C^2(\mathbb{R}^m, \mathrm{MAT}(n,n,\mathbb{R})) \quad \text{symmetric}$$

$$\rho(A_\mu(x)) \quad \text{bounded}$$

$$\|A_\mu(x)-A_\mu(\tilde{x})\| \le L\,\|x-\tilde{x}\|_2, \quad x,\tilde{x} \in \mathbb{R}^m, \quad \mu = 1(1)m.$$

Method:

$$C(h) = I + \tfrac{1}{2}\lambda \sum_{\mu=1}^{m} A_\mu(x)(T_{k\mu}-T_{k\mu}^{-1}) + \frac{r}{2m}\sum_{\mu=1}^{m}(T_{k\mu}-2I+T_{k\mu}^{-1})$$

with $r \in \mathbb{R}$.

Amplification matrix:

$$G(h,y,x) = (1-r)I + \lambda i \sum_{\mu=1}^{m} A_\mu(x)\sin\omega_\mu + \frac{r}{m}\,I\sum_{\mu=1}^{m}\cos\omega_\mu$$

where $\omega_\mu = ky_\mu$ for $\mu = 1(1)m$. The differential equation constitutes a symmetric hyperbolic system. The theory can be found in Mizohata (1973). The case $m = 2$ was covered in the preceding example.

The m-dimensional wave equation

$$v_{tt}(x,t) = \sum_{\mu=1}^{m} b_\mu(x)\partial_\mu[b_\mu(x)\partial_\mu v(x,t)]$$

$$b_\mu \in C^2(\mathbb{R}^m, \mathbb{R}_+)$$

can be reduced to such a system by means of the substitution

$$u(x,t) = (v_t(x,t),\ b_1(x)\partial_1 v(x,t),\dots,b_m(x)\partial_m v(x,t)).$$

In this special case, the coefficients of the system are elements of $MAT(m+1,m+1,\mathbb{R})$:

$$A_\mu(x) = (a_{\sigma\tau}^{(\mu)}(x))$$

where

$$a_{\sigma\tau}^{(\mu)}(x) = \begin{cases} b_\mu(x) & \text{for } \sigma = 1 \text{ and } \tau = \mu+1 \\ b_\mu(x) & \text{for } \tau = 1 \text{ and } \sigma = \mu+1 \\ 0 & \text{otherwise.} \end{cases}$$

For $m = r = 1$, this obviously is the Friedrichs method previously considered. For $m = 2$, this is simpler than the product method of Example 10.6. For $m > 2$, the m-dimensional Friedrichs method is substantially simpler than the product methods which can be created for these cases.

C is consistent for all $\lambda \in \mathbb{R}_+$ and all $r \in \mathbb{R}$, but we skip the proof. C is positive definite exactly when

$$r \in (0,1] \quad \text{and} \quad \lambda \max_{\mu=1(1)m} \sup_{x\in\mathbb{R}^m} \rho(A_\mu(x)) \le r/m,$$

for it is exactly under these conditions that all the matrices

$$(1-r)I, \quad \lambda A_\mu + \frac{r}{m}I \quad \text{and} \quad -\lambda A_\mu + \frac{r}{m}I$$

are positive semidefinite. For $A_\mu(x) = cI$, $\mu = 1(1)m$, $c \in \mathbb{R}$ and $r = 1$ it follows for $\omega_\mu = \pi/2$ that

$$||G(h,y,x)||_2 = \lambda cm.$$

By Theorem 9.31, the stability condition, at least in this special case, agrees with the condition under which the method is positive definite. However, there are also cases in which the method is stable but not positive definite.

We want to compare the above condition on the Friedrichs method for $m = 2$, $r = 1$ with our stability condition for the product method. The former is

$$h \max_{\mu=1(1)2} \sup_{x\in\mathbb{R}^2} \rho(A_\mu(x)) \le k/2$$

and the latter,

$$h \max_{\mu=1(1)2} \sup_{x\in\mathbb{R}^2} \rho(A_\mu(x)) \le k.$$

However, one also has to take the separation of the lattice points into account. They are k and $k\sqrt{2}$ respectively (see Figure 10.8). For the product method, the ratio of the

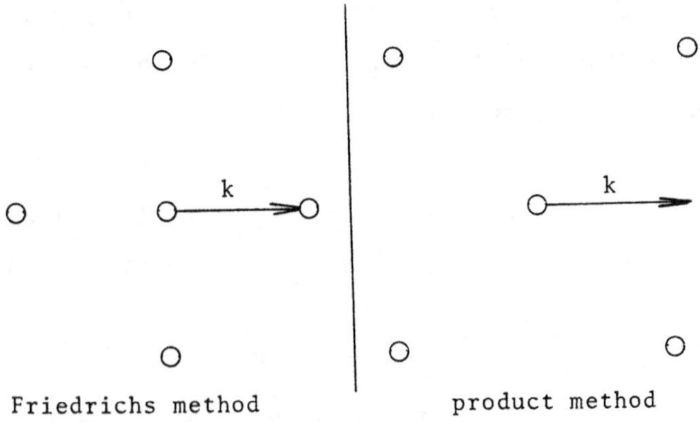

Figure 10.8

maximum possible time increment to this separation nevertheless is greater by a factor of $\sqrt{2}$. The product method provides a better approximation for the domain of determinancy of the differential equation. That is the general advantage of the product method, and guarantees it some attention. It is also called optimally stable, which is a way of saying that its stability condition is the Courant-Friedrichs-Lewy condition. □

Example 10.9: *Lax-Wendroff-Richtmyer method.*
Differential equation as in Example 10.7, with the additional conditions

$$A_\mu \in C^3(\mathbb{R}^m, \mathrm{MAT}(n,n,\mathbb{R}))$$

$$\|\partial_\sigma \partial_\tau A_\mu(x)\|_2 \text{ bounded}, \quad \mu = 1(1)m, \quad \sigma = 1(1)m, \quad \tau = 1(1)m.$$

Method:

$$C(h) = I + S\circ[I + \tfrac{1}{2}S + \tfrac{r}{2m} \sum_{\mu=1}^{m} (T_{k\mu} - 2I + T_{k\mu}^{-1})]$$

with $r \in \mathbb{R}$ and

$$S = \tfrac{1}{2}\lambda \sum_{\mu=1}^{m} A_\mu(x)(T_{k\mu} - T_{k\mu}^{-1}).$$

For $m = r = 1$ and $A_\mu(x) = A \equiv$ *constant*, we have the ordinary Lax-Wendroff method (cf. Example 9.26), for then, with $T_k = T_{k1}$,

$$S = \tfrac{1}{2}\lambda A(T_k - T_k^{-1})$$

$$C(h) = I + \tfrac{1}{2}\lambda A(T_k - T_k^{-1}) + \tfrac{1}{8}\lambda^2 A^2 (T_k^2 - 2I + T_k^{-2}) + \tfrac{1}{4}\lambda A(T_k^2 - T_k^{-2}) - \tfrac{1}{2}\lambda A(T_k - T_k^{-1})$$

$$C(h) = I + \tfrac{1}{4}\lambda A(T_k^2 - T_k^{-2}) + \tfrac{1}{8}\lambda^2 A^2 (T_k^2 - 2I + T_k^{-2}).$$

Replacing $2k$ by $\tilde{k}$ and $\lambda/2$ by $\tilde{\lambda}$ yields the expression

$$I + \frac{1}{2}\tilde{\lambda}A(T_{\tilde{k}} - T_{\tilde{k}}^{-1}) + \frac{1}{2}\tilde{\lambda}^2 A^2 (T_{\tilde{k}} - 2I + T_{\tilde{k}}^{-1}).$$

In any case, when $r = 1$, C only contains powers T_k^s with even sums $\sum_{\mu=1}^{m} s_\mu$. Figure 10.10 shows which lattice points

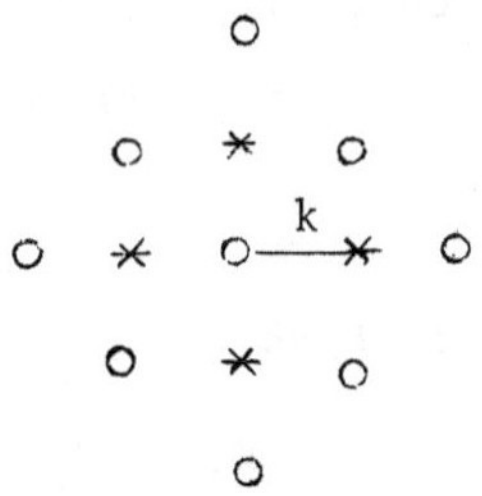

Figure 10.10

are used to compute C for $m = 2$. The points $*$ are used only when $r \neq 1$. The Lax-Wendroff-Richtmyer method has order of consistency $O(h^2)$. It is perhaps the most important method for dealing with symmetric hyperbolic systems. The choice $r \in (0,1)$ is sometimes to be recommended for generalizations to nonlinear problems.

We present a short sketch of the consistency proof. It follows from

$$u_t(x,t) = \sum_{\mu=1}^{m} A_\mu(x)\partial_\mu u(x,t)$$

that

$$u_{tt}(x,t) = \sum_{\mu=1}^{m} A_\mu(x)\partial_\mu\Big[\sum_{\nu=1}^{m} A_\nu(x)\partial_\nu u(x,t)\Big].$$

For $u \in C_0^3(\mathbb{R}^m, \mathbb{C}^n)$, one shows sequentially that

$$Su(x,t) = h\sum_{\mu=1}^{m} A_\mu(x)\partial_\mu u(x,t) + O(h^3)$$

$$\tfrac{1}{2}S^2 u(x,t) = \tfrac{1}{2}h^2 \sum_{\mu=1}^{m} A_\mu(x)\partial_\mu\Big[\sum_{\nu=1}^{m} A_\nu(x)\partial_\nu u(x,t)\Big] + O(h^3)$$

$$\frac{1}{2m} S \sum_{\mu=1}^{m} (T_{k\mu} - 2I + T_{k\mu}^{-1}) u(x,t)$$

$$= \frac{1}{2m} hk^2 \sum_{\mu=1}^{m} \sum_{\nu=1}^{m} A_\mu(x) \partial^3_{\mu\nu\nu} u(x,t) + O(h^3).$$

Altogether, this yields

$$C(h)u(x,t) = u(x,t) + hu_t(x,t) + \tfrac{1}{2}h^2 u_{tt}(x,t) + O(h^3)$$

$$= u(x,t+h) + O(h^3).$$

We want to derive sufficient stability criteria with the aid of the Lax-Nirenberg Theorem 9.34 (cf. Meuer 1972). However, the theorem is not directly applicable. In reducing C to the normal form

$$C(h) = \sum_s B_s(x,h) T_k^s$$

one ordinarily obtains coefficients $B_s(x,h)$ which actually depend on h. For example, for $m = 1$, we have

$$S^2 = \tfrac{1}{4}\lambda^2 A(x)A(x+g)T_k^2 - [\tfrac{1}{4}\lambda^2 A(x)A(x+g) + \tfrac{1}{4}\lambda^2 A(x)A(x-g)]I$$

$$+ \tfrac{1}{4}\lambda^2 A(x)A(x-g)T_k^{-2}$$

where $A = A_1$, $g = ke_1$, and $T_k = T_{k1}$. But the operator

$$C^*(h) = \sum_s B_s(x,0) T_k^s$$

has coefficients which are independent of h. One easily shows:

(1) $\|C(h) - C^*(h)\|_2 = O(h)$. Thus C and C^* are both stable or both unstable.

(2) For every $u \in C_o^3(\mathbb{R}^m, \mathbb{C}^n)$ we have

$$\|[C(h) - C^*(h)](u)\|_2 = O(h^2).$$

Hence C^* is at least first order consistent.

(3) C^* has amplification matrix

$$G^*(h,y,x) = I + \hat{S}[(1-r)I + \tfrac{1}{2}\hat{S} + \tfrac{r}{m}I \sum_{\mu=1}^{m} \cos \omega_\mu]$$

where $\omega_\mu = ky_\mu$ $(\mu = 1(1)m)$ and

$$\hat{S} = i\lambda \sum_{\mu=1}^{m} A_\mu(x)\sin \omega_\mu.$$

For $m = 1$, we have

$$\begin{aligned} C(h) - C^*(h) &= \tfrac{1}{8}\lambda^2 A(x)[A(x+g) - A(x)]T_k^2 \\ &\quad - \tfrac{1}{8}\lambda^2 A(x)[A(x+g) + A(x-g) - 2A(x)]I \\ &\quad + \tfrac{1}{8}\lambda^2 A(x)[A(x-g) - A(x)]T_k^{-2}. \end{aligned}$$

(1) follows immediately. The proof of (2) depends on the differences

$$C(h) - C^*(h) = \tfrac{1}{4}\lambda^2 gA(x)A'(x)[T_k^2 - T_k^{-2}] + O(h^2).$$

For $m > 1$, we leave this to the reader.

Now we can apply the Lax-Nirenberg Theorem 9.34 to C^*. Then it suffices for stability that

$$||G^*(h,y,x)||_2 \le 1.$$

By Theorem 9.31 this condition is also necessary. Now let H be the product of $G^*(h,y,x)$ with the Hermite transposed matrix $(G^*)^H$,

$$P = \sum_{\mu=1}^{m} A_\mu(x) \sin \omega_\mu$$

and let

$$\eta = \frac{1}{m} \sum_{\mu=1}^{m} \cos \omega_\mu.$$

We have $-1 \le \eta \le 1$. η assumes all these values independently of λ, r, and h. It follows from the Schwartz inequality that

$$\eta^2 \leq \frac{1}{m} \sum_{\mu=1}^{m} \cos^2 \omega_\mu .$$

H may be represented as follows:

$$H = [I - \tfrac{1}{2}\lambda^2 P^2 - i\lambda(1-r+r\eta)P][I - \tfrac{1}{2}\lambda^2 P^2 + i\lambda(1-r+r\eta)P]$$

$$H = (I - \tfrac{1}{2}\lambda^2 P^2)^2 + \lambda^2(1-r+r\eta)^2 P^2 .$$

P is real and symmetric. The eigenvalues α of P are real. To every eigenvalue α of P there corresponds an eigenvalue $\tilde{\alpha}$ of H. Thus it is both necessary and sufficient for the stability of C^* and C that $\tilde{\alpha}$ never be greater than 1. Hence we must examine the following inequality:

$$\tilde{\alpha} = (1 - \tfrac{1}{2}\lambda^2\alpha^2)^2 + \lambda^2(1-r+r\eta)^2\alpha^2 \leq 1.$$

For $\alpha = 0$, this is always satisfied. α is always zero only if all the matrices $A_\mu(x)$ are zero everywhere. In this trivial case, one has stability for all λ and r. In all other cases, we can restrict ourselves to those combinations of x and ω_μ with $\rho(P) > 0$, and consider an equivalent set of inequalities:

$$\tfrac{1}{4}\lambda^2\rho(P)^2 + (1-r+r\eta)^2 \leq 1.$$

For $r \leq 0$ or $r > 1$, $\eta < -1/r$, this inequality is self-contradictory. In the nontrivial cases, then, $r \in (0,1]$ is necessary. For these r, the inequalities can be converted to the equivalent inequalities

$$\frac{1}{4r}\lambda^2\rho(P)^2 + \eta^2 - (1-r)(1-\eta)^2 \leq 1. \qquad (10.11)$$

We now set

$$K = \max_{\mu=1(1)m} \sup_{x \in \mathbb{R}^m} \rho(A_\mu(x))$$

and assert that

$$r \in (0,1] \quad \text{and} \quad \tfrac{1}{2}\lambda K \leq \frac{\sqrt{r}}{m}. \tag{10.12}$$

is sufficient for the stability of C^* and C. To see this, let w be an arbitrary eigenvector of a matrix P with $\|w\|_2 = 1$ and $P(w) = \alpha w$.

$$\alpha = \sum_{\mu=1}^{m} [w^T A_\mu(x) w] \sin \omega_\mu.$$

Again we apply the Schwartz inequality to obtain

$$\alpha^2 \leq \sum_{\mu=1}^{m} [w^T A_\mu(x) w]^2 \cdot \sum_{\mu=1}^{m} \sin^2 \omega_\mu$$

$$\alpha^2 \leq m\, K^2 \sum_{\mu=1}^{m} \sin^2 \omega_\mu$$

$$\rho(P)^2 \leq m\, K^2 \sum_{\mu=1}^{m} \sin^2 \omega_\mu$$

$$\frac{1}{4r}\lambda^2 \rho(P)^2 \leq \frac{1}{m} \sum_{\mu=1}^{m} \sin^2 \omega_\mu$$

$$\frac{1}{4r}\lambda^2 \rho(P)^2 + \eta^2 \leq 1.$$

This inequality is somewhat stronger than (10.11).

There remains the question whether stability condition (10.12) is at all realistic. The answer is that whenever the matrices $A_\mu(x)$ have some special structure, it is worthwhile to refer back to the necessary and sufficient condition (10.11). A well-known example is the generalized wave equation. As noted in Example 10.7, for this equation we have

$$A_\mu(x) = (a_{\sigma\tau}^{(\mu)}(x))$$

where

$$a_{\sigma\tau}^{(\mu)}(x) = b_\mu(x) \quad \text{for} \quad \sigma = 1,\ \tau = \mu+1 \text{ and } \tau = 1,\ \sigma = \mu+1$$

$$a_{\sigma\tau}^{(\mu)}(x) = 0 \qquad \text{otherwise.}$$

Letting

$$K = \max_{\mu=1(1)m} \sup_{x \in \mathbb{R}^m} b_\mu(x)$$

we also have

$$K = \max_{\mu=1(1)m} \sup_{x \in \mathbb{R}^m} \rho(A_\mu(x)).$$

But in contrast to the above,

$$\rho(P)^2 \le K^2 \sum_{\mu=1}^{m} \sin^2\omega_\mu.$$

With the help of (10.11), one obtains a condition which is better by a factor of $\sqrt{m}$:

$$r \in (0,1] \quad \text{and} \quad \tfrac{1}{2}\lambda K \le \sqrt{\frac{r}{m}}\,.$$

The same weakening of the stability condition (factor $\sqrt{m}$) is also possible for the m-dimensional Friedrichs method, in the case of the generalized wave equation.

So far we have ignored general methods for which there are different spacings k_μ of the lattice points in the directions e_μ. Instead of $\lambda = h/k$ or $\lambda = h/k^2$, one could then have possibly m different step increment ratios $\lambda_\mu = h/k_\mu$ or $\lambda_\mu = h/k_\mu^2$. Such methods have definite practical significance. Now one can obtain $k_1 = k_2 = \dots = k_m$ with the coordinate transformation

$$\tilde{x}_\mu = \sigma_\mu x_\mu \quad \text{where} \quad \sigma_\mu > 0,\ \mu = 1(1)m$$

This transformation changes the coefficients of the differential equation. They are multiplied by σ_μ or σ_μ^2 or

$\sigma_\mu \sigma_\nu$. In many cases, the following approach has proved useful. First transform the coordinates so that the coefficients mapped into each other by the change of variables are nearly the same. For a symmetric hyperbolic system this means

$$\sup_{x \in \mathbb{R}^m} \rho(A_1(x)) = \ldots = \sup_{x \in \mathbb{R}^m} \rho(A_m(x)).$$

Then choose the increments independent of μ. This corresponds to a method with $k_\mu = k/\sigma_\mu$ in the original coordinate system.

11. Extrapolation methods

All of the concrete examples of difference methods which we have discussed so far have been convergent of first or second order. Such simple methods are actually of great significance in practice. This will come as a great surprise to anyone familiar with the situation for ordinary differential equations, for there in practice one doesn't consider methods of less than fourth order convergence.

High precision can only be achieved with methods of high order convergence. This is especially true for partial differential equations. Consider a method with m space variables, of k-th order, and with $h/\Delta x = \lambda \equiv$ *constant*. Then the computational effort for a fixed time interval $[0,T]$ is $O(h^{-m-1-\varepsilon})$. For explicit methods, $\varepsilon = 0$, while for implicit methods, $\varepsilon > 0$ at times. The latter depends on the amount of effort required to solve the system of equations. In any case, $m+1+\varepsilon \geq 2$. To improve the precision by a factor of q thus is to multiply the computational effort by a factor of $\tilde{q} = q^{(m+1+\varepsilon)/k}$.

In solving a parabolic differential equation we have as a rule that $h/(\Delta x)^2 = \lambda \equiv$ *constant*. The growth law $O(h^{-m/2-1-\varepsilon})$ for the computational effort appears more favorable. However, a remainder of $O(h^k) + O((\Delta x)^k) = O(h^{k/2})$ implies $\tilde{q} = q^{(m+2+2\varepsilon)/k}$. $\tilde{q} = q^{(m+2+2\varepsilon)/2k}$ is only achieved with a remainder $O(h^k) + O((\Delta x)^{2k}) = O(h^k)$.

How then is one to explain the preference for simpler methods in practice? There are in fact a number of important reasons for this, which we will briefly discuss.

(1) In many applications, a complicated geometry is involved. The boundary conditions (and sometimes, insufficiently smooth coefficients for the differential equations) lead to solutions which are only once or twice differentiable. Then methods of higher order carry no advantage. For ordinary differential equations, there is no influence of geometry or of boundary conditions in this sense; with several space variables, however, difficulties of this sort become dominant.

(2) The stability question is grounds enough to restrict oneself to those few types of methods for which there is sufficient experience in hand. A method which is stable for a pure initial value problem with equations with arbitrarily often differentiable coefficients, may well lose this stability in the face of boundary conditions, less smooth coefficients, or nonlinearities. In addition, stability is a conclusion based on incrementations $h \le h_o$. It is often quite unclear how h_o depends on the above named influences. In this complicated theoretical situation, practical experience becomes a decisive factor.

(3) The precision demanded by engineers and physicists

is often quite modest. This fact is usually unnoticed in the context of ordinary differential equations, since the computing times involved are quite insignificant. As a result, the question of precision demanded is barely discussed. As with the evaluation of simple transcendental functions, one simply uses the mantissa length of the machine numbers as a basis for precision. The numerical solution of partial differential equations, however, quickly can become so expensive, that the engineer or physicist would rather reduce the demands for precision. This cost constraint may well be relaxed with future technological progress in hardware.

These arguments should not be taken to mean that higher order convergence methods have no future. Indeed one would hope that their significance would gradually increase. The derivation of such methods is given a powerful assist by extrapolation methods. We begin with an explanation of the basic procedure of these methods. In order to keep the formulas from getting too long, we will restrict ourselves to problems in $\mathbb{R}^2$, with one space and one time variable, x and t.

The starting point is a properly posed problem and a corresponding consistent and stable difference method. Only solutions for s-times differentiable initial functions are considered. The step size of the difference method is denoted by h. The foundation of all extrapolation methods is the following assumption:

Assumption: The solutions $w(x,t,h)$ of the difference method have an *asymptotic expansion*

$$w(x,t,h) = \sum_{\nu=0}^{r-1} \tau_\nu(x,t)h^{\gamma_\nu} + p(x,t,h), \quad (x,t) \in G,$$
$$h \in (0,h_o]$$

where $r \geq 2$ and

$$\|p(x,t,h)\| = O(h^{\gamma_r}), \qquad (x,t) \in G,\ h \in (0,h_o]$$
$$\tau_\nu : G \to \mathbb{C}^n, \qquad \nu = 0(1)r-1$$
$$0 = \gamma_o < \gamma_1 \quad \dots < \gamma_r.$$

τ_o is the desired exact solution of the problem. □

We begin with a discussion of what is called *global extrapolation*. For this, one carries out the difference method for r different incrementations h_j, $j = 1(1)r$, each for the entire time interval. The r computations are independent of each other. For each level $t = t_k$, where $t_k/h_j \in \mathbb{Z}$ for all $j = 1(1)r$, one can now form a linear combination $\tilde{w}(x,t_k,h_1,\dots,h_r)$ of the quantities $w(x,t_k,h_j)$ so that

$$\tilde{w}(x,t_k,h_1,\dots,h_r) = \tau_o(x,y) + R.$$

Letting $h_\nu = q_\nu h$, $\nu = 1(1)r$, and letting h converge to zero, we get

$$R = O(h_1^{\gamma_r}).$$

$\tilde{w}$ is computed recursively:

$$T_{j,o} = w(x,t_k,h_{j+1}), \quad j = 0(1)r-1$$
$$T_{j,\nu} = T_{j,\nu-1} - \beta_{j\nu}[T_{j-1,\nu-1} - T_{j,\nu-1}], \quad \nu = 1(1)r-1,$$
$$j = \nu(1)r-1$$
$$\tilde{w}(x,t_k,h_1,\dots,h_r) = T_{r-1,r-1}.$$

In general the coefficients $\beta_{j\nu} \in \mathbb{R}$ depend in complicated ways on the step sizes h_j and the exponents γ_ν. In the following two important special cases, however, the computation is relatively simple.

Case 1: $h_j = \frac{1}{2}h_{j-1}$, $j = 2(1)r$, γ_ν arbitrary

$$\beta_{j\nu} = \frac{1}{2^{\gamma_\nu}-1}$$

Case 2: $\gamma_\nu = \nu\gamma$, $\gamma > 0$, $\nu = 1(1)r$, h_j arbitrary

$$\beta_{j\nu} = \frac{1}{\left(\frac{h_{j-\nu}}{h_j}\right)^\gamma - 1}$$

The background can be found in Stoer-Bulirsch, 1980, Chapter 2, and Grigorieff (1972), Chapter 5. This procedure, by the way, is well-known for Romberg and Bulirsch quadrature and midpoint rule extrapolation for ordinary differential equations (cf. Stoer-Bulirsch 1980).

In practice, the difference method is only carried out for finitely many values of x. Extrapolation is then possible for those x which occur for all increments h_j. The case $h_j/(\Delta x)_j^2 \equiv$ *constant* presents extra difficulties. The ratios of the h_j's are very important, both for the size of the remainder and the computational effort. For solving hyperbolic differential equations one can also use the Romberg or the Bulirsch sequence.

Romberg sequence:

$$h_j = h/2^{j-1}, \quad j = 1(1)r.$$

Bulirsch sequence:

$$h_1 = h, \ h_{2j} = h/2^j, \quad h_{2j+1} = h/(3\cdot 2^{j-1}), \quad j \geq 1.$$

Because of the difficulties associated with the case $h_j/(\Delta x)_j^2 \equiv$ *constant*, it is wise to use a spacing of the $(\Delta x)_j$

based on these sequences for solving parabolic differential equations. In principle, one could use other sequences for global extrapolation, however.

Before applying an extrapolation method, we ask ourselves two decisive questions: Does there exist an asymptotic expansion? What are the exponents γ_ν? Naturally $\gamma_\nu = 2\nu$ would be optimal. Usually one must be satisfied with $\gamma_\nu = \nu$. In certain problems, nonintegral exponents can occur. In general the derivation of an asymptotic expansion is a very difficult theoretical problem. This is true even for those cases where practical experience speaks for the existence of such expansions. However, the proofs are relatively simple for linear initial value problems without boundary conditions.

As an example we use the problem

$$u_t(x,t) = A(x)u_x(x,t) + q(x,t), \quad x \in \mathbb{R},\ t \in (0,T)$$
$$u(x,0) = \phi(x), \quad x \in \mathbb{R}.$$

The conditions on the coefficient matrix have to be quite strict. We demand

$$A \in C^\infty(\mathbb{R}, \mathrm{MAT}(n,n,\mathbb{R}))$$
$$A(x) \text{ real and symmetric}, \quad ||A(x)|| \leq L_1$$
$$||A(x)-A(\tilde{x})|| \leq L_2|x-\tilde{x}|, \quad x,\tilde{x} \in \mathbb{R}.$$

Let the $w(x,t,h)$ be the approximate values obtained with the *Friedrichs* method. Let a fixed $\lambda = h/\Delta x > 0$ be chosen and let

$$\lambda \sup_{x\in\mathbb{R}} \rho(A(x)) \leq 1.$$

The method is consistent and stable in the Banach space $L^2(\mathbb{R},\mathbb{C}^n)$ (cf. Example 8.9). In the case of an inhomogeneous equation, we use the formula

$$w(x,t+h,h) = \tfrac{1}{2}[I+\lambda A(x)]w(x+\Delta x,t,h) + \tfrac{1}{2}[I-\lambda A(x)]w(x-\Delta x,t,h) + hq(x,t).$$

Theorem 11.1: Let $r \in \mathbb{N}$, $\phi \in C_0^\infty(\mathbb{R}, \mathbb{R}^n)$ and $q \in C_0^\infty(\mathbb{R} \times [0,T], \mathbb{R}^n)$. Then it is true for all $h \in (0,h_0]$ that

$$w(x,t,h) = \sum_{\nu=0}^{r-1} \tau_\nu(x,t)h^\nu + p(x,t,h), \quad x \in \mathbb{R},\ t \in [0,T],\ t/h \in \mathbb{Z}$$

$$\tau_\nu \in C_0^\infty(\mathbb{R} \times [0,T], \mathbb{R}^n)$$

$$\|p(\cdot,t,h)\|_2 = O(h^r) \quad \text{uniformly in } t.$$

Proof: Since there is nothing to prove for $r = 1$, we suppose that $r > 1$. We use the notation

$$V = C_0^\infty(\mathbb{R}, \mathbb{R}^n), \qquad W = C_0^\infty(\mathbb{R} \times [0,T], \mathbb{R}^n).$$

The most important tool for the proof is the fact that for $\phi \in V$ and $q \in W$, the solution u of the above problem belongs to W. This is a special case of the existence and uniqueness theorems for linear hyperbolic systems (cf., e.g., Mizohata 1973). For arbitrary $v \in W$, we examine the difference quotients

$$Q_1(v)(x,t,h) = h^{-1}\{v(x,t+h) - \tfrac{1}{2}[v(x+\Delta x,t)+v(x-\Delta x,t)]\}$$

$$Q_2(v)(x,t,h) = (2\Delta x)^{-1}\{v(x+\Delta x,t)-v(x-\Delta x,t)\}$$

$$Q(v) = Q_1(v) - A(x)Q_2(v).$$

Although $w(x,t,h)$ is only defined for $t/h \in \mathbb{Z}$, one can apply Q to w:

$$Q(w(\cdot,\cdot,h))(x,t,h) = q(x,t),\ x \in \mathbb{R},\ t \in [0,T],\ t/h \in \mathbb{Z},\ h \in (0,h_0].$$

For $v \in W$, $Q_1(v)$ and $Q_2(v)$ can be expanded separately with Taylor's series

$$Q(v)(x,t,h) = v_t(x,t) - A(x)v_x(x,t)$$
$$+ \sum_{\nu=2}^{s} h^{\nu-1} D_\nu(v)(x,t) + h^s Z(x,t,h).$$

Here $s \in \mathbb{N}$ is arbitrary. For $s = 1$, the sum from 2 to s vanishes. The operators D_ν, $\nu = 2(1)\infty$, are differential operators containing $A(x)$ as well as partial derivatives of order ν. We have $D_\nu(v) \in W$. The support of Z is bounded. For fixed h, $Z(\cdot,\cdot,h) \in W$. $Z(x,t,h)$ is bounded for all x, t, and h.

The quantities $\tau_\nu \in W$, $\nu = 0(1)r-1$ are defined recursively:

$$\nu=0: \quad \frac{\partial}{\partial t}\tau_0(x,t) = A(x)\frac{\partial}{\partial x}\tau_0(x,t)+q(x,t) \qquad x \in \mathbb{R},\ t \in [0,T]$$
$$\tau_0(x,0) = \phi(x)$$

$$\nu>0: \quad \frac{\partial}{\partial t}\tau_\nu(x,t) = A(x)\frac{\partial}{\partial x}\tau_\nu(x,t) - \sum_{\mu=0}^{\nu-1} D_{\nu+1-\mu}(\tau_\mu)(x,t) \qquad x \in \mathbb{R},\ t \in [0,T]$$

$$\tau_\nu(x,0) = 0$$

It follows that $\tau_\nu \in W$, $\nu = 0(1)r-1$. The difference quotients $Q(\tau_\nu)$ yield

$$Q(\tau_0)(x,t,h) = q(x,t)+\sum_{\mu=2}^{2r-1} h^{\mu-1}D_\mu(\tau_0)(x,t)+h^{2r-1}Z_0(x,t,h)$$

$$Q(\tau_\nu)(x,t,h) = -\sum_{\mu=0}^{\nu-1} D_{\nu+1-\mu}(\tau_\mu)(x,t)+\sum_{\mu=2}^{2r-2\nu-1} h^{\mu-1}D_\mu(\tau_\nu)(x,t)$$
$$+ h^{2r-2\nu-1}Z_\nu(x,t,h), \qquad \nu = 1(1)r-1.$$

In the last equation, the sum from 2 to $2r-2\nu-1$ vanishes when $\nu = r-1$. Next the ν-th equation is multiplied by h^ν

and all the equations are added. Letting $\tau = \Sigma\tau_\nu h^\nu$ we get

$$Q(\tau)(x,t,h) = q(x,t) - \sum_{\nu=1}^{r-1} h^\nu \sum_{\mu=0}^{\nu-1} D_{\nu+1-\mu}(\tau_\mu)(x,t)$$

$$+ \sum_{\nu=0}^{r-2} h^\nu \sum_{\mu=2}^{r-\nu} h^{\mu-1} D_\mu(\tau_\nu)(x,t)$$

$$+ \sum_{\nu=0}^{r-2} h^\nu \sum_{\mu=r-\nu+1}^{2r-2\nu-1} h^{\mu-1} D_\mu(\tau_\nu)(x,t)$$

$$+ \sum_{\nu=0}^{r-1} h^{2r-\nu-1} Z_\nu(x,t,h).$$

The first two double sums are actually the same, except for sign. To see this, substitute $\tilde{\mu} = \nu+\mu-1$ in the second, obtaining

$$\sum_{\nu=0}^{r-2} \sum_{\tilde{\mu}=\nu+1}^{r-1} h^{\tilde{\mu}} D_{\tilde{\mu}-\nu+1}(\tau_\nu)(x,t).$$

Then change the order of summation:

$$\sum_{\tilde{\mu}=1}^{r-1} h^{\tilde{\mu}} \sum_{\nu=0}^{\tilde{\mu}-1} D_{\tilde{\mu}+1-\nu}(\tau_\nu)(x,t).$$

Now the substitution $(\tilde{\mu},\nu) \to (\nu,\mu)$ yields the first double sum.

While the first two terms in this representation of $Q(\tau)$ cancel, the last two contain a common factor of h^r. Thus we get

$$Q(\tau)(x,t,h) = q(x,t) + h^r Z(x,t,h),$$

$$x \in \mathbb{R},\ t \in [0,T],\ t+h \in [0,T],\quad h \in (0,h_o].$$

Z has the same properties as Z_ν: bounded support, continuous for fixed h, bounded for all $x \in \mathbb{R}$, $t \in [0,T]$, and $h \in (0,h_o]$. The quanity $\tau-w$ satisfies the difference equation

$$Q(\tau)(x,t,h) - Q(w)(x,t,h) = h^r Z(x,t,h)$$
$$\tau(x,0,h) - w(x,0,h) = 0.$$

Thus, $\tau - w$ is a solution of the Friedrichs method with initial function 0 and inhomogeneity $h^r Z(x,t,h)$. It follows from the stability of the method and from $\|Z(\cdot,t,h)\|_2 \leq L$ for $t/h \in \mathbb{Z}$ and $h \in (0,h_o]$, that for these t and h,

$$\|\tau(\cdot,t,h) - w(\cdot,t,h)\|_2 \leq \tilde{L}\cdot h^r. \quad \square$$

From the practical point of view, the restriction to functions ϕ and q with compact support is inconsequential because of the finite domain of dependence of the differential equation *and* the difference method. Only the differentiability conditions are of significance. Parabolic equations do not have a finite dependency domain. The vector spaces V and W are therefore not suitable for these differential equations. However, they can be replaced by vector spaces of those functions for which

$$\sup_{x \in \mathbb{R}} |\phi^{(j)}(x)x^k| < \infty, \qquad j = 0(1)s,\ k = 1(1)\infty$$
$$\sup_{x \in \mathbb{R}} |(\tfrac{\partial}{\partial x})^j q(x,t)x^k| < \infty, \quad j = 0(1)s,\ k = 1(1)\infty,\ t = [0,T].$$

$s \in \mathbb{N}$ suitable but fixed.

These spaces could also have been used in Theorem 11.1. The proof of a similar theorem for the Courant-Isaacson-Rees method would founder, for the splitting $A(x) = A_+(x) - A_-(x)$ is not differentiable in x, i.e., just because $A(x)$ is arbitrarily often differentiable, it does not follow that this is necessarily so for $A_+(x)$ and $A_-(x)$.

Global extrapolation does not correspond exactly to

the model of midpoint rule extrapolation for ordinary differential equations, for there one has a case of *local extrapolation*. Although the latter can be used with partial differential equations only in exceptional cases, we do want to present a short description of the method here. Let

$$h = n_1h_1 = n_2h_2 = \dots = n_rh_r, \quad n_j \in \mathbb{N}, \ j = 1(1)r.$$

At first the difference method is only carried out for the interval $[0,h]$. For $t = h$, there are then r approximations for τ_0 available. With the aid of the Neville scheme, a higher order approximation for $t = h$ is computed. The quantities obtained through this approximation then become the initial values for the interval $[h,2h]$. There are two difficulties with this:

(1) When the computation is based on finitely many points x, the extrapolation is only possible for those x which are used in *all* r computations. Practically, this means that for $j = 1(1)r$, the same x-values must be used. Since $\lambda = h_j/(\Delta x)_j \equiv$ *constant* or $\lambda = h_j/(\Delta x)_j^2 \equiv$ *constant*, for the larger increments h_j the method has to be carried out repeatedly, with the lattice shifted in the x-direction. This leads to additional difficulties except for pure initial value problems. In any case, the computational effort is increased by this.

(2) Local extrapolation of a difference method is a new difference method. Its stability does not follow from the stability of the method being extrapolated. Frequently the new method is not stable, and then local extrapolation is not applicable. Occasionally so-called weakly stable methods

arise, which yield useful results with h values that are not too small. Insofar as stability is present, this must be demonstrated independently of the stability of the original method. Local extrapolation therefore is a heuristic method in the search for higher order methods.

The advantages of local over global extrapolation, however, are obvious. For one thing, not as many intermediate results have to be stored, so that the programming task is simplified. For another, the step size h can be changed in the interval $[0,T]$. The Neville scheme yields good information for the control of the step size. In this way the method attains a greater flexibility, which can be exploited to shorten the total computing time.

As an example of local extrapolation, we again examine the Friedrichs method $C(h)$ for the problem considered above. The asymptotic expansion begins with $\tau_0(x,y) + h\tau_1(x,y) + h^2\tau_2(x,y)$. Let $r = 2$, $h_1 = h$, and $h_2 = h/2$. Then

$$E_2(h) = 2(C(h/2))^2 - C(h)$$

is a second order method. We check to see if it is stable. Let $\Delta x = h/\lambda$ and $g = \Delta x/2$. Then

$$C(h) = \tfrac{1}{2}[I+\lambda A(x)]T_g^2 + \tfrac{1}{2}[I-\lambda A(x)]T_g^{-2}$$

$$C(h/2) = \tfrac{1}{2}[I+\lambda A(x)]T_g + \tfrac{1}{2}[I-\lambda A(x)]T_g^{-1}$$

$$\begin{aligned} 2(C(h/2))^2 = {} & \tfrac{1}{2}[I+\lambda A(x)][I+\lambda A(x+g)]T_g^2 \\ & + \tfrac{1}{2}[I+\lambda A(x)][I-\lambda A(x+g)]T_g^0 \\ & + \tfrac{1}{2}[I-\lambda A(x)][I+\lambda A(x-g)]T_g^0 \\ & + \tfrac{1}{2}[I-\lambda A(x)][I-\lambda A(x-g)]T_g^{-2} \end{aligned}$$

$$\begin{aligned} E_2(h) = &\tfrac{1}{2}\lambda[I+\lambda A(x)]A(x+g)T_g^2 \\ &+ I - \tfrac{1}{2}\lambda[I+\lambda A(x)]A(x+g) + \tfrac{1}{2}\lambda[I-\lambda A(x)]A(x-g) \\ &- \tfrac{1}{2}\lambda[I-\lambda A(x)]A(x-g)T_g^{-2}. \end{aligned}$$

By Theorem 5.13, terms of order $O(h)$ have no influence on stability. Therefore $E_2(h)$ is stable exactly when the method $\tilde{E}_2(h)$, created by replacing $A(x+g)$ and $A(x-g)$ with $A(x)$, is stable:

$$\begin{aligned} \tilde{E}_2(h) &= \tfrac{1}{2}\lambda[I+\lambda A(x)]A(x)T_g^2 + I - \lambda^2 A(x)^2 - \tfrac{1}{2}\lambda[I-\lambda A(x)]A(x)T_g^{-2} \\ &= I + \tfrac{1}{2}\lambda A(x)(T_{\Delta x} - T_{\Delta x}^{-1}) + \tfrac{1}{2}\lambda^2 A(x)^2(T_{\Delta x} - 2I + T_{\Delta x}^{-1}). \end{aligned}$$

For $A(x) \equiv$ *constant*, $\tilde{E}_2(h)$ is the Lax-Wendroff method (cf. Example 9.26). This method is stable for $\lambda\rho(A(x)) \leq 1$. If $A(x)$ is real, symmetric, and constant, it even follows that

$$\|\tilde{E}_2(h)\|_2 \leq 1.$$

With the help of Theorem 9.34 (Lax-Nirenberg) we obtain a sufficient stability condition for nonconstant A. $\tilde{E}_2(h)$ and $E_2(h)$ are stable under the following conditions:

(1) $A \in C^2(\mathbb{R}, \mathrm{MAT}(n,n,\mathbb{R}))$

(2) $A(x)$ is always symmetric

(3) The first and second derivatives of A are bounded

(4) $\lambda\rho(A(x)) \leq 1$, $x \in \mathbb{R}$.

By Theorem 9.31, Condition (4) is also necessary for stability.

In the constant coefficient case, $E_2(h)$ coincides with the special case $m = 1$, $r = 1$ of method $C(h)$ of Example 10.9. Both methods have the same order of consistency and the same stability condition, but they are different for nonconstant A.

The difference $E_2(h)-(C(h/2))^2 = (C(h/2))^2-C(h)$ gives a good indication of order of magnitude of the local error. One can use it for stepwise control. In this respect, local extrapolation of the Friedrichs method has an advantage over direct application of the Lax-Wendroff method.

The derivation of $\tilde{E}_2(h)$ can also be carried through the amplification matrix of $C(h)$. $C(h/2)$ has amplification matrix

$$G(h/2,y,x) = \cos\omega\cdot I+i\lambda\sin\omega\cdot A(x)$$

where $\omega = yg = \frac{1}{2}y\Delta x$. It follows that

$$\begin{aligned} H_2(h,y,x) &= 2G(h/2,y,x)^2 - G(h,y,x) \\ &= 2\cos^2\omega\cdot I - 2\lambda^2\sin^2\omega\cdot A(x)^2 \\ &\quad +2i\lambda\sin 2\omega\cdot A(x) - \cos 2\omega\cdot I-i\lambda\sin 2\omega\cdot A(x) \\ &= I-2\lambda^2\sin^2\omega\cdot A(x)^2 + i\lambda\sin 2\omega\cdot A(x). \end{aligned}$$

That is the amplification matrix of $\tilde{E}_2$. Through further extrapolation, we will now try to derive a method E_3 of third order consistency:

$$E_3(2h) = \frac{4}{3}E_2(h)^2 - \frac{1}{3}E_2(2h).$$

Consistency is obvious, since there exists an asymptotic expansion. We have to investigate the amplification matrix

$$H_3(2h,y,x) = \frac{4}{3}H_2(h,y,x)^2 - \frac{1}{3}H_2(2h,y,x)$$

Let μ be an eigenvalue of $\lambda A(x)$, and $\eta_2,\hat{\eta}_2,\eta_3$ the corresponding eigenvalues of $H_2(h,y,x)$, $H_2(2h,y,x)$, and $H_3(2h,y,x)$. Then

$$\eta_2 = 1-2\omega^2\mu^2 + \frac{2}{3}\omega^4\mu^2 + i[2\omega\mu - \frac{4}{3}\omega^3\mu] + O(|\omega|^5)$$

$$\eta_2^2 = 1-8\omega^2\mu^2 + \frac{20}{3}\omega^4\mu^2 + 4\omega^4\mu^4 + i[4\omega\mu - \frac{8}{3}\omega^3\mu - 8\omega^3\mu^3] + O(|\omega|^5)$$

$$\hat{\eta}_2 = 1-8\omega^2\mu^2 + \frac{32}{3}\omega^4\mu^2 + i[4\omega\mu - \frac{32}{3}\omega^3\mu] + O(|\omega|^5)$$

$$\eta_3 = 1-8\omega^2\mu^2 + \frac{16}{3}\omega^4\mu^2 + \frac{16}{3}\omega^4\mu^4 + i[4\omega\mu - \frac{32}{3}\omega^3\mu^3] + O(|\omega|^5)$$

$$|\eta_3|^2 = 1 + \frac{32}{3}\omega^4(\mu^2-\mu^4) + O(|\omega|^5).$$

For stability it is necessary that $|\eta_3| \le 1$, that is, $|\mu| \ge 1$. On the other hand, for $\omega = \pi/2$ we have

$$\begin{aligned} H_2(2h,y,x) &= I \\ H_2(h,y,x) &= I-2\lambda^2A(x)^2 \\ \eta_3 &= 1 + \frac{16}{3}(\mu^4-\mu^2) \end{aligned}$$

and hence the condition $|\mu| \le 1$. Thus E_3 is stable only if by chance all of the eigenvalues of $\lambda A(x)$ are $+1$ or 0 or -1, for all $x \in \mathbb{R}$. In this exceptional case, the Friedrichs method turns into a characteristic method, and thus need not concern us here.

For characteristic methods, local extrapolation is almost always possible as with ordinary differential tions. This is mostly true even if boundary conditions are present. The theoretical background can be found in Hackbusch (1973), (1977).

PART II.
BOUNDARY VALUE PROBLEMS FOR ELLIPTIC DIFFERENTIAL EQUATIONS

12. Properly posed boundary value problems

Boundary value problems for elliptic differential equations are of great significance in physics and engineering. They arise, among other places, in the areas of fluid dynamics, electrodynamics, stationary heat and mass transport (diffusion), statics, and reactor physics (neutron transport). In contrast to boundary value problems, initial value problems for elliptic differential equations are not properly posed as a rule (cf. Example 1.14).

Within mathematics itself the theory of elliptic differential equations appears in numerous other areas. For a long time the theory was a by-product of the theory of functions and the calculus of variations. To this day variational methods are of great practical significance for the numerical solution of boundary value problems for elliptic differential equations. Function theoretical methods can frequently be used to find a closed solution for, or at least greatly simplify, planar problems.

The following examples should clarify the relationship

between boundary value problems and certain questions of function theory and the calculus of variations. Throughout, G will be a simply connected bounded region in $\mathbb{R}^2$ with a continuously differentiable boundary ∂G.

Example 12.1: *Euler differential equation* from the calculus of variations. Find a mapping $u: \overline{G} \to \mathbb{R}$ which satisfies the following conditions:

(1) u is continuous on $\overline{G}$ and continuously differentiable on G.

(2) $u(x,y) = \psi(x,y)$ for all $(x,y) \in \partial G$.

(3) u minimizes the integral

$$I[w] = \iint_{\overline{G}} [a_1(x,y)w_x(x,y)^2 + a_2(x,y)w_y(x,y)^2 + c(x,y)w(x,y)^2 - 2q(x,y)w(x,y)]dxdy$$

in the class of all functions w satisfying (1) and (2). Here $a_1, a_2 \in C^1(\overline{G}, \mathbb{R})$, $c,q \in C^1(\overline{G}, \mathbb{R})$, and $\psi \in C^1(\partial G, \mathbb{R})$ with

$$\begin{aligned} a_1(x,y) &\geq \alpha > 0 \\ a_2(x,y) &\geq \alpha > 0 \qquad (x,y) \in \overline{G} \\ c(x,y) &\geq 0. \end{aligned}$$

It is known from the calculus of variations that this problem has a uniquely determined solution (cf., e.g., Gilbarg-Trudinger 1977, Ch. 10.5). In addition it can be shown that u is twice continuously differentiable on G and solves the following boundary value problem:

$$\begin{aligned} -[a_1(x,y)u_x]_x - [a_2(x,y)u_y]_y + c(x,y)u &= q(x,y), \quad (x,y) \in G \\ u(x,y) &= \psi(x,y), \qquad (x,y) \in \partial G. \end{aligned} \tag{12.2}$$

The differential equation is called the *Euler differential equation* for the variational problem. Its principal part is

$$-a_1u_{xx} - a_2u_{yy}.$$

The differential operator

$$\Delta = \frac{\partial^2}{\partial x^2} + \frac{\partial^2}{\partial y^2}$$

is called the *Laplace operator* (Laplacian). In *polar coordinates,*

$$x = r\cos\phi$$

$$y = r\sin\phi$$

it looks like

$$\Delta = \frac{\partial^2}{\partial r^2} + \frac{1}{r}\frac{\partial}{\partial r} + \frac{1}{r^2}\frac{\partial^2}{\partial \phi^2}\ .$$

The equation

$$-\Delta u(x,y) = q(x,y)$$

is called the *Poisson equation* and

$$-\Delta u(x,y) + cu(x,y) = q(x,y), \qquad c \equiv \textit{constant}$$

is called the *Helmholtz equation*. □

With boundary value problems, as with initial value problems, there arises the question of whether the given problem is uniquely solvable and if this solution depends continuously on the preconditions. In Equation (12.2) the preconditions are the functions q and ψ. Strictly speaking, one should also examine the effect of "small deformations" of the boundary curve. Because of the special problems this entails, we will avoid this issue. For many boundary value problems, both the uniqueness of the solution and its continuous dependence on the preconditions follows from the *maximum-minimum principle* (extremum principle).

Theorem 12.3: *Maximum-minimum principle.* If $c(x,y) \geq 0$ and $q(x,y) \geq 0$ ($q(x,y) \leq 0$) for all $(x,y) \in \bar{G}$, then every nonconstant solution u of differential equation (12.2) assumes its minimum, if it is negative (its maximum, if it is positive) on ∂G and not in G.

A proof may be found in Hellwig 1977, Part 3, Ch. 1.1.

Theorem 12.4: Let boundary value problem (12.2) with $c(x,y) \geq 0$ for all $(x,y) \in \bar{G}$ be given. Then

(1) It follows from

$$q(x,y) \geq 0, \quad (x,y) \in \bar{G}$$

and

$$\psi(x,y) \geq 0, \quad (x,y) \in \partial G$$

that

$$u(x,y) \geq 0, \quad (x,y) \in \bar{G}.$$

(2) There exists a constant $K > 0$ such that

$$|u(x,y)| \leq \max_{(\tilde{x},\tilde{y})\in\partial G} |\psi(\tilde{x},\tilde{y})| + K \cdot \max_{(\tilde{x},\tilde{y})\in\bar{G}} |q(\tilde{x},\tilde{y})|, \quad (x,y) \in \bar{G}.$$

The first assertion of the theorem is a reformulation of the maximum minimum principle which in many instances is more easily applied. The second assertion shows that the boundary value problem is properly posed in the maximum norm.

Proof: (1) follows immediately from Theorem 12.3. To prove (2), we begin by letting

$$w(x,y) = \Psi + [\exp(\beta\xi) - \exp(\beta x)]Q$$

where

$$\Psi = \max_{(\tilde{x},\tilde{y})\in\partial G} |\psi(\tilde{x},\tilde{y})|, \quad Q = \max_{(\tilde{x},\tilde{y})\in\overline{G}} |q(\tilde{x},\tilde{y})|$$

$$\beta \equiv \textit{const.} > 0, \qquad \xi \equiv \textit{const.} > \max_{(\tilde{x},\tilde{y})\in\overline{G}} \tilde{x}.$$

Further, let

$$M = \max_{(\tilde{x},\tilde{y})\in\overline{G}} \{|\tfrac{\partial}{\partial x} a_1(\tilde{x},\tilde{y})|, c(\tilde{x},\tilde{y})\}.$$

Without loss of generality, we may suppose that the first component, x, is always nonnegative on $\overline{G}$. Since $a_1(x,y) \geq \alpha$, we have

$$\begin{aligned} r(x,y) &= -[a_1(x,y)w_x(x,y)]_x - [a_2(x,y)w_y(x,y)]_y \\ &\qquad + c(x,y)w(x,y) \\ &= Q\exp(\beta x)[a_1(x,y)\beta^2 + \beta\tfrac{\partial}{\partial x} a_1(x,y) - c(x,y)] \\ &\quad + c(x,y)[Q\exp(\beta\xi) + \Psi] \\ &\geq Q\exp(\beta x)[\alpha\beta^2 - M(\beta+1)]. \end{aligned}$$

Now choose β so large that

$$\alpha\beta^2 - M(\beta+1) \geq 1.$$

It follows that

$$r(x,y) \geq Q, \qquad (x,y) \in \overline{G}.$$

In addition,

$$w(x,y) \geq \Psi, \quad (x,y) \in \partial G.$$

From this it follows that

$$\left.\begin{aligned} q(x,y) + r(x,y) &\geq 0 \\ q(x,y) - r(x,y) &\leq 0 \end{aligned}\right\} \quad (x,y) \in \overline{G}$$

$$\left.\begin{aligned} u(x,y) + w(x,y) &= \psi(x,y) + w(x,y) \geq 0 \\ u(x,y) - w(x,y) &= \psi(x,y) - w(x,y) \leq 0 \end{aligned}\right\} \quad (x,y) \in \partial G.$$

Together with (1) we obtain

$$\left.\begin{aligned} u(x,y) + w(x,y) &\ge 0 \\ u(x,y) - w(x,y) &\le 0 \end{aligned}\right\} \quad (x,y) \in \overline{G}$$

which is equivalent to

$$|u(x,y)| \le w(x,y), \quad (x,y) \in \overline{G}. \qquad \square$$

To check the uniqueness of the solution u, and its continuous dependence on the preconditions ψ and q, pick a different solution $\tilde{u}$ for preconditions $\tilde{\psi}$ and $\tilde{q}$. From Theorem 12.4(2), for $(x,y) \in \overline{G}$, we obtain the inequality

$$|u(x,y) - \tilde{u}(x,y)| \le \max_{(\tilde{x},\tilde{y})\in\partial G} |\psi(\tilde{x},\tilde{y}) - \tilde{\psi}(\tilde{x},\tilde{y})|$$

$$+ K \cdot \max_{(\tilde{x},\tilde{y})\in\overline{G}} |q(\tilde{x},\tilde{y}) - \tilde{q}(\tilde{x},\tilde{y})|.$$

This implies that the solution u is uniquely determined and depends continuously on the preconditions ψ and q.

Example 12.5: *Potential equation, harmonic functions.* Boundary value problem:

$$\Delta u(x,y) = 0, \qquad (x,y) \in G$$
$$u(x,y) = \psi(x,y), \qquad (x,y) \in \partial G.$$

Here $\psi \in C^0(\partial G, \mathbb{R})$. As a special case of (12.2), this problem has a uniquely determined solution which depends continuously on the boundary condition ψ. The homogeneous differential equation

$$\Delta u(x,y) = 0$$

is called the *potential equation*. Its solutions are called *harmonic functions.* Harmonic functions are studied carefully in classical function theory (cf. Ahlfors 1966, Ch. 4.6). Many of these function theoretical results were extended later and by different methods to more general differential equations and to higher dimensions. In this, the readily visualized classical theory served as a model. We will now review the most important results of the classical theory.

(1) Let $f(z)$ be a holomorphic mapping. Then $f(z)$, $\overline{f(z)}$, $\mathrm{Re}(f(z))$, and $\mathrm{Im}(f(z))$ are all harmonic functions.

(2) Every function which is harmonic on an open set is real analytic, i.e., at every interior point of the set it has a local expansion as a uniformly convergent power series in x and y.

(3) When the set G is the unit disk, the solution of the boundary value problem for the potential equation can be given by means of the *Poisson integral formula*

$$u(x,y) = \begin{cases} \dfrac{1}{2\pi}\displaystyle\int_0^{2\pi} \psi(\cos\tilde{\phi},\sin\tilde{\phi})\frac{1-r^2}{1-2r\cos(\phi-\tilde{\phi})+r^2}\,d\tilde{\phi} & \text{for } r<1 \\ \psi(x,y) & \text{for } r=1. \end{cases}$$

Here (r,ϕ) are the polar coordinates of (x,y). The Poisson integral formula is a simple consequence of the Cauchy integral formula.

(4) The Poisson integral formula leads to the expression

$$u(x,y) = \frac{\alpha_0}{2} + \sum_{\nu=1}^{\infty} r^\nu[\alpha_\nu \cos(\nu\phi) + \beta_\nu \sin(\nu\phi)]$$

where

$$\alpha_\nu = \frac{1}{\pi}\int_0^{2\pi} \psi(\cos\tilde{\phi},\ \sin\tilde{\phi})\cos(\nu\tilde{\phi})\,d\tilde{\phi}$$

$$\beta_\nu = \frac{1}{\pi}\int_0^{2\pi} \psi(\cos\tilde{\phi},\ \sin\tilde{\phi})\sin(\nu\tilde{\phi})\,d\tilde{\phi}.$$

The functions $r^\nu\cos(\nu\phi)$, $r^\nu\sin(\nu\phi)$ are the simplest harmonic functions. Thus the above expansion of $u(x,y)$ is analogous to the power series expansion for holomorphic functions.

(5) The potential equation is invariant with respect to one-to-one holomorphic transformations. Thus one need consider the boundary value problem for the potential equation only on the unit disk, since by the Riemann mapping theorem, every simply connected region with at least two boundary points can be mapped onto the unit disk conformally (i.e., globally one-to-one and holomorphically).

(6) It follows from the Schwarz reflection principle that at every boundary point where the boundary curve ∂G and the boundary function ψ are both real analytic, the solution u is also real analytic. At these points, u can be continued across the border.

The conformal mappings of a simply connected region onto the unit circle can be given in closed or almost closed form for a great number of regions. As a result, conclusion (5) is of considerable practical significance. It is frequently worthwhile to map regions with a complicated border onto the unit disk or onto some other simple region, such as a rectangle. Unfortunately, the Riemann mapping theorem has no generalization to higher dimensions. The exploitation of conformal mappings is thus restricted to the plane. Differential equations differing from the potential equation are not in general invariant with respect to conformal maps. However, it is usually easy to specify the differential equation for the transformed function. In executing the transformation, the *Wirtinger calculus* has proved itself to be of use, and we briefly describe it now.

Instead of the (mutually independent) coordinates x and y, we consider the (mutually dependent) complex coordinates z and $\bar{z}$, where

$$z = x + iy \quad , \qquad \bar{z} = x - iy,$$

$$x = \frac{1}{2}(z+\bar{z}) \quad , \qquad y = \frac{1}{2i}(z-\bar{z}).$$

The differential operators $\partial/\partial z$ and $\partial/\partial\bar{z}$ are defined by

$$\frac{\partial}{\partial z} = \frac{1}{2}\frac{\partial}{\partial x} + \frac{1}{2i}\frac{\partial}{\partial y},$$

$$\frac{\partial}{\partial \bar{z}} = \frac{1}{2}\frac{\partial}{\partial x} - \frac{1}{2i}\frac{\partial}{\partial y}.$$

Conversely, we have

$$\frac{\partial}{\partial x} = \frac{\partial}{\partial z} + \frac{\partial}{\partial \bar{z}},$$

$$\frac{\partial}{\partial y} = i\left(\frac{\partial}{\partial z} - \frac{\partial}{\partial \bar{z}}\right).$$

The potential equation now assumes the form

$$\Delta u(x,y) = 4\,\frac{\partial^2 u(z,\bar{z})}{\partial z\,\partial\bar{z}} = 0.$$

A function

$$f(z) = f(z,\bar{z}) = a(z,\bar{z}) + ib(z,\bar{z})$$

is holomorphic exactly when it satisfies the differential equation

$$\frac{\partial f}{\partial \bar{z}}(z,\bar{z}) = 0$$

on an open set. This equation is just another form of the Cauchy-Riemann differential equations

$$a_x(x,y) = b_y(x,y),$$
$$a_y(x,y) = -b_x(x,y).$$

For a holomorphic function $f(z)$ it is further true that

$$\frac{\partial f(z)}{\partial z} = f'(z)$$
$$\frac{\partial \overline{f(z)}}{\partial z} = \frac{\partial f(z)}{\partial \bar{z}} = \frac{\partial f(\bar{z})}{\partial z} = 0$$
$$\frac{\partial \overline{f(z)}}{\partial \bar{z}} = \overline{f'(z)}.$$

Now let $w = f(z)$ be a conformal mapping of the region G onto the region G^*. Then it follows from

$$\frac{\partial}{\partial z} = \frac{\partial f(z)}{\partial z}\frac{\partial}{\partial w} + \frac{\partial \overline{f(z)}}{\partial z}\frac{\partial}{\partial \bar{w}} = f'(z)\frac{\partial}{\partial w}$$

$$\frac{\partial^2}{\partial z \partial \bar{z}} = \frac{\partial f'(z)}{\partial \bar{z}}\frac{\partial}{\partial w} + f'(z)\left[\frac{\partial f(z)}{\partial \bar{z}}\frac{\partial^2}{\partial w \partial w} + \frac{\partial \overline{f(z)}}{\partial \bar{z}}\frac{\partial^2}{\partial w \partial \bar{w}}\right]$$

$$= f'(z)\overline{f'(z)}\frac{\partial^2}{\partial w \partial \bar{w}}$$

that

$$\frac{\partial^2}{\partial w \partial \bar{w}} = \frac{1}{f'(z)\overline{f'(z)}}\frac{\partial^2}{\partial z \partial \bar{z}}.$$

With the help of this equation one easily transforms differential equations of the form

$$-\Delta u(x,y) = H(x,y,u)$$

or

$$-4\frac{\partial^2 u(z,\bar{z})}{\partial z \partial \bar{z}} = H(z,\bar{z},u).$$

Example 12.6: *First boundary value problem for the Poisson equation.*

$$-\Delta u(x,y) = q(x,y), \qquad (x,y) \in G$$
$$u(x,y) = \psi(x,y), \qquad (x,y) \in \partial G.$$

In many algorithms it is assumed that either $\psi(x,y) \equiv 0$ or $q(x,y) \equiv 0$. The general case can usually be reduced to these special cases by means of a substitution: let ψ be extendable to a function $\hat{\psi} \in C^2(\overline{G}, \mathbb{R})$ and let

$$\tilde{u}(x,y) = u(x,y) - \hat{\psi}(x,y)$$
$$\tilde{q}(x,y) = q(x,y) + \Delta\hat{\psi}(x,y).$$

We then obtain the new problem

$$-\Delta\tilde{u}(x,y) = \tilde{q}(x,y), \qquad (x,y) \in G$$
$$\tilde{u}(x,y) = 0\ , \qquad (x,y) \in \partial G.$$

If, on the other hand,

$$q(x,y) = P(z,\overline{z}) = \sum_{\mu,\nu=0}^{k} a_{\mu\nu}\, z^{\mu}\overline{z}^{\nu},$$

then one can define

$$\approx u(x,y) = u(x,y) + \frac{1}{4}\sum_{\mu,\nu=0}^{k} \frac{a_{\mu\nu}}{(\mu+1)(\nu+1)}\, z^{\mu+1}\overline{z}^{\nu+1}.$$

$$\approx\psi(x,y) = \psi(x,y) + \frac{1}{4}\sum_{\mu,\nu=0}^{k} \frac{a_{\mu\nu}}{(\mu+1)(\nu+1)}\, z^{\mu+1}\overline{z}^{\nu+1}.$$

$\overset{\approx}{u}$ is the solution of the problem

$$-\Delta\overset{\approx}{u}(x,y) = 0, \qquad (x,y) \in G$$
$$\overset{\approx}{u}(x,y) = \overset{\approx}{\psi}(x,y), \quad (x,y) \in \partial G. \qquad \square$$

Example 12.7: *Third boundary value problem.*

$$-\Delta u(x,y) + \alpha u(x,y) = q(x,y), \qquad (x,y) \in G$$

$$\frac{\partial u(x,y)}{\partial n} + \beta u(x,y) = \psi(x,y), \qquad (x,y) \in \partial G.$$

Here $\alpha,\beta \in \mathbb{R}$, $\psi \in C^0(\partial G, \mathbb{R})$, $q \in C^0(\overline{G}, \mathbb{R})$ and $\frac{\partial}{\partial n}$ is the derivative in the direction of the outward normal of ∂G. We know, from the theory of partial differential equations (cf., e.g., Walter 1970, Appendix), that:

(1) Whenever the real numbers α,β satisfy the relations

$$\alpha \geq 0, \quad \beta \geq 0, \quad \alpha+\beta > 0$$

the problem has a unique solution. The solution depends continuously on the preconditions $q(x,y)$ and $\psi(x,y)$. There is a valid monotone principle: $q(x,y) \geq 0$ and $\psi(x,y) \geq 0$ implies $u(x,y) \geq 0$.

(2) If $\alpha = \beta = 0$, then $u(x,y) + c$, $c \equiv$ *constant*, is a solution whenever $u(x,y)$ is. Therefore the problem is not uniquely solvable. However, in certain important cases, it can be reduced to a properly posed boundary value problem of the first type. To this end, we choose $q_1(x,y)$ and $q_2(x,y)$ so that

$$\frac{\partial}{\partial x} q_1(x,y) + \frac{\partial}{\partial y} q_2(x,y) = q(x,y).$$

The differential equation can then be written as a first order system:

$$-u_x(x,y) + v_y(x,y) = q_1(x,y),$$

$$-u_y(x,y) - v_x(x,y) = q_2(x,y).$$

v is called the *conjugate function* for u. If $q \in C^1(\overline{G}, \mathbb{R})$,

v satisfies the differential equation

$$-\Delta v(x,y) = \tilde{q}(x,y) = \frac{\partial}{\partial x} q_2(x,y) - \frac{\partial}{\partial y} q_1(x,y).$$

We now compute the tangential derivative of v at a boundary point. Let (ω_1,ω_2) be the unit vector in the direction of the outward normal. Then $(-\omega_2,\omega_1)$ is the corresponding tangential unit vector, with the positive sense of rotation.

$$\begin{aligned} &-\omega_2 v_x(x,y) + \omega_1 v_y(x,y) \\ &= -\omega_2[-u_y(x,y)-q_2(x,y)] + \omega_1[u_x(x,y)+q_1(x,y)] \\ &= \psi(x,y) + \omega_1 q_1(x,y) + \omega_2 q_2(x,y) = \tilde{\psi}(x,y). \end{aligned}$$

$\tilde{\psi}(x,y)$ thus is computable for all boundary points (x,y), given $\psi(x,y)$, $q_1(x,y)$, and $q_2(x,y)$. Since the function v is unique, we obtain the integral condition

$$\int_{\partial G} \tilde{\psi}(x,y)ds = 0, \qquad ds = \text{arc length along } \partial G.$$

If the integrability condition is not satisfied, the original problem is not solvable. Otherwise, one can integrate $\tilde{\psi}$ to obtain a $\tilde{\tilde{\psi}} \in C^1(\partial G, \mathbb{R})$ with

$$\frac{d\tilde{\tilde{\psi}}(x,y)}{ds} = \tilde{\psi}(x,y).$$

$\tilde{\tilde{\psi}}$ is only determined up to a constant. Finally we obtain the following boundary value problem of the first type for v:

$$\begin{aligned} -\Delta v(x,y) &= \tilde{q}(x,y), \qquad (x,y) \in G \\ v(x,y) &= \tilde{\tilde{\psi}}(x,y), \qquad (x,y) \in \partial G. \end{aligned}$$

One recomputes u from v through the above first order system. However, this is not necessary in most practical instances (e.g., problems in fluid dynamics) since our interest is only in the derivatives of u.

(3) For $\alpha < 0$ or $\beta < 0$, the problem has unique solutions in some cases and not in others. For example, for $\alpha = 0$, $-\beta = \nu \in \mathbb{N}$, $q \equiv 0$, and $\psi \equiv 0$, one obtains the family of solutions

$$u(x,y) = \gamma r^{\nu}\sin(\nu\phi), \qquad \gamma \in \mathbb{R}$$
$$r^2 = x^2+y^2, \quad x = r\cos\phi,\ y = r\sin\phi.$$

Thus the problem is not uniquely solvable. In particular, there is no valid maximum-minimum principle. □

Example 12.8: *Biharmonic equation;* load deflection of a homogeneous plate. The differential equation

$$\Delta\Delta u(x,y) = u_{xxxx} + 2u_{xxyy} + u_{yyyy} = 0$$

is called the *biharmonic* equation. As with the harmonic equation, its solutions are real analytic on every open set. The deflection of a homogeneous plate is described by the differential equation

$$\Delta\Delta u(x,y) = q(x,y), \qquad (x,y) \in G$$

with boundary conditions

$$\left.\begin{aligned} u(x,y) &= \psi_1(x,y) \\ -\Delta u(x,y) &= \psi_2(x,y) \end{aligned}\right\} \quad (x,y) \in \partial G \tag{1}$$

or

$$\left.\begin{aligned} u(x,y) &= \psi_3(x,y) \\ \frac{\partial u(x,y)}{\partial n} &= \psi_4(x,y) \end{aligned}\right\} \quad (x,y) \in \partial G. \tag{2}$$

Here $q \in C^0(\overline{G}, \mathbb{R})$, $\psi_1 \in C^2(\partial G, \mathbb{R})$, $\psi_2, \psi_4 \in C^0(\partial G, \mathbb{R})$, and

$\psi_3 \in C^1(\partial G, \mathbb{R})$. The boundary conditions (1) and (2) depend on the type of stress at the boundary. In the first case, the problem can be split into two second-order subproblems:

$$\text{(a)} \quad \begin{aligned} -\Delta v(x,y) &= q(x,y), & (x,y) &\in G \\ v(x,y) &= \psi_2(x,y), & (x,y) &\in \partial G \end{aligned}$$

and

$$\text{(b)} \quad \begin{aligned} -\Delta u(x,y) &= v(x,y), & (x,y) &\in G \\ u(x,y) &= \psi_1(x,y), & (x,y) &\in \partial G. \end{aligned}$$

As special cases of (12.2), these problems are both properly posed, since the maximum minimum principle applies. All properties--especially the monotone principle--carry over immediately to the fourth-order equation with boundary conditions (1). To solve the split system (a), (b), it suffices to have $\psi_1 \in C^0(\partial G, \mathbb{R})$ instead of $\psi_1 \in C^2(\partial G, \mathbb{R})$. Boundary value problem (2) is also properly posed, but unfortunately it cannot be split into a problem with two second-order differential equations. Thus both the theoretical and the numerical treatment are substantially more complicated. There is no simple monotone principle comparable to Theorem 12.4(1).

The variation integral belonging to the differential equation

$$\Delta\Delta u(x,y) = q(x,y)$$

is

$$I[w] = \iint_{\overline{G}} [(\Delta w(x,y))^2 - 2q(x,y)w(x,y)]dx\,dy.$$

The boundary value problem is equivalent to the variation problem

$$I[u] = \min\{I[w] \mid w \in W\}$$

with

$$W = \{w \in C^2(\overline{G}, \mathbb{R}) \mid w \text{ satisfies boundary cond. (1)}\}$$

or

$$W = \{w \in C^1(\overline{G}, \mathbb{R}) \cap C^2(G, \mathbb{R}) \mid w \text{ satisfies boundary cond. (2)}\}.$$

It can be shown that u is actually four times continuously differentiable in G. □

Error estimates for numerical methods typically use higher derivatives of the solution u of the boundary value problem. Experience shows that the methods may converge extremely slowly whenever these derivatives do not exist or are unbounded. This automatically raises the question of the existence and behavior of the higher derivatives of u. Matters are somewhat simplified by the fact that the solution will be sufficiently often differentiable in $\overline{G}$ if the boundary of the region, the coefficients of the differential equation, and the boundary conditions are sufficiently often differentiable. In practice one often encounters regions with corners, such as rectangles

$$G = (a,b) \times (c,d)$$

or L-shaped regions

$$G = (-a,a) \times (0,b) \cup (0,a) \times (-b,b).$$

The boundaries of these regions are not differentiable, and therefore the remark just made is not relevant. We must first define continuous differentiability for a function ψ defined on the boundary of such a region. There should be an open set $U \subset \mathbb{R}^2$ and a function $f \in C^1(U, \mathbb{R})$ with the following properties: (1) $\partial G \subset U$, and (2) ψ = restriction of f to ∂G. Higher order differentiability is defined analogously.

For the two cornered regions mentioned above, this definition is equivalent to the requirement that the restriction of ψ to each closed side of the region be sufficiently often continuously differentiable.

Example 12.9: *Poisson equation on the square.*

$$-\Delta u(x,y) = q(x,y), \qquad (x,y) \in G = (0,1) \times (0,1)$$
$$u(x,y) = \psi(x,y), \qquad (x,y) \in \partial G.$$

Whenever $u \in C^{2k}(\overline{G}, \mathbb{R})$, then for $\nu = 1(1)k$

$$(\frac{\partial}{\partial x})^{2\nu}u(x,y) + (-1)^{\nu-1}(\frac{\partial}{\partial y})^{2\nu}u(x,y)$$

$$= \Big[\sum_{j=0}^{\nu-1} (-1)^{\nu-j-1} (\frac{\partial}{\partial x})^{2j}(\frac{\partial}{\partial y})^{2\nu-2j-2}\Big]\Delta u(x,y).$$

Let (x_0,y_0) be one of the corner points of the square and let ψ be 2k-times continuously differentiable. Then the left side of the equation at the point (x_0,y_0) is determined by ψ alone. We have the following relations between ψ and q:

$$\psi_{xx}(x_0,y_0) + \psi_{yy}(x_0,y_0) = -q(x_0,y_0)$$

$$\psi_{xxxx}(x_0,y_0) - \psi_{yyyy}(x_0,y_0) = -q_{xx}(x_0,y_0) + q_{yy}(x_0,y_0)$$

etc.

When these equations are false, u does not belong to $C^{2k}(\overline{G}, \mathbb{R})$. On the other hand a more careful analysis will show that u does belong to $C^{2k}(\overline{G}, \mathbb{R})$ if the above equations are satisfied and q and ψ are sufficiently often differentiable.

The validity of the equations can be enforced through

the addition of a function with the "appropriate singularity". For $\nu = 1(1)\infty$, let

$$v_\nu(x,y) = \frac{2(-1)^\nu}{\pi} \operatorname{Im}(z^{2\nu}\log z)$$

$$\log z = \log r + i\phi \quad \text{where} \quad r = |z|,\ \phi = \arg|z|,\quad -\pi < \phi \leq \pi.$$

For $x > 0$ and $y > 0$ we have

$$v_\nu(x,0) = 0$$
$$v_\nu(0,y) = y^{2\nu}.$$

Set

$$c_{\mu\nu} = \psi_{xx}(\mu,\nu)+\psi_{yy}(\mu,\nu)+q(\mu,\nu),\quad \mu = 0,1 \quad\text{and}\quad \nu = 0,1$$

$$\tilde{u}(x,y) = u(x,y) + \frac{1}{\pi}\sum_{\mu=0}^{1}\sum_{\nu=0}^{1} c_{\mu\nu}\operatorname{Im}(z_{\mu\nu}^2\log z_{\mu\nu})$$

$$\tilde{\psi}(x,y) = \psi(x,y) + \frac{1}{\pi}\sum_{\mu=0}^{1}\sum_{\nu=0}^{1} c_{\mu\nu}\operatorname{Im}(z_{\mu\nu}^2\log z_{\mu\nu})$$

where

$$z_{00} = z,\quad z_{10} = -i(z-1),\quad z_{01} = i(z-i),\quad z_{11} = -(z-i-1).$$

The new boundary value problem reads

$$-\Delta\tilde{u}(x,y) = q(x,y),\qquad (x,y)\in G$$
$$\tilde{u}(x,y) = \tilde{\psi}(x,y),\qquad (x,y)\in \partial G.$$

We have $\tilde{u} \in C^2(\overline{G}, \mathbb{R})$.

The problem

$$-\Delta u(x,y) = 1,\qquad (x,y)\in G$$
$$u(x,y) = 0,\qquad (x,y)\in \partial G$$

has been solved twice, with the simplest of difference methods (cf. Section 13), once directly, and once by means of $\tilde{\psi}$ and $\tilde{u}$. Table 12.10 contains the results for increments h at the points (a,a). The upper numbers were computed directly with the difference method, and the lower numbers with the given boundary correction.

h \ a	1/2	1/8	1/32	1/128
1/16	0.7344577(-1) 0.7370542(-1)	0.1808965(-1) 0.1821285(-1)		
1/64	0.7365719(-1) 0.7367349(-1)	0.1819750(-1) 0.1820544(-1)	0.1993333(-2) 0.1999667(-2)	
1/256	0.7367047(-1) 0.7367149(-1)	0.1820448(-1) 0.1820498(-1)	0.1999212(-2) 0.1999622(-2)	0.1784531(-3) 0.1788425(-3)

Table 12.10

h \ a	1/2	1/8	1/32	1/128
1/64	0.736713349(-1) 0.736713549(-1)	0.182048795(-1) 0.182049484(-1)	0.199888417(-2) 0.199961973(-2)	
1/256	0.736713532(-1) 0.736713533(-1)	0.182049475(-1) 0.182049478(-1)	0.199961516(-2) 0.199961941(-2)	0.178796363(-3) 0.178842316(-3)

Table 12.11

Table 12.11 contains the values extrapolated from the preceding computations. Extrapolation proceeded in the sense of a pure h^2-expansion:

$$w_h(a,a) = \frac{1}{3}[4\ u_h(a,a) - u_{2h}(a,a)].$$

With the exception of the point (1/128,1/128), the last line is accurate to within one unit in the last decimal place. At the exceptional point, the error is less than 100 units of the last decimal. The values in the vicinity of the corners are particularly difficult to compute. It is clear that the detour via $\tilde{\psi}$ and $\tilde{u}$ is worthwhile. Incidentally, these numerical results provide a good example of the kind of accuracy which can be achieved on a machine with a mantissa length of 48 bits. With boundary value problems, rounding error hardly plays a role, because the systems of equations are solved with particularly nice algorithms.

Example 12.12: *Poisson equation on a nonconvex region with corners.*

$$-\Delta u(x,y) = q(x,y), \qquad (x,y) \in G_\alpha$$
$$u(x,y) = \psi(x,y), \qquad (x,y) \in \partial G_\alpha$$

$$G_\alpha = \{(x,y) \in \mathbb{R}^2 \mid x^2+y^2 < 1 \text{ and } |y| > x \tan \tfrac{\alpha}{2}\}$$
$$\text{for } \alpha \in (\pi, 2\pi).$$

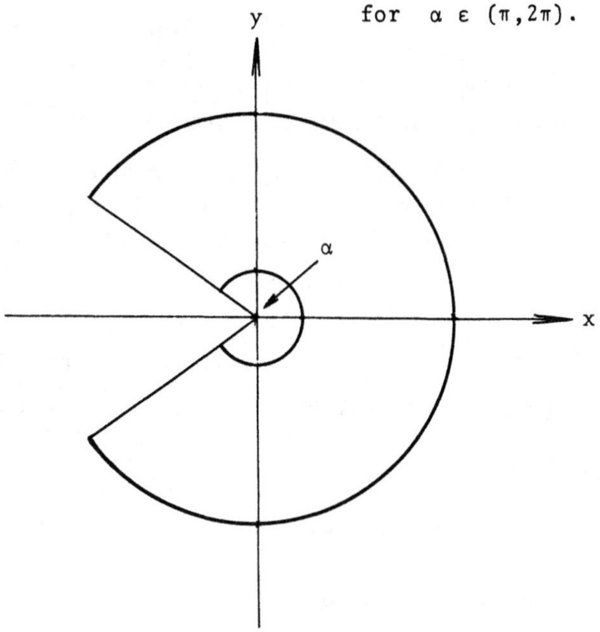

Figure 12.13

The region (Figure 12.13) has three corners

$$(0,0),\ (\cos \alpha/2,\ \sin \alpha/2),\ (\cos \alpha/2,\ -\sin \alpha/2).$$

The interior angles are

$$\alpha,\quad \pi/2,\quad \pi/2.$$

The remarks at 12.9 apply to the right angles. But at the interior angle $\alpha > \pi$ other singularities in the derivatives of u arise. Let

$$\psi(x,y) = \mathrm{Re}(z^{\pi/\alpha}) = \mathrm{Re}\ \exp[(\pi/\alpha)\log z]$$

$$\log z = \log r + i\phi, \qquad -\pi < \phi \le \pi$$

$$q(x,y) = 0.$$

Then

$$u(x,y) = \mathrm{Re}(z^{\pi/\alpha}),$$

and for $\alpha = 3\pi/2$, this is $u(x,y) = \mathrm{Re}(z^{2/3})$. Obviously not even the first derivatives of u are bounded in G. Here $\psi(x,y) \equiv 0$ on the intervals from $(0,0)$ to $(\cos \alpha/2, \sin \alpha/2)$ and from $(0,0)$ to $(\cos \alpha/2, -\sin \alpha/2)$. Since $q(x,y) \equiv 0$ also, the singularity has nothing to do with the derivatives of ψ or q at the point $(0,0)$. It arises from the global behavior of the functions. It is not possible to subtract a function with the "appropriate singularity" in advance. Problems of this type are of great practical significance. In the Ritz method (cf. §14) and the collocation methods (cf. §16) one should use special initial functions to take account of these types of solutions. □

The following two examples should demonstrate that boundary value problems for parabolic and hyperbolic differential equations are either not solvable or not uniquely solvable.

Example 12.14: *Boundary value problem for the heat equation.*

$$u_y(x,y) = u_{xx}(x,y), \qquad (x,y) \in G$$
$$u(x,y) = \psi(x,y), \qquad (x,y) \in \partial G$$

where $\psi \in C^0(\partial G, \mathbb{R})$. The boundary value problem is over-determined. For example, let $G = (0,1) \times (0,1)$. Then the initial boundary value problem already is properly posed. Therefore the set of all boundary values for which the problem is solvable cannot lie entirely in the set of all boundary values. For regions with continuously differentiable boundary there are similar consequences which we will not enter into here. □

Example 12.15: *Boundary value problem for the wave equation.*

$$u_{xx}(x,y) - u_{yy}(x,y) = 0, \qquad (x,y) \in G$$
$$u(x,y) = \psi(x,y), \qquad (x,y) \in \partial G$$

where $\psi \in C^0(\partial G, \mathbb{R})$. This problem also is not properly posed. We restrict ourselves to two simple cases. Let

$$G = Q_1 = (0,1) \times (0,1)$$

or

$$G = Q_2 = \{(x,y) \in \mathbb{R}^2 \mid \sqrt{2} - |x| > y > |x|\}.$$

The two regions differ in that the boundary of Q_2 consists of characteristics while the boundary of Q_1 nowhere coincides with the characteristics. According to Example 1.9, the general solution for the wave equation has the representation $r(x+y) + s(x-y)$. If $u(x,y)$ is a solution for $G = Q_1$, then so is

$$u(x,y) + \cos[2\pi(x+y)] - \cos[2\pi(x-y)]$$
$$= u(x,y) - 2\sin(2\pi x)\sin(2\pi y).$$

The problem therefore is not uniquely solvable. In case $G = Q_2$, r and s can be determined merely from the conditions on two neighboring sides of the square (*characteristic initial value problem*) and therefore the problem is overdetermined. □

13. Difference methods

In composing difference methods for initial value problems, the major problem lies in finding a consistent method (of higher order, preferably) which is also stable. For boundary value problems, this problem is of minor significance, since the obvious consistent difference methods are stable as a rule. In particular, with boundary value problems one does not encounter difficulties of the sort corresponding to the limitations on the step size ratio $h/\Delta x$ or $h/(\Delta x)^2$ encountered with initial value problems.

We consider boundary value problems on bounded regions. Such regions are not invariant under applications of the translation operators. The difference operators are defined, therefore, only on a discrete subset of the region--the lattice. In practice one proceeds in the same manner with initial value problems, but here, even in theory we will dispense with the distinctions, and start with the assumption that the difference operators are defined on the same Banach space as the differential operators.

From the practical point of view, the real difficulty with boundary value problems lies in the necessity of solving large systems of linear or even nonlinear equations for each problem. We will consider this subject extensively in the third part of this book. The systems of equations which arise with boundary value problems are rather specialized in the main. But they barely differ from the systems which arise with implicit methods for the solution of initial value problems.

Error estimation is the other major area of concern in a treatment of boundary value problems.

In this chapter, G will always be a bounded region (an open, bounded, and connected set) in $\mathbb{R}^2$. We denote the boundary of G by Γ. Let

$$r_\Gamma\colon C^0(\overline{G}, \mathbb{R}) \to C^0(\Gamma, \mathbb{R})$$

be the natural map which assigns to each function $u \in C^0(\overline{G}, \mathbb{R})$ its restriction to the boundary Γ, called the *boundary restriction map*. In $C^0(\overline{G}, \mathbb{R})$ and $C^0(\Gamma, \mathbb{R})$ we use the norms

$$\|u\|_\infty = \max_{(x,y)\in\overline{G}} |u(x,y)|$$

and

$$\|\psi\|_\infty = \max_{(x,y)\in\Gamma} |\psi(x,y)|.$$

Both spaces are Banach spaces, and r_Γ is a continuous linear map with $\|r_\Gamma\| = 1$.

<u>Definition 13.1</u>: A finite set $M \subset G$ is called a *lattice* in G. It has *mesh size*

$$|M| = 2 \max_{(x,y)\in\overline{G}} \min_{(u,v)\in\Gamma\cup M} \|(x,y) - (u,v)\|_\infty.$$

The space of *all* lattice functions $f: M \to \mathbb{R}$ we denote by $C^0(M, \mathbb{R})$. With the norm

$$\|f\|_\infty = \max_{(x,y)\varepsilon M} |f(x,y)|,$$

$C^0(M, \mathbb{R})$ becomes a finite dimensional Banach space. The natural map

$$r_M: C^0(\overline{G}, \mathbb{R}) \to C^0(M, \mathbb{R})$$

is called the *lattice restriction map*. The lattice

$$\{(x,y) \in G \mid x = \mu h,\ y = \nu h \text{ with } \mu,\nu \in \mathbb{Z}\},\ 0 < h \leq h_0$$

is called the *standard lattice* in G. It has mesh size h if h_0 is chosen sufficiently small. □

Obviously r_M is linear, continuous, and surjective, and $\|r_M\| = 1$. If the points of M are numbered arbitrarily, the space $C^0(M, \mathbb{R})$ can be identified with $\mathbb{R}^n$ (n = number of points in M) by means of the isomorphism

$$f \longleftrightarrow (f(x_1,y_1),\ldots,f(x_n,y_n)).$$

Thus it is possible to consider *differentiable* maps

$$F : C^0(M, \mathbb{R}) \to C^0(M, \mathbb{R}).$$

In this chapter we will consider only the following problem together with a few special cases.

Problem 13.2:

$$Lu(x,y) = q(x,y), \quad (x,y) \in G$$
$$u(x,y) = \psi(x,y), \quad (x,y) \in \Gamma.$$

Here L is always a semilinear uniformly elliptic second-order differential operator of the form

$$Lu = -a_{11}u_{xx} - 2a_{12}u_{xy} - a_{22}u_{yy}$$
$$-b_1u_x - b_2u_y + H(x,y,u),$$

where

$$a_{11},\ a_{12},\ a_{22},\ b_1,\ b_2 \in C^\infty(\overline{G}, \mathbb{R}),$$
$$H \in C^\infty(\overline{G} \times \mathbb{R}, \mathbb{R}), \quad q \in C^0(\overline{G}, \mathbb{R}), \quad \psi \in C^0(\Gamma, \mathbb{R}).$$

Furthermore, for all $(x,y) \in \overline{G}$ and all $z \in \mathbb{R}$, let

$$H(x,y,0) = 0, \quad H_z(x,y,z) \geq 0,$$
$$a_{11} > 0, \quad a_{11}a_{22} - a_{12}^2 > 0.$$

If $u \in C^0(\overline{G}, \mathbb{R}) \cap C^2(G, \mathbb{R})$, u is called the *classical solution* of the problem. □

The next definition contains the general conditions on a difference method for solving 13.2.

Definition 13.3: A sequence

$$D = \{(M_j, F_j, R_j) \mid j = 1(1)\infty\}$$

is called a *difference method* for Problem 13.2 if the following three conditions are satisfied:

(1) The M_j are lattices in G with mesh sizes $h_j = |M_j|$ converging to zero.

(2) The F_j are continuous maps

$$C^0(\Gamma, \mathbb{R}) \times C^0(M_j, \mathbb{R}) \to C^0(M_j, \mathbb{R}).$$

For each fixed $\psi \in C^0(\Gamma, \mathbb{R})$, all $F_j(\psi,\cdot)$ are continuously differentiable maps of $C^0(M_j, \mathbb{R})$ to $C^0(M_j, \mathbb{R})$.

(3) The R_j are continuous linear maps

$$C^0(M_j, \mathbb{R}) \to C^0(M_j, \mathbb{R}).$$

The method D is called *consistent* if the following condition is satisfied:

(4) There exists an $m \geq 2$ with the property that for all $u \in C^m(\overline{G}, \mathbb{R})$,

$$\lim_{j\to\infty} \|F_j(\psi, r_j(u)) - R_j(r_j(q))\|_\infty = 0.$$

Here $r_j = r_{M_j}$, $\psi = r_\Gamma(u)$, and $q(x,y) = Lu(x,y)$ for all $(x,y) \in G$.

The method D is called *stable* if the following condition is satisfied:

(5) There exist $K > 0$, $\tilde{K} > 0$ and $j_0 \in \mathbb{N}$ with the following properties:

$$\|F_j(\psi,w_j)-F_j(\psi,\tilde{w}_j)\|_\infty \geq K\|w_j-\tilde{w}_j\|_\infty$$

$$\|R_j(w_j)-R_j(\tilde{w}_j)\|_\infty \leq \tilde{K}\|w_j-\tilde{w}_j\|_\infty$$

$$\psi \in C^0(\Gamma, \mathbb{R}),\ j = j_0(1)\infty,\ w_j,\tilde{w}_j \in C^0(M_j, \mathbb{R}).$$ □

Example 13.4: *The standard discretization of the model problem.* We consider a consistent and stable difference method for the *model problem*

$$-\Delta u(x,y) = q(x,y), \quad (x,y) \in G = (0,1)^2$$

$$u(x,y) = \psi(x,y), \quad (x,y) \in \Gamma.$$

For $j = 1(1)\infty$ we set

M_j : standard lattice with mesh size 2^{-j}

$$F_j(\psi,w_j)(x,y) = \frac{1}{h_j^2}\{4w_j(x,y)-\hat{w}_j(x+h_j,y)-\hat{w}_j(x-h_j,y) - \hat{w}_j(x,y+h_j)-\hat{w}_j(x,y-h_j)\}$$

$$R_j(w_j)(x,y) = w_j(x,y)$$
$$w_j \in C^o(M_j, \mathbb{R}), \quad (x,y) \in M_j.$$

Here

$$\hat{w}_j(x,y) = \begin{cases} w_j(x,y) & \text{when} \quad (x,y) \in M_j \\ \psi(x,y) & \text{when} \quad (x,y) \in \Gamma. \end{cases}$$

The proof of the consistency condition, (4), we leave to the reader. Stability, (5), follows from Theorem 13.16 below.

The eigenvalues and eigenfunctions of the linear maps

$$F_j(0,\cdot)\colon C^o(M_j, \mathbb{R}) \to C^o(M_j, \mathbb{R})$$

can be given in closed form. One easily checks that the functions

$$v_{\mu\nu}(x,y) = \sin(\mu\pi x)\cdot\sin(\nu\pi y) \qquad (x,y) \in M_j$$
$$\mu,\nu = 1(1)2^j-1$$

are linearly independent eigenfunctions. The corresponding eigenvalues are

$$\lambda_{\mu\nu} = \frac{2}{h^2}[2 - \cos(\mu\pi h) - \cos(\nu\pi h)], \quad h = h_j.$$

Since lattice M_j consists of $(2^j-1)^2$ points, we have a complete system of eigenfunctions. All eigenvalues are positive and lie in the interval $[\lambda_{11},\lambda_{mm}]$ where $m = 2^j-1$. We have

$$\lambda_{11} = \frac{4}{h^2}[1 - \cos(\pi h)] = 2\pi^2 - \frac{1}{6}\pi^4 h^2 + O(h^4)$$
$$\lambda_{mm} = \frac{4}{h^2}[1 + \cos(\pi h)] = \frac{8}{h^2} - 2\pi^2 + \frac{1}{6}\pi^4 h^2 + O(h^4).$$

With an arbitrary numbering of the lattice points, there are real symmetric matrices A_j for the maps $F_j(0,\cdot)$. With respect to the spectral norm $\|\cdot\|_2$ they satisfy the conditions

$$\frac{\lambda_{mm}}{\lambda_{11}} = \frac{1+\cos(\pi h)}{1-\cos(\pi h)} = \frac{4}{(\pi h)^2} - \frac{2}{3} + O(h^4).$$

The functions $v_{\mu\nu}$, regarded as functions in $\mathbb{R}^2$, are also eigenfunctions of the differential operator $-\Delta$. For example,

$$-\Delta v_{11}(x,y) = 2\pi^2 v_{11}(x,y), \quad (x,y) \in \mathbb{R}^2.$$

Since the functions $v_{\mu\nu}$ vanish on the boundary of $(0,1)^2$, they are also eigensolutions of the boundary value problem. □

Now let D be an arbitrary difference method for solving Problem 13.2. An approximation $w_j \in C^0(M_j, \mathbb{R})$ for the exact solution u of 13.2 is obtained, when possible, from the finitely many *difference equations*

$$F_j(\psi, w_j)(x,y) = R_j(r_j(q))(x,y), \quad (x,y) \in M_j.$$

Thus our first question is: Does the system of equations in the finitely many unknowns $w_j(x,y)$ have a unique solution? For stable methods, a positive answer is supplied by the following theorem.

Theorem 13.5: Let $F \in C^1(\mathbb{R}^n, \mathbb{R}^n)$ and $K > 0$. Then the following two conditions are equivalent:

(1) $\|F(x)-F(\tilde{x})\| \geq K\|x-\tilde{x}\|, \quad x,\tilde{x} \in \mathbb{R}^n$

(2) F is bijective. The inverse map Q is continuously differentiable and

$$\|Q(x)-Q(\tilde{x})\| \leq \frac{1}{K}\|x-\tilde{x}\|, \quad x,\tilde{x} \in \mathbb{R}^n.$$

Proof that (1) implies (2): Let $F'(x)$ be the Jacobian of

F. We show that $F'(x)$ is regular for all x. For if not, then there would exist $x_0, y_0 \in \mathbb{R}^n$ with $F'(x_0)y_0 = 0$ and $y_0 \neq 0$. This means that the directional derivative at the point x_0 in the direction y_0 is zero:

$$\lim_{|h|\to 0} \frac{1}{|h|} \|F(x_0+hy_0)-F(x_0)\| = 0.$$

Thus there exists an $h_0 > 0$ such that

$$\frac{1}{h_0} \|F(x_0+h_0y_0) - F(x_0)\| < K\|y_0\|$$

or

$$\|F(x_0+h_0y_0) - F(x_0)\| < K\|h_0y_0\| .$$

This contradicts (1). Therefore $F'(x)$ is regular everywhere.

F is injective. $F(x) = F(\tilde{x})$ implies $x = \tilde{x}$ at once by virtue of (1). Since $F'(x)$ is always regular, it follows from the implicit function theorem that the inverse map Q is continuously differentiable and that F is an open mapping. It maps open sets to open sets. In particular, $F(\mathbb{R}^n)$ is an open set.

We must still show that F is surjective. Let $x_0 \in \mathbb{R}^n$ be an arbitrary but fixed vector. By (1) we have

$$\|F(x) - F(0)\| \geq K\|x\|$$

$$\|F(x)-x_0\| + \|x_0-F(0)\| \geq K\|x\| .$$

For all x outside the ball

$$E = \{x \in \mathbb{R}^n \mid \|x\| \leq 2\|x_0-F(0)\|/K\}$$

we have

$$d(x) = \|F(x)-x_0\| > \|F(0)-x_0\| .$$

Therefore there exists an $x_1 \in E$ with $d(x_1) \le d(x)$, $x \in \mathbb{R}^n$. On the other hand,

$$d(x_1) = \inf_{y \in F(\mathbb{R}^n)} \|y - x_0\| .$$

Since $F(\mathbb{R}^n)$ is open, it follows that $x_0 \in F(\mathbb{R}^n)$. Thus F is surjective. It also follows from (1) that

$$\|x-\tilde{x}\| = \|F(Q(x))-F(Q(\tilde{x}))\| \ge K\|Q(x)-Q(\tilde{x})\| .$$

This completes the proof of (2).

Proof that (2) implies (1): Let $x, \tilde{x} \in \mathbb{R}^n$. It follows by virtue of (2) that

$$\|x-\tilde{x}\| = \|Q(F(x))-Q(F(\tilde{x}))\| \le \frac{1}{K}\|F(x)-F(\tilde{x})\| . \qquad \square$$

Theorem 13.6: Let D be a consistent and stable difference method for Problem 13.2 and let $m, j_0 \in \mathbb{N}$, and $K > 0$ be constants as in Definition 13.3. For arbitrary $u \in C^2(\overline{G}, \mathbb{R})$ we define the lattice functions w_j, $j = j_0(1)\infty$, to be the solutions of the difference equations

$$F_j(\psi, w_j) = R_j(r_j(q)).$$

Here $\psi = r_\Gamma(u)$ and $q = Lu$. Then we have:

(1) $\|r_j(u)-w_j\|_\infty \le \frac{1}{K}\|F_j(\psi, r_j(u))-R_j(r_j(q))\|_\infty$, $\quad j = j_0(1)\infty$.

(2) If $u \in C^m(\overline{G}, \mathbb{R})$, then

$$\lim_{j\to\infty} \|r_j(u)-w_j\|_\infty = 0.$$

Proof: ψ depends only on u and not on j. By Theorem 13.5, the maps $F_j(\psi,\cdot)$ have differentiable inverses Q_j. We have

$$r_j(u) = Q_j(F_j(\psi, r_j(u)))$$

$$w_j = Q_j(R_j(r_j(q)))$$

$$\|r_j(u)-w_j\|_\infty \le \|Q_j(F_j(\psi, r_j(u)))-Q_j(R_j(r_j(q)))\|_\infty.$$

(1) follows from Theorem 13.5(2) and (2) follows from (1) and Definition 13.3(4). □

In Problem 13.2, q and ψ are given. All convergence conditions which take account of the properties of the exact solution u are of only relative utility. Unfortunately, it is very difficult to decide the convergence question simply on the basis of a knowledge of q and ψ. Nevertheless one knows that for fixed $\psi \in C^0(\Gamma, \mathbb{R})$, the set of q for which the difference method converges is closed in $C^0(\overline{G}, \mathbb{R})$.

<u>Theorem 13.7</u>: Let D be a consistent and stable difference method for Problem 13.2, let $\psi \in C^0(\Gamma, \mathbb{R})$ and let

$$S_\psi = \{q \in C^0(\overline{G}, \mathbb{R}) \mid \text{there exists a } u \in C^2(\overline{G}, \mathbb{R}) \text{ such that } r_\Gamma(u) = \psi,$$
$$q = Lu \text{ and } \lim_{j\to\infty} \|r_j(u)-w_j\|_\infty = 0\}.$$

Further let $\tilde{q} \in \overline{S}_\psi$ and

$$F_j(\psi, \tilde{w}_j) = R_j(r_j(\tilde{q})), \quad j = j_0(1)\infty.$$

Then there exists a $\tilde{u} \in C^0(\overline{G}, \mathbb{R})$ such that

$$\lim_{j\to\infty} \|r_j(\tilde{u})-\tilde{w}_j\|_\infty = 0.$$

Note that the function $\tilde{u}$ need not necessarily be the classical solution of the boundary value problem.

Proof: Let

$$q^{(1)},q^{(2)} \in S_\psi, \quad q^{(1)} = Lu^{(1)}, q^{(2)} = Lu^{(2)},$$

$$F_j(\psi,w_j^{(1)}) = R_j(r_j(q^{(1)})), \quad j = j_0(1)\infty$$

$$F_j(\psi,w_j^{(2)}) = R_j(r_j(q^{(2)})), \quad j = j_0(1)\infty.$$

Then:

$$\|r_j(u^{(1)})-r_j(u^{(2)})\|_\infty$$

$$\leq \|r_j(u^{(1)})-w_j^{(1)}\|_\infty + \|w_j^{(1)}-w_j^{(2)}\|_\infty + \|w_j^{(2)}-r_j(u_j^{(2)})\|_\infty.$$

Let Q_j, $j = j_0(1)\infty$, again be the inverse functions of $F_j(\psi,\cdot)$ and K and $\tilde{K}$ the constants from stability condition 13.3(5). It follows from Theorem 13.5 that:

$$\|r_j(u^{(1)}-u^{(2)})\|_\infty$$

$$\leq \|r_j(u^{(1)})-w_j^{(1)}\|_\infty + \|r_j(u^{(2)})-w_j^{(2)}\|_\infty + \frac{\tilde{K}}{K}\|r_j(q^{(1)}-q^{(2)})\|_\infty$$

$$\leq \|r_j(u^{(1)})-w_j^{(1)}\|_\infty + \|r_j(u^{(2)})-w_j^{(2)}\|_\infty + \frac{\tilde{K}}{K}\|q^{(1)}-q^{(2)}\|_\infty.$$

In passing to the limit $j \to \infty$, the left side of the inequality converges to $\|u^{(1)} - u^{(2)}\|$, while the mesh $|M_j|$ converges to zero. On the right, the first two summands converge to zero by hypothesis. All this means that

$$\|u^{(1)} - u^{(2)}\|_\infty \leq \frac{\tilde{K}}{K}\|q^{(1)} - q^{(2)}\|_\infty.$$

Thus corresponding to the Cauchy sequence

$$\{q^{(\nu)} \in S_\psi \mid \nu = 1(1)\infty\}$$

there is a Cauchy sequence

$$\{u^{(\nu)} \in C^o(\overline{G}, \mathbb{R}) \mid \nu = 1(1)\infty\}.$$

Let

$$\tilde{q} = \lim_{\nu\to\infty} q^{(\nu)}, \qquad \tilde{u} = \lim_{\nu\to\infty} u^{(\nu)}.$$

Then for $\nu = 1(1)\infty$ we have the inequalities

$$\begin{aligned} &\|r_j(\tilde{u})-\tilde{w}_j\| \\ &\leq \|r_j(\tilde{u}-u^{(\nu)})\|_\infty + \|r_j(u^{(\nu)})-w_j^{(\nu)}\|_\infty + \|w_j^{(\nu)}-\tilde{w}_j\|_\infty \\ &\leq \|\tilde{u}-u^{(\nu)}\|_\infty + \|r_j(u^{(\nu)})-w_j^{(\nu)}\|_\infty + \frac{\tilde{K}}{K}\|q^{(\nu)}-\tilde{q}\|_\infty. \end{aligned}$$

For $\varepsilon > 0$ there is a $\nu_0 \in \mathbb{N}$ with

$$\|\tilde{u}-u^{(\nu_0)}\|_\infty < \frac{\varepsilon}{3}$$

$$\|\tilde{q}-q^{(\nu_0)}\|_\infty < \frac{\varepsilon}{3}\frac{K}{\tilde{K}}.$$

For this ν_0 we choose a $j_1 \in \mathbb{N}$ such that

$$\|r_j(u^{(\nu_0)})-w_j^{(\nu_0)}\|_\infty \leq \frac{\varepsilon}{3}, \qquad j = j_1(1)\infty.$$

Altogether then, for $j \geq j_1$ we have

$$\|r_j(\tilde{u})-\tilde{w}_j\|_\infty \leq \varepsilon. \qquad \square$$

For the most important of the difference methods for elliptic differential equations, stability follows from a monotone principle. The first presentation of this relationship may be found in Gerschgorin 1930. The method was then expanded extensively by Collatz 1964 and others.

The monotone principle just mentioned belongs to the theory of semi-ordered vector spaces. We recall some basic concepts. Let Ω be an arbitrary set and V a vector space of elements $f:\Omega \to \mathbb{R}$. In V there is a *natural semiorder*

$$f \leq g \iff [f(x) \leq g(x), \quad x \in \Omega].$$

The following computational rules hold:

$$\begin{aligned}
&f \leq f \\
&f \leq g,\ g \leq f \quad \Rightarrow \quad f = g \\
&f \leq g,\ g \leq h \quad \Rightarrow \quad f \leq h \\
&f < g,\ \lambda \in \mathbb{R}_+ \quad \Rightarrow \quad \lambda f \leq \lambda g \\
&f \leq g \quad \Rightarrow \quad -g \leq -f \\
&0 \leq f,\ 0 \leq g \quad \Rightarrow \quad 0 \leq f+g.
\end{aligned}$$

We further define

$$|f|(x) = |f(x)|.$$

From this it follows that

$$0 \leq |f|, \quad f \leq |f|.$$

When Ω is a finite set or when Ω is compact and all $f \in V$ are continuous,

$$\|f\|_\infty = \max_{x\in\Omega} |f(x)|$$

exists. Obviously,

$$\| |f| \|_\infty = \|f\|_\infty.$$

We use this semiorder for various basic sets Ω, including

Ω	V
$\{1,2,\ldots,n\}$	$\mathbb{R}^n$
$\{1,2,\ldots,m\}\times\{1,2,\ldots,n\}$	$MAT(m,n,\mathbb{R})$
Lattice M	$C^0(M,\mathbb{R})$
$\overline{G}$	$C^0(\overline{G},\mathbb{R})$.

Definition 13.8: $A \in MAT(n,n,\mathbb{R})$ is called an *M-matrix* if there exists a splitting $A = D - B$ with the following properties:

(1) D is a regular, diagonal matrix; the diagonal of B is identically zero.

(2) $D \geq 0$, $B \geq 0$.

(3) $A^{-1} \geq 0$.

Theorem 13.9: Let $A = D - B$ be a splitting of $A \in MAT(n,n,\mathbb{R})$, where D and B satisfy 13.8(1) and (2). Then A is an M-matrix if and only if $\rho(D^{-1}B) < 1$ (ρ = *spectral radius*).

Proof: Let $\rho(D^{-1}B) < 1$. Then the series

$$S = \sum_{\nu=0}^{\infty} (D^{-1}B)^{\nu}$$

converges and $S \geq 0$. Obviously,

$$(I-D^{-1}B)S = S(I-D^{-1}B) = I, \quad A^{-1} = SD^{-1} \geq 0.$$

Conversely, let $A^{-1} \geq 0$ and let λ be an eigenvalue of $D^{-1}B$ with x the corresponding eigenvector. Then we have the following inequalities:

$$|\lambda||x| = |D^{-1}Bx| \leq D^{-1}B|x|$$

$$(I-D^{-1}B)|x| \leq (1-|\lambda|)|x|$$

$$(D-B)|x| \leq (1-|\lambda|)D|x|$$

$$|x| \leq (1-|\lambda|)A^{-1}D|x|.$$

Since $x \neq 0$, $A^{-1} \geq 0$, and $D \geq 0$, the last inequality implies that $|\lambda| < 1$ and $\rho(D^{-1}B) < 1$. □

The eigenvalues of $D^{-1}B$ can be estimated with the help of Gershgorin circles (cf. Stoer-Bulirsch 1980). For this let

$$A = \{a_{ij} \mid i = 1(1)n,\ j = 1(1)n\}.$$

One obtains the following sufficient conditions for $\rho(D^{-1}B) < 1$:

Condition 13.10: A is *diagonal dominant*, i.e.

$$\sum_{\substack{j=1\\ j\neq i}}^{n} |a_{ij}| < |a_{ii}|, \quad i = 1(1)n.$$

Condition 13.11: A is *irreducible diagonal dominant*, i.e.,

$$\sum_{\substack{j=1\\ j\neq i}}^{n} |a_{ij}| \le |a_{ii}|, \quad i = 1(1)n,$$

A is irreducible and there exist $r \in \{0,1,\ldots,n\}$ such that

$$\sum_{\substack{j=1\\ j\neq r}}^{n} |a_{rj}| < |a_{rr}|.$$

Definition 13.12: Let V_1 and V_2 be semiordered. A mapping $F:V_1 \to V_2$ is called

isotonic	if $f \le g \Rightarrow F(f) \le F(g)$
antitonic	if $f \le g \Rightarrow F(g) \le F(f)$
inverse isotonic	if $F(f) \le F(g) \Rightarrow f \le g$

for all $f,g \in V_1$.

Definition 13.13: Let V be the vector space whose elements consist of mappings $f:\Omega \to \mathbb{R}$. Then $F:V \to V$ is called *diagonal* if for all $f,g \in V$ and all $x \in \Omega$ it is true that:

$$f(x) = g(x) \Rightarrow F(f)(x) = F(g)(x). \quad \square$$

In order to give substance to these concepts, we consider the affine maps $F:x \to Ax+c$, where $A \in MAT(n,n,\mathbb{R})$ and

$c \in \mathbb{R}^n$. Then we have:

$A \geq 0 \iff F$ isotonic

$-A \geq 0 \iff F$ antitonic

A an M-matrix $\Rightarrow$ F inverse isotonic

A diagonal matrix $\iff$ F diagonal

$A \geq 0$, regular diagonal matrix $\iff$ F diagonal, isotonic, and inverse isotonic.

A mapping $F: \mathbb{R}^n \to \mathbb{R}^n$ is diagonal if it can be written as follows:

$$y_i = f_i(x_i), \quad i = 1(1)n.$$

The concepts of isotonic, antitonic, inverse isotonic, and diagonal were originally defined in Ortega-Rheinboldt 1970. Equations of the form $F(f) = g$ with F inverse isotonic are investigated thoroughly in Collatz 1964, where he calls them of *monotone type*.

Theorem 13.14: Let $A \in \mathrm{MAT}(n,n,\mathbb{R})$ be an M-matrix and let $F: \mathbb{R}^n \to \mathbb{R}^n$ be diagonal and isotonic. Then the mapping $\hat{F}: \mathbb{R}^n \to \mathbb{R}^n$ defined by

$$\hat{F}(x) = Ax + F(x), \qquad x \in \mathbb{R}^n$$

is inverse isotonic and furthermore

$$\|\hat{F}(x) - \hat{F}(\tilde{x})\|_\infty \geq \frac{\|x-\tilde{x}\|_\infty}{\|A^{-1}\|_\infty}.$$

Proof: Since F is diagonal, one can write the equation $y = F(x)$ componentwise as follows:

$$y_i = f_i(x_i), \quad i = 1(1)n.$$

For fixed but arbitrary $x = (x_1,\ldots,x_n) \in \mathbb{R}^n$ and

$\tilde{x} = (\tilde{x}_1,\ldots,\tilde{x}_n) \in \mathbb{R}^n$ we define, for $i = 1(1)n$,

$$e_{ii} = \begin{cases} \dfrac{f_i(x_i) - f_i(\tilde{x}_i)}{x_i - \tilde{x}_i} & \text{if } \tilde{x}_i \neq x_i \\ 1 & \text{otherwise} \end{cases}$$

$$E = \operatorname{diag}(e_{ii}).$$

F isotonic implies $E \geq 0$. In addition,

$$F(x) - F(\tilde{x}) = E(x-\tilde{x}).$$

Let $A = D - B$ be a splitting of A as in Definition 13.8. It follows from

$$\hat{F}(\tilde{x}) = A\tilde{x} + F(\tilde{x}) = \tilde{y}$$
$$\hat{F}(x) = Ax + F(x) = y$$

that

$$\hat{F}(x) - \hat{F}(\tilde{x}) = (D+E-B)(x-\tilde{x}).$$

Since

$$S = \sum_{\nu=0}^{\infty} (D^{-1}B)^{\nu} \geq 0$$

converges,

$$T = \sum_{\nu=0}^{\infty} [(D+E)^{-1}B]^{\nu} \geq 0$$

certainly converges. The elements in the series are certainly no greater than the elements in preceding series. Therefore,

$$I - (D+E)^{-1}B$$

is regular, and

$$[I-(D+E)^{-1}B]^{-1} = T \geq 0.$$

$D+E-B$ is also an M-matrix. We have $x \leq \tilde{x}$ for $\hat{F}(x) \leq \hat{F}(\tilde{x})$, and this holds for all $x,\tilde{x} \in \mathbb{R}^n$. This shows that $\hat{F}$ is inverse monotone.

In addition we have

$$\begin{aligned}||x-\tilde{x}||_\infty &= ||(D+E-B)^{-1}[\hat{F}(x)-\hat{F}(\tilde{x})]||_\infty \\ &= ||T(D+E)^{-1}[\hat{F}(x)-\hat{F}(\tilde{x})]||_\infty \\ &\le ||T(D+E)^{-1}||_\infty\, ||\hat{F}(x)-\hat{F}(\tilde{x})||_\infty\end{aligned}$$

or

$$||\hat{F}(x)-\hat{F}(\tilde{x})||_\infty \ge \frac{||x-\tilde{x}||_\infty}{||T(D+E)^{-1}||_\infty} .$$

The row sum norm of the matrix

$$T(D+E)^{-1} = \{\sum_{\nu=0}^{\infty} [(D+E)^{-1}B]^\nu\}(D+E)^{-1}$$

is obviously no greater than the norm of

$$SD^{-1} = \{\sum_{\nu=0}^{\infty} [D^{-1}B]^\nu\}D^{-1} = A^{-1}.$$

This implies that

$$||T(D+E)^{-1}||_\infty \le ||A^{-1}||_\infty$$

$$||\hat{F}(x)-\hat{F}(\tilde{x})||_\infty \ge \frac{||x-\tilde{x}||_\infty}{||A^{-1}||_\infty} . \qquad \square$$

<u>Theorem 13.15</u>: Hypotheses:

(a) $A \in MAT(n,n,\mathbb{R})$ is an M-matrix

(b) $F: \mathbb{R}^n \to \mathbb{R}^n$ is diagonal and isotonic, $\hat{F}(x) = Ax + F(x)$

(c) $v \in \mathbb{R}^n$, $v \ge 0$, $Av \ge z = (1,\dots,1) \in \mathbb{R}^n$

(d) $w \in \mathbb{R}^n$, $||\hat{F}(w)||_\infty \le 1$.

Conclusions:

(1) It is true for all $x,\tilde{x} \in \mathbb{R}^n$ that

$$||\hat{F}(x)-\hat{F}(\tilde{x})||_\infty \ge \frac{||x-\tilde{x}||_\infty}{||v||_\infty} .$$

(2) $F(0) = 0$ implies $|w| \leq v$.

Proof: For all $x \in \mathbb{R}^n$ it follows from $Av \geq z$ that

$$|x| \leq \|x\|_\infty z \leq \|x\|_\infty Av.$$

Since $A^{-1} \geq 0$ it follows that

$$A^{-1}|x| \leq \|x\|_\infty v$$

$$\|A^{-1}x\|_\infty \leq \|A^{-1}|x|\|_\infty \leq \|x\|_\infty \|v\|_\infty$$

$$\|A^{-1}\|_\infty \leq \|v\|_\infty .$$

Combining this with Theorem 13.14 yields conclusion (1):

$$\|\hat{F}(x)-\hat{F}(\tilde{x})\|_\infty \geq \frac{\|x-\tilde{x}\|_\infty}{\|v\|_\infty}, \qquad x,\tilde{x} \in \mathbb{R}^n.$$

For the proof of (2) we need to remember that $F(x)$ is isotonic and $F(-x)$ is antitonic. $F(0) = 0$ implies that

$$-z \leq \hat{F}(w) \leq z$$

$$-Av \leq \hat{F}(w) \leq Av$$

$$-Av+F(-v) \leq \hat{F}(w) \leq Av+F(v)$$

$$-\hat{F}(-v) \leq \hat{F}(w) \leq \hat{F}(v).$$

Since $\hat{F}$ is inverse isotonic, it follows that

$$-v \leq w \leq v. \qquad \square$$

We conclude our generalized considerations of functions on semiordered vector spaces with this theorem, and return to the topic of difference methods. In order to lend some substance to the subject, we assume that the points of lattice M_j have been enumerated in *some* way from 1 to n_j. We will not distinguish between a lattice function $w_j \in C^0(M_j, \mathbb{R})$ and the vector

$$[w_j(x_1,y_1),\ldots,w_j(x_{n_j},y_{n_j})] \in \mathbb{R}^{n_j}.$$

Thus, for each linear mapping $F:C^0(M_j, \mathbb{R}) \to C^0(M_j, \mathbb{R})$ there is a matrix $A \in \mathrm{MAT}(n_j,n_j, \mathbb{R})$ and vice versa. This matrix depends naturally on the enumeration of the lattice points. However, properties such as "$A \geq 0$" or "A is a diagonal matrix" or "A is an M-matrix" either hold for every enumeration or for none. The primary consequence of these monotonicity considerations is the following theorem.

<u>Theorem 13.16</u>: Let $D = \{(M_j,F_j,R_j) \mid j = 1(1)\infty\}$ be a difference method satisfying the following properties:

(1) $F_j(\psi,w_j) = F_j^{(1)}(w_j) + F_j^{(2)}(w_j) - F_j^{(3)}(\psi)$,

$\psi \in C^0(\Gamma, \mathbb{R})$, $w_j \in C^0(M_j, \mathbb{R})$.

(2) $F_j^{(1)}$ is a linear mapping having an M-matrix.

(3) $F_j^{(2)}$ is diagonal and isotonic, and $F_j^{(2)}(0) = 0$.

(4) $F_j^{(3)}$ and R_j are linear and isotonic, and $\|R_j\| \leq \tilde{K}$. Also it is true for all $\psi \in C^0(\Gamma, \mathbb{R})$ and $w_j \in C^0(M_j, \mathbb{R})$ with $\psi \geq 1$ and $w_j \geq 1$ that

$$F_j^{(3)}(\psi) + R_j(w_j) \geq (1,\ldots,1).$$

(5) The method $\{(M_j,F_j^{(1)}-F_j^{(3)},R_j) \mid j = 1(1)\infty\}$ is consistent if the function H in Problem 13.2 is identically zero.

<u>Conclusion</u>: D is stable.

<u>Remark 13.17</u>: The individual summands of F_j, R_j as a rule correspond to the following terms of a boundary value problem:

Difference method	Boundary value problem
$F_j^{(1)}$	$Lu(x,y) - H(x,y,u(x,y))$
$F_j^{(2)}$	$H(x,y,u(x,y))$
$F_j^{(3)}$	$\psi(x,y)$
R_j	$q(x,y)$.

Since $F_j^{(2)}$ must be isotonic, $H_z(x,y,z)$ can never be negative. If this fails, the theorem is not applicable.

Consistency as in 13.3(4) can almost always be obtained by multiplying the difference equations with a sufficiently high power of $h_j = |M_j|$. The decisive question is whether stability survives this approach. Condition (4) of Theorem 13.16 is a normalization condition which states precisely when such a multiplication is permissible. At most points (x,y) of the lattice it is the rule that $F_j^{(3)}(\psi)(x,y) = 0$. Such points we call *boundary-distant points*. Among other things, 13.16(4) implies that it follows from $w_j \geq 1$ that for all boundary-distant points (x,y),

$$R_j(w_j)(x,y) \geq 1.$$

In practice, one is interested only in the consistent methods D. But stability follows from consistency alone for $H \equiv 0$ and $F_j^{(2)}(w_j) \equiv 0$. In general, it suffices to have $F_j^{(2)}$ isotonic.

In Example 13.4 we have $F_j^{(2)}(w_j) \equiv 0$, R_j the identity and

$$F_j^{(1)}(w_j)-F_j^{(3)}(\psi) = \frac{1}{h_j^2}\{4w_j(x,y)-\hat{w}_j(x+h_j,y)-\hat{w}_j(x-h_j,y) -\hat{w}_j(x,y+h_j)-\hat{w}_j(x,y-h_j)\}$$

$$\hat{w}_j(x,y) = \begin{cases} w_j(x,y) & \text{if } (x,y) \in M_j \\ \psi(x,y) & \text{if } (x,y) \in \Gamma. \end{cases}$$

$F_j^{(1)}$ has a corresponding symmetric, irreducible, diagonal dominant matrix (cf. Condition 13.11). Thus condition (2) of Theorem 13.16 is satisfied. Since $F_j^{(2)} \equiv 0$, condition (3) is satisfied. Since R_j is the identity, one can choose $\tilde{K} = 1$ in (4). Also when $w_j \geq 1$ and $\psi \geq 1$, then obviously

$$F_j^{(3)}(\psi) \geq 0, \quad R_j(w_j) \geq 1.$$

Therefore (4) is also satisfied. Consistency (5) is easily obtained for $m = 4$ from a Taylor expansion of the difference equations. For $u \in C^4(\overline{G}, \mathbb{R})$ one obtains

$$||F_j(\psi, r_j(u)) - r_j(q)||_\infty \leq \hat{K} h_j^2$$

where

$$\hat{K} = \frac{1}{6} \max_{(x,y)\in\overline{G}} (|u_{xxxx}|, |u_{yyyy}|).$$

Thus the method is stable by Theorem 13.16. □

Theorem 13.16 is reduced to Theorem 13.15 with the aid of two lemmas.

Lemma 13.18: There exists an $s \in C^\infty(\overline{G}, \mathbb{R})$ with $s(x,y) \geq 0$ and

$$\tilde{L}s(x,y) = Ls(x,y) - H(x,y,s(x,y)) \geq 1, \quad (x,y) \in \overline{G}.$$

Proof: For all $(x,y) \in \overline{G}$ let

$$a_{11}(x,y) \geq K_1 > 0, \quad |b_1(x,y)| \leq K_2, \quad \beta_1 \leq x \leq \beta_2.$$

We set

$$\alpha = \frac{3K_2}{K_1}, \quad \beta = \frac{\beta_1+\beta_2}{2}$$

and show that

$$s(x,y) = \{\cosh[\alpha(\beta_2-\beta)] - \cosh[\alpha(x-\beta)]\}/(2\alpha K_2)$$

is a function with the desired properties. First of all, it follows from $|x-\beta| \leq \beta_2-\beta$ that

$$\cosh[\alpha(x-\beta)] \leq \cosh[\alpha(\beta_2-\beta)],$$

and from this, that $s(x,y) \geq 0$. Since s depends only on x, we have

$$\begin{aligned}\tilde{L}s &= -a_{11}s_{xx} - b_1 s_x \\ &= \frac{3}{2K_1} a_{11}\cosh[\alpha(x-\beta)] + \frac{1}{2K_2} b_1 \sinh[\alpha(x-\beta)].\end{aligned}$$

Since it is always the case that

$$|\sinh[\alpha(x-\beta)]| < \cosh[\alpha(x-\beta)],$$

it is also true that

$$\tilde{L}s(x,y) \geq \cosh[\alpha(x-\beta)] \geq 1. \qquad \square$$

Remark 13.19: The function s plays a definite role in error estimates, since s should then be as small as possible. Our approach was meant to cover the most general case. In many specific cases there are substantially smaller functions of this type, as the following three examples demonstrate.

$$\begin{aligned}
&b_1(x,y) = 0: && s(x,y) = \frac{1}{2K_1}(x-\beta_1)(\beta_2-x)\\
&b_1(x,y) \geq 0: && s(x,y) = \frac{1}{2K_1}(x-\beta_1)(2\beta_2-\beta_1-x)\\
&L = -\Delta,\ G = \text{unit circle}: && s(x,y) = \tfrac{1}{4}(1-x^2-y^2).
\end{aligned}$$

In the last example, the choice of s is optimal, since there are no smaller functions with the desired properties. In the other two examples, one may possibly obtain more advantageous constants K_1, β_1, and β_2 by exchanging x and y. □

Lemma 13.20: There exists a $v \in C^0(\overline{G}, \mathbb{R})$ and a $j_0 \in \mathbb{N}$ such that

$$v(x,y) \geq 0, \qquad (x,y) \in \overline{G}$$

$$F_j^{(1)}(r_j(v))(x,y) \geq (1,\ldots,1), \quad j = j_0(1)\infty.$$

Proof: We choose the s of Lemma 13.18 and define $v = 2s + 2$. It is obviously true that $v \in C^\infty(\overline{G}, \mathbb{R})$, and

$$v(x,y) \geq 2, \quad \tilde{L}v(x,y) \geq 2 \quad \text{for} \quad (x,y) \in \overline{G}.$$

The function v is a solution of the boundary value problem 13.2 with

$$H \equiv 0, \quad \psi = r_\Gamma(v) \geq 2, \quad q(x,y) = \tilde{L}v(x,y) \geq 2.$$

Insofar as the method $\{(M_j, F_j^{(1)} - F_j^{(3)}, R_j) \mid j = 1(1)\infty\}$ is consistent with respect to this problem, we have

$$\lim_{j\to\infty} \| F_j^{(1)}(r_j(v)) - F_j^{(3)}(\psi) - R_j(r_j(q)) \|_\infty = 0.$$

We now choose j_0 so large that for all $j \geq j_0$ we have

$$\| F_j^{(1)}(r_j(v)) - F_j^{(3)}(\psi) - R_j(r_j(q)) \|_\infty \leq 1.$$

$F_j^{(3)}$ and R_j are linear and isotonic. For $\psi \geq 1$ and $q \geq 1$ we have

$$F_j^{(3)}(\psi) + R_j(r_j(q)) \geq (1,\ldots,1).$$

Since we actually have $\psi \geq 2$ and $q \geq 2$, it follows that

$$F_j^{(3)}(\psi) + R_j(r_j(q)) \geq (2,\ldots,2)$$

and hence that

$$F_j^{(1)}(r_j(v)) \geq (1,\ldots,1). \qquad \square$$

Remark 13.21: Instead of $v \in C^0(\overline{G}, \mathbb{R})$ and $v \geq 0$, we actually proved $v \in C^\infty(\overline{G}, \mathbb{R})$ and $v \geq 2$. However, the conditions of the lemma are sufficient for the sequel. Since one is again interested in the smallest possible functions of this type, constructions other than the one of our proof could be used. These other methods need only yield a continuous function $v \geq 0$. $\square$

Proof of Theorem 13.16: We choose v as in Lemma 13.20. Then we can apply Theorem 13.15. The quantities are related as follows:

Theorem 13.16	Theorem 13.15
$F_j^{(1)}$	A
$F_j^{(2)}$	F
$r_j(v)$	v
$F_j^{(1)} + F_j^{(2)}$	$\hat{F}$
0	w .

For $j \geq j_0$ it follows from Theorem 13.15(1) that:

$$\|F_j^{(1)}(w_j)+F_j^{(2)}(w_j)-F_j^{(1)}(\tilde{w}_j)-F_j^{(2)}(\tilde{w}_j)\|_\infty$$
$$\geq \frac{\|w_j-\tilde{w}_j\|_\infty}{\|r_j(v)\|_\infty} \geq \frac{\|w_j-\tilde{w}_j\|_\infty}{\|v\|_\infty}, \quad w_j,\tilde{w}_j \in C^0(M_j, \mathbb{R}).$$

$\|v\|_\infty$ does not depend on j. This proves the first inequality in 13.3(5) with $K = 1/\|v\|_\infty$. The second inequality is equivalent to $\|R_j\| \leq \tilde{K}$. $\square$

In view of the last proof, one may choose $K = 1/\|v\|_\infty$ in Definition 13.3(5). Here v is an otherwise arbitrary function satisfying the properties given in Lemma 13.20.

Conclusion (1) of Theorem 13.6 yields the error estimate

$$\|r_j(u)-w_j\|_\infty \le \|v\|_\infty \|F_j(\psi,r_j(u))-R_j(r_j(q))\|_\infty, \quad j = j_0(1)\infty.$$

Here

u is the exact solution of the boundary value problem

w_j is the solution of the difference equation

$F_j(\psi,r_j(u)) - R_j(r_j(q))$ is the *local error*

v is a *bounding function* (which depends only on $F_j^{(1)}$).

The inequality can be sharpened to a pointwise estimate with the help of conclusion (2) of Theorem 13.15. For all lattice points $(x,y) \in M_j$ and $j = j_0(1)\infty$ we have

$$|u(x,y)-w_j(x,y)| \le v(x,y)\|F_j(\psi,r_j(u))-R_j(r_j(q))\|_\infty.$$

In many important special cases, e.g., the model problem (Example 13.4), R_j is the identity. A straightforward modification of the proof of Lemma 13.20 then leads to the following result: let $\varepsilon > 0$ and let $s \in C^m(\overline{G}, \mathbb{R})$ with $s \ge 0$ and $\tilde{L}s(x,y) \ge 1$ (cf. Lemma 13.18). Then there exists a $j_1 \in \mathbb{N}$ such that

$$|u(x,y)-w_j(x,y)| \le (1+\varepsilon)s(x,y)\|F_j(\psi,r_j(u))-r_j(q))\|_\infty,$$
$$j = j_1(1)\infty.$$

In the model problem

$$s(x,y) = \frac{1}{4}\,x(1-x) + \frac{1}{4}\,y(1-y)$$

is such a function. Here one can actually choose $j_1 = 1$, independently of ε. It therefore follows that

$$|u(x,y)-w_j(x,y)| \le s(x,y)\|F_j(\psi,r_j(u))-r_j(q)\|_\infty, \quad j = 1(1)\infty.$$

We will now construct several concrete difference methods. Let

$$e^{(1)} = \begin{pmatrix}1\\0\end{pmatrix}, \quad e^{(2)} = \begin{pmatrix}0\\1\end{pmatrix}, \quad e^{(3)} = \begin{pmatrix}-1\\0\end{pmatrix}, \quad e^{(4)} = \begin{pmatrix}0\\-1\end{pmatrix}.$$

(cf. Figure 13.22). With each point $(x,y) \in G$ we associate four neighboring points $N_\nu(x,y,h) \in \overline{G}$ with $\nu = 1(1)4$ and $h > 0$. Now for $\nu = 1(1)4$ let:

$$\lambda_\nu = \begin{cases} h \text{ if } & (x,y)+\lambda e^{(\nu)} \in G \text{ for all } \lambda \in [0,h] \\ \min\,\{\lambda > 0 \mid (x,y)+\lambda e^{(\nu)} \in \Gamma\} & \text{otherwise} \end{cases}$$

$$N_\nu(x,y,h) = (x,y) + \lambda_\nu e^{(\nu)}$$

$$d_\nu(x,y,h) = \|(x,y) - N_\nu(x,y,h)\|_2 = \lambda_\nu \|e^{(\nu)}\|_2.$$

Obviously we have

$$0 < d_\nu(x,y,h) \le h, \qquad \nu = 1(1)4.$$

By Definition 13.1, the standard lattice with mesh size h is

$$M_h = \{(x,y) \in G \mid x = \mu h,\ y = \nu h \text{ where } \mu,\nu \in \mathbb{Z}\},$$
$$0 < h \le h_o.$$

For $(x,y) \in M_h$ all the neighboring points $N_\nu(x,y,h)$ belong to M_h or Γ.

For brevity, we introduce the notation $\mathrm{Lip}^{(\ell)}(\overline{G}, \mathbb{R})$. This is a subspace of $C^\ell(\overline{G}, \mathbb{R})$ defined as follows: for every $f \in \mathrm{Lip}^{(\ell)}(\overline{G}, \mathbb{R})$ there exists an $L > 0$ such that

$$\left|\frac{\partial^{\mu+\nu} f}{\partial x^\mu \partial y^\nu}(x,y) - \frac{\partial^{\mu+\nu} f}{\partial x^\mu \partial y^\nu}(\tilde{x},\tilde{y})\right| \le L\|(x,y)-(\tilde{x},\tilde{y})\|_\infty$$

$$(x,y) \in \overline{G}, \quad (\tilde{x},\tilde{y}) \in \overline{G}, \quad \mu+\nu = \ell.$$

Obviously,

$$C^{\ell+1}(\overline{G}, \mathbb{R}) \subset \mathrm{Lip}^{(\ell)}(\overline{G}, \mathbb{R}).$$

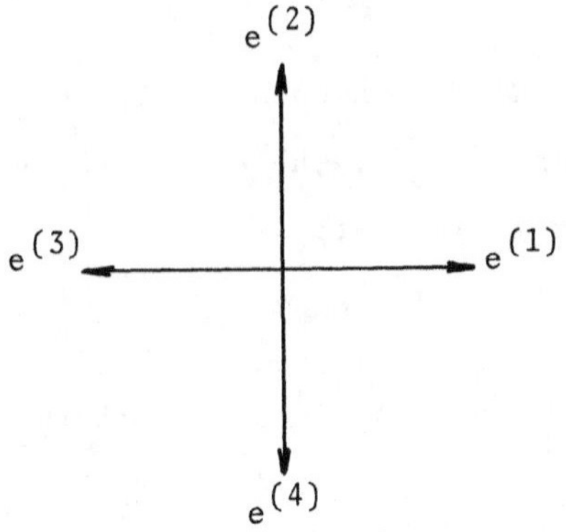

Figure 13.22. Direction vectors for the difference method

The next lemma contains a one-dimensional difference equation which we will use as the basis for the difference methods in G.

Lemma 13.23: Let $\beta > 0$ and $a,u \in C^2([-\beta,\beta], \mathbb{R}) \cap C^3((-\beta,\beta), \mathbb{R})$. Suppose further that there is a positive constant L such that for all $t,s \in (-\beta,\beta)$ and $\nu = 0(1)3$ the following inequalities are valid:

$$|a^{(\nu)}(t)| \le L, \quad |u^{(\nu)}(t)| \le L,$$

$$|a^{(\nu)}(t)-a^{(\nu)}(s)| \le L|t-s|,$$

$$|u^{(\nu)}(t)-u^{(\nu)}(s)| \le L|t-s|.$$

Then it is true for all $h_1,h_2 \in (0,\beta]$ that

$$\frac{2}{h_1h_2(h_1+h_2)}\{h_2a(\tfrac{1}{2}h_1)[u(h_1)-u(0)]+h_1a(-\tfrac{1}{2}h_2)[u(-h_2)-u(0)]\}$$
$$= a(0)u''(0)+a'(0)u'(0) + \frac{h_1-h_2}{12}[4a(0)u'''(0)+6a'(0)u''(0) + 3a''(0)u'(0)] + R$$

where

$$|R| \le \frac{41}{24} L^2 \frac{h_1^3+h_2^3}{h_1+h_2}.$$

Proof: We examine the function

$$f(s) = \frac{2}{h_1h_2(h_1+h_2)} [h_2a(\tfrac{s}{2}h_1)u(sh_1) + h_1a(-\tfrac{s}{2}h_2)u(-sh_2)$$

$$- h_2a(\tfrac{s}{2}h_1)u(0) - h_1a(-\tfrac{s}{2}h_2)u(0)], \qquad s \in [0,1].$$

The ν-th derivatives, $\nu = 0(1)3$, are

$$f^{(\nu)}(s) = \frac{2}{h_1h_2(h_1+h_2)} \sum_{\mu=0}^{\nu} \frac{1}{2^\mu}\binom{\nu}{\mu}\Big[h_1^\nu h_2a^{(\mu)}(\tfrac{s}{2}h_1)u^{(\nu-\mu)}(sh_1)$$

$$+ (-1)^\nu h_1h_2^\nu a^{(\mu)}(-\tfrac{s}{2}h_2)u^{(\nu-\mu)}(-sh_2)\Big]$$

$$- \frac{2}{h_1h_2(h_1+h_2)} \frac{u(0)}{2^\nu}\Big[h_1^\nu h_2a^{(\nu)}(\tfrac{s}{2}h_1)$$

$$+ (-1)^\nu h_1h_2^\nu a^{(\nu)}(-\tfrac{s}{2}h_2)\Big].$$

It follows that

$$f(0) = f'(0) = 0$$

$$f''(0) = 2a(0)u''(0) + 2a'(0)u'(0)$$

$$f'''(0) = (h_1-h_2)[2a(0)u'''(0)+3a'(0)u''(0) + \tfrac{3}{2}a''(0)u'(0)].$$

By Taylor's Theorem,

$$f(1) = f(0) + f'(0) + \tfrac{1}{2}f''(0) + \tfrac{1}{6}f'''(0) + \tfrac{1}{6}[f'''(\theta)-f'''(0)],$$

$$0 < \theta < 1.$$

The conclusion follows, with $R = \frac{1}{6}[f'''(\theta)-f'''(0)]$, once we show that

$$|f'''(\theta)-f'''(0)| \le \frac{41}{24} L^2 \frac{h_1^3+h_2^3}{h_1+h_2}.$$

But this inequality follows from

$$|a^{(\mu)}(\tfrac{s}{2}h_1)u^{(\nu-\mu)}(sh_1) - a^{\mu}(0)u^{(\nu-\mu)}(0)|$$

$$\leq |a^{(\mu)}(\tfrac{s}{2}h_1)|\ |u^{(\nu-\mu)}(sh_1) - u^{(\nu-\mu)}(0)|$$

$$+ |u^{(\nu-\mu)}(0)|\ |a^{(\mu)}(\tfrac{s}{2}h_1) - a^{(\mu)}(0)| \leq \tfrac{3}{2} L^2 h_1$$

and, similarly,

$$|a^{\mu}(- \tfrac{s}{2}h_2)u^{(\nu-\mu)}(-sh_2) - a^{\mu}(0)u^{(\nu-\mu)}(0)| \leq \tfrac{3}{2} L^2 h_2. \qquad \square$$

Remark 13.24: Whenever $a,u \in C^4([-\beta,\beta], \mathbb{R})$, a constant L with the desired properties always exists. A convenient choice is

$$L = \max_{\substack{\nu=0(1)4 \\ t\in[-\beta,\beta]}} (|a^{(\nu)}(t)|,\ |u^{(\nu)}(t)|).$$

Example 13.25: *Standard Five Point Method.*

Differential operator:

$$Lu = -[a_1u_x]_x - [a_2u_y]_y + H(x,y,u)$$

where $a_1,a_2 \in C^\infty(\overline{G}, \mathbb{R})$, $H \in C^\infty(\overline{G} \times \mathbb{R}, \mathbb{R})$, and

$$\left.\begin{array}{ll} a_1(x,y) > 0, & a_2(x,y) > 0 \\ H(x,y,0) = 0, & H_z(x,y,z) \geq 0 \end{array}\right. \quad (x,y) \in \overline{G},\ z \in \mathbb{R}.$$

Lattice:

$h_j = 2^{-(j+\ell)}$, $j = 1(1)\infty$, ℓ sufficiently large, but fixed

M_j: standard lattice with mesh size h_j.

A point $(x,y) \in M_j$ is called *boundary-distant* if all neighboring points $N_\nu(x,y,h_j)$ belong to G; otherwise it is called *boundary-close.*

Derivation of the difference equations: At the boundary-distant lattice points, the first two terms of $Lu(x,y)$ are replaced, one at a time, with the aid of Lemma 13.23. We abbreviate h_j, w_j, and n_j, merely writing h, w, and n:

$$\frac{1}{h^2}\{a_1(x+\tfrac{1}{2}h,y)[w(x,y) - w(x+h,y)]$$
$$+ a_1(x-\tfrac{1}{2}h,y)[w(x,y) - w(x-h,y)]$$
$$+ a_2(x,y+\tfrac{1}{2}h)[w(x,y) - w(x,y+h)]$$
$$+ a_2(x,y-\tfrac{1}{2}h)[w(x,y) - w(x,y-h)]\}$$
$$+ H(x,y,w(x,y)) = q(x,y).$$

If one replaces w by the exact solution u of the boundary value problem, where $u \in \mathrm{Lip}^{(3)}(\overline{G}, \mathbb{R})$, the local error will be $O(h^2)$. An analogous procedure at the boundary-close lattice points yields

$$\Sigma_1 K_\nu(x,y)[w(x,y) - w(N_\nu(x,y,h))]$$
$$+ \Sigma_2 K_\nu(x,y)w(x,y) + H(x,y,w(x,y))$$
$$= q(x,y) + \Sigma_2 K_\nu(x,y)\psi(N_\nu(x,y,h)).$$

where

$$K_\nu(x,y) = \frac{2a_\mu(\tilde{x}_\nu,\tilde{y}_\nu)}{d_\nu(x,y,h)[d_\mu(x,y,h)+d_{\mu+2}(x,y,h)]},$$

$$\mu = \begin{cases} 1 & \nu = 1,3 \\ 2 & \nu = 2,4 \end{cases}$$

$$(\tilde{x}_\nu,\tilde{y}_\nu) = (x,y) + \tfrac{1}{2}d_\nu(x,y,h)e^{(\nu)}.$$

In the sums Σ_1 and Σ_2, ν runs through the subsets of $\{1,2,3,4\}$:

Σ_1: all ν with $N_\nu(x,y,h) \in G$

Σ_2: all ν with $N_\nu(x,y,h) \in \Gamma$.

Formally, the equations for the boundary-distant points are special cases of the equations for the boundary-close points. However, they differ substantially with respect to the local error. In applying Lemma 13.23 at the boundary-close points, one must choose

$$h_1 = d_1(x,y,h) \quad \text{and} \quad h_2 = d_3(x,y,h)$$

for the first summand of $Lu(x,y)$, and

$$h_1 = d_2(x,y,h) \quad \text{and} \quad h_2 = d_4(x,y,h)$$

for the second. The local error contains the remainder R and also the additional term

$$\frac{h_1-h_2}{12}\,[4a(0)u'''(0) + 6a'(0)u''(0) + 3a''(0)u'(0)].$$

Altogether there results an error of $O(h)$. However, this may be reduced to $O(h^3)$ by a trick (cf. Gorenflo 1973). Divide the difference equations at the boundary-close points by

$$b(x,y) = \Sigma_2 K_\nu(x,y).$$

The new equations now satisfy the normalization condition (4) of Theorem 13.16, since for $\psi \geq 1$ and $q \geq 1$ it is obviously true that

$$[q(x,y) + \Sigma_2 K_\nu(x,y)\psi(x,y)]/\Sigma_2 K_\nu(x,y) \geq 1.$$

At the boundary-distant points such an "optical" improvement of the local error is not possible. Therefore the maximum is $O(h^2)$.

We can now formally define (cf. Theorem 13.16) the *difference operators:*

$$b(x,y) = \begin{cases} 1 & \text{whenever } (x,y) \text{ is boundary-distant} \\ \Sigma_2 K_\nu(x,y) & \text{whenever } (x,y) \text{ is boundary-close} \end{cases}$$

$$F_j^{(1)}(w)(x,y) = \Big[\sum_{\nu=1}^{4} K_\nu(x,y)w(x,y) - \Sigma_2 K_\nu(x,y)w(N_\nu(x,y,h))\Big]/b(x,y)$$

$$F_j^{(2)}(w)(x,y) = H(x,y,w(x,y))/b(x,y)$$

$$F_j^{(3)}(w)(x,y) = [\Sigma_2 K_\nu(x,y)\psi(N_\nu(x,y,h))]/b(x,y)$$

$$R_j(r_j(q))(x,y) = q(x,y)/b(x,y).$$

For $F_j^{(1)}$ there is a matrix $B^{-1}A$; B is a diagonal matrix with diagonal elements $b(x,y)$, whereas the particular A naturally also depends on the enumeration of the lattice points. In practice, there are two methods of enumeration which have proven themselves to be of value:

(1) *Enumeration by columns and rows:* (x,y) precedes $(\tilde{x},\tilde{y})$ if one of the following conditions is satisfied:

(a) $x < \tilde{x}$, (b) $x = \tilde{x}$ and $y < \tilde{y}$.

With this enumeration, the matrix A becomes *block tridiagonal:*

$$A = \begin{pmatrix} D_1 & -S_1 & & & \mathcal{O} \\ -\tilde{S}_1 & D_2 & -S_2 & & \\ & -\tilde{S}_2 & D_3 & \ddots & \\ & & \ddots & \ddots & -S_k \\ \mathcal{O} & & & -\tilde{S}_k & D_{k+1} \end{pmatrix}$$

The matrices D_μ are quadratic and tridiagonal. Their diagonal is positive, and all other elements are nonpositive. The matrices S_μ and $\tilde{S}_\mu$ are nonnegative.

(2) *Enumeration by the checkerboard pattern:* Divide the lattice M_j into two disjoint subsets (the white and black squares of a checkerboard):

$$M_j^{(1)} = \{(\mu h, \nu h) \in M_j \mid \mu+\nu \text{ even}\}$$

$$M_j^{(2)} = \{(\mu h, \nu h) \in M_j \mid \mu+\nu \text{ odd}\}.$$

The elements of $M_j^{(1)}$ are enumerated first, and the elements of $M_j^{(2)}$ second. In each of these subsets, we use the column and row ordering of (1). The result is a matrix of the form

$$A = \begin{pmatrix} D_1 & -S \\ -\tilde{S} & D_2 \end{pmatrix} .$$

D_1 and D_2 are quadratic diagonal matrices with positive diagonals. S and $\tilde{S}$ are nonnegative matrices.

In Figures 13.26 and 13.27 we have an example of the two enumerations.

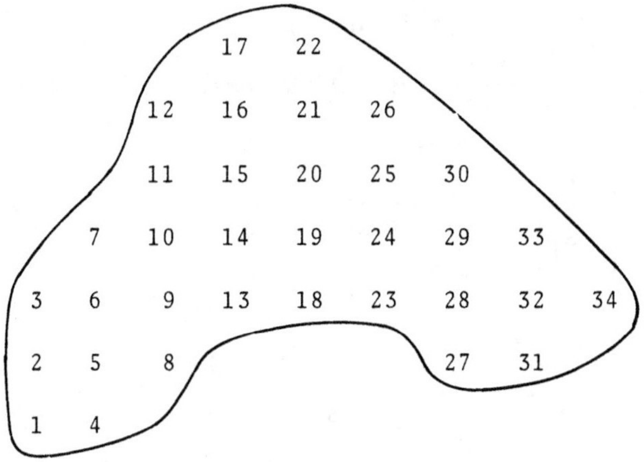

Figure 13.26. Enumeration by columns and rows

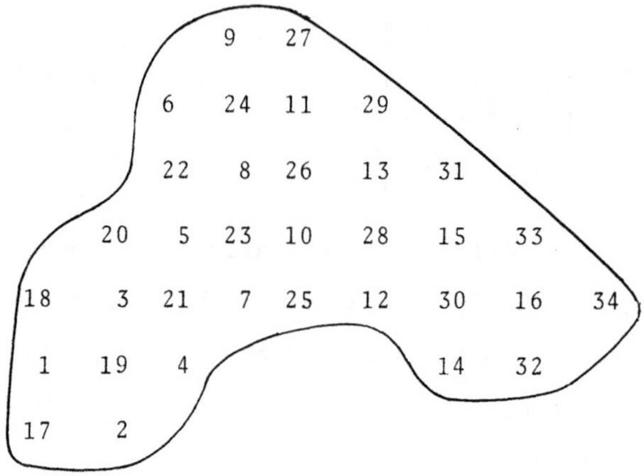

Figure 13.27. Enumeration on the checkerboard pattern

We will now show that A and $B^{-1}A$ are M-matrices. It is obvious that:

(1) $a_{\rho\rho} = \sum_{\nu=1}^{4} K_\nu(x,y) > 0, \quad \rho = 1(1)n, \quad (x,y) \in M_j.$

(2) $a_{\rho\sigma} = -K_\nu(x,y) < 0$ or $a_{\rho\sigma} = 0$ for all $\rho,\sigma = 1(1)n$ with $\sigma \neq \rho$.

(3) $a_{\rho\rho} \geq \sum_{\substack{\sigma=1 \\ \sigma\neq\rho}}^{n} |a_{\rho\sigma}|, \quad \rho = 1(1)n.$

(4) For each row $(a_{\rho 1},\ldots,a_{\rho n})$, belonging to a boundary-close point,

$$a_{\rho\rho} > \sum_{\substack{\sigma=1 \\ \sigma\neq\rho}}^{n} |a_{\rho\sigma}|.$$

(5) $a_{\rho\sigma} = 0$ implies $a_{\sigma\rho} = 0$ for $\rho,\sigma = 1(1)n$.

In case A is irreducible, it is even irreducible diagonal dominant, by (1) through (4) (cf. condition 13.11). Otherwise, A is reducible, and by (5) there exists a permutation matrix P such that

$$PAP^{-1} = \begin{pmatrix} A_1 & & & \mathcal{O} \\ & A_2 & & \\ & & \ddots & \\ \mathcal{O} & & & A_\ell \end{pmatrix}$$

The matrices A_ν, $\nu = 1(1)\ell$ are quadratic and irreducible. Each matrix A_ν has at least one row which belongs to a boundary-close point. Hence all of these matrices are irreducible diagonal dominant, and thus M-matrices. Consequently, A is also an M-matrix.

For certain h and certain simple regions [e.g. $G = (0,1) \times (0,1)$ or $G = \{(x,y) \in (0,1) \times (0,1) \mid x+y < 1\}$, $h = 1/m$] it will be the case that $d_\nu(x,y,h) = h$. When this condition is met, we have the additional results:

(6) $K_\nu(x,y,h) = K_\mu(N_\nu(x,y,h),h)$ where $\mu-1 = (\nu+1)\bmod 4$, $(x,y) \in M_j$.

(7) $a_{\rho\sigma} = a_{\sigma\rho}$ for $\rho,\sigma = 1(1)n$.

(8) A is positive definite.

(9) $B^{-1}A$ is similar to $B^{-1/2}AB^{-1/2}$ and therefore has positive eigenvalues only.

Of the conditions of Theorem 13.16 we have shown so far that (2) ($B^{-1}A$ is an M-matrix), (4) (normalization condition), and (5) (consistency) are satisfied.

$$F_j^{(2)}(w)(x,y) = H(x,y,w(x,y))/b(x,y)$$

is trivially diagonal and isotonic. Thus condition (3) is also satisfied. Therefore, the method is stable.

In the following examples we restrict ourselves to the region $G = (0,1) \times (0,1)$; for the lattice M_j we always choose the standard lattice with mesh width $h = h_j = 2^{-j}$. In this way we avoid all special problems related to proximity to the boundary. In principle, however, they could be solved with methods similar to those in Example 13.25. For brevity's sake, we also consider only linear differential operators without the summand $H(x,y,u(x,y))$. Then the summand $F_j^{(2)}$ drops out of the difference operator. When $(x,y) \in \Gamma$, we use $w(x,y)$ for $\psi(x,y)$.

Example 13.28: *Differential operator:*

$$Lu = -a_{11}u_{xx} - a_{22}u_{yy} - b_1u_x - b_2u_y.$$

Coefficients as in Problem 13.2.

Difference equations:

$$\begin{aligned}
&\frac{1}{h^2}\{[a_{11}(x,y)+ch][2w(x,y)-w(x+h,y)-w(x-h,y)]\\
&\qquad + [a_{22}(x,y)+ch][2w(x,y)-w(x,y+h)-w(x,y-h)]\}\\
&- \frac{1}{2h}\{b_1(x,y)[w(x+h,y)-w(x-h,y)]\\
&\qquad + b_2(x,y)[w(x,y+h)-w(x,y-h)]\}\\
&= q(x,y).
\end{aligned}$$

Here $c \geq 0$ is an arbitrary, but fixed, constant. When $u \in \mathrm{Lip}^{(3)}(\bar{G}, \mathbb{R})$, we obtain a local error of $O(h^2)$ when $c = 0$, and $O(h)$ when $c > 0$. For small h, $F_j^{(1)}$ can be given by an M-matrix. The necessary and sufficient conditions for this are

$$\frac{h}{2}|b_1(x,y)| \leq a_{11}(x,y) + ch, \qquad (x,y) \in M_j$$

$$\frac{h}{2}|b_2(x,y)| \leq a_{22}(x,y) + ch, \qquad (x,y) \in M_j$$

which is equivalent to

$$\frac{h}{2}[|b_1(x,y)|-2c] \le a_{11}(x,y), \qquad (x,y) \in M_j$$

$$\frac{h}{2}[|b_2(x,y)|-2c] \le a_{22}(x,y), \qquad (x,y) \in M_j.$$

If one of the above conditions is not met, the matrix may possibly be singular. Therefore these inequalities must be satisfied in every case. For $c = 0$, one obtains the smaller local error, and for $c > 0$, the larger stability interval $h \in (0,h_0]$. In the problems of fluid dynamics, $|b_1|$ or $|b_2|$ are often substantially larger than a_{11} and a_{22}. For $c > 0$, we introduce a *numerical viscosity* (as with the Friedrichs method, cf. Ch. 6). This could be accomplished in many other ways as well. One can then improve the global error by extrapolation. □

<u>Example 13.29</u>: *Differential operator:* as in Example 13.28. *Difference equations:*

$$\frac{1}{h^2}\{a_{11}(x,y)[2w(x,y)-w(x+h,y)-w(x-h,y)]$$
$$+ a_{22}(x,y)[2w(x,y)-w(x,y+h)-w(x,y-h)]\}$$
$$- \frac{1}{h}\{D_1(x,y) + D_2(x,y)\} = q(x,y).$$

Here D_1 and D_2 are defined as follows, where $(x,y) \in M_j$,

$$D_1(x,y) = \begin{cases} b_1(x,y)[w(x+h,y)-w(x,y)] & \text{for } b_1(x,y) \ge 0 \\ b_1(x,y)[w(x,y)-w(x-h,y)] & \text{for } b_1(x,y) < 0 \end{cases}$$

$$D_2(x,y) = \begin{cases} b_2(x,y)[w(x,y+h)-w(x,y)] & \text{for } b_2(x,y) \ge 0 \\ b_2(x,y)[w(x,y)-w(x,y-h)] & \text{for } b_2(x,y) < 0. \end{cases}$$

$F_j^{(1)}$ is given by an M-matrix for arbitrary $h > 0$. This is the advantage of this method with one-sided difference quotients to approximate the first derivatives. The local error is $O(h)$ for $u \in \mathrm{Lip}^{(3)}(\bar{G}, \mathbb{R})$. Extrapolation is possible only if b_1 and b_2 do not change in sign. Note the similarity with the method of Courant, Isaacson, and Rees (cf. Ch. 6). □

Example 13.30: *Differential operator:*

$$Lu = -a\Delta u - 2bu_{xy}$$

where $a,b \in C^{\infty}(\bar{G}, \mathbb{R})$ satisfy

$$\left.\begin{array}{l} a(x,y) > 0, \\ a(x,y)^2 - b(x,y)^2 > 0. \end{array}\right. \qquad (x,y) \in \bar{G}$$

Difference equations:

$$\begin{aligned} \frac{1}{2h^2}\{&a(x,y)[2w(x,y)-w(x+h,y+h)-w(x-h,y-h)] \\ &+ a(x,y)[2w(x,y)-w(x+h,y-h)-w(x-h,y+h)] \\ &- b(x,y)[w(x+h,y+h)-w(x-h,y+h)-w(x+h,y-h)+w(x-h,y-h)]\} \\ &= q(x,y). \end{aligned}$$

When

$$|b(x,y)| \leq a(x,y), \qquad (x,y) \in M_j$$

one obtains an M-matrix independent of h. However, the differential operator is uniformly elliptic only for

$$|b(x,y)| < a(x,y), \qquad (x,y) \in \bar{G}.$$

When $b(x,y) \equiv 0$, the system of difference equations splits into two linear systems of equations, namely for the points

$$(\mu h, \nu h) \quad \text{where} \quad \mu + \nu \quad \text{is even}$$

and

$$(\mu h, \nu h) \quad \text{where} \quad \mu + \nu \quad \text{is odd.}$$

One can then restrict oneself to solving one of the systems. The local error is of order $O(h^2)$ for $u \in \mathrm{Lip}^{(3)}(\bar{G}, \mathbb{R})$. □

Example 13.31: *Multiplace method.*

Differential operator:

$$Lu(x,y) = -\Delta u(x,y).$$

Difference equations:

$$\frac{1}{h^2}\{5w(x,y) - [w(x+h,y)+w(x,y+h)+w(x-h,y)+w(x,y-h)]$$
$$- \frac{1}{4}[w(x+h,y+h)+w(x-h,y+h)+w(x-h,y-h)+w(x+h,y-h)]\}$$
$$= q(x,y) + \frac{1}{8}[q(x+h,y)+q(x,y+h)+q(x-h,y)+q(x,y-h)].$$

The local error is $O(h^4)$ for $u \in \mathrm{Lip}^{(5)}(\bar{G}, \mathbb{R})$. Theorem 13.16 is applicable because $F_j^{(1)}$ always has an M-matrix. The natural generalization to more general regions leads to a method with a local error of $O(h^3)$. More on other methods of similar type may be found in Collatz 1966. □

So far we have only considered boundary value problems of the first type, i.e., the functional values on Γ were given. Nevertheless, the method also works with certain other boundary value problems.

Example 13.32: *Boundary value problem:*

$$-\Delta u(x,y) = q(x,y), \qquad (x,y) \in G = (0,1) \times (0,1)$$
$$u(x,y) = \psi(x,y), \qquad (x,y) \in \Gamma \quad \text{and} \quad x \neq 0$$
$$u(0,y) - \alpha u_x(0,y) = \phi(y), \quad y \in (0,1)$$

where ψ and ϕ are continuous and bounded and $\alpha > 0$ is fixed.

Lattice:

$\hat{M}_j$: the standard lattice M_j with mesh width $h=h_j=2^{-j}$ combined with the points $(0,\mu h)$, $\mu = 1(1)2^j-1$.

Difference equations:

For the points in $\hat{M}_j \cap (0,1) \times (0,1)$, we use the same equations as for the model problem (see Example 13.4). For $y = \mu h$, $\mu = 1(1)2^j-1$, and $u \in \mathrm{Lip}^{(3)}(\overline{G}, \mathbb{R})$ we have

$$u(h,y) = u(0,y) + hu_x(0,y) + \tfrac{1}{2}h^2 u_{xx}(0,y) + O(h^3)$$

$$= u(0,y) + hu_x(0,y) - \tfrac{1}{2}h^2[q(0,y)+u_{yy}(0,y)] + O(h^3).$$

If we replace $-h^2u_{yy}(0,y)$ by

$$2u(0,y) - u(0,y+h) - u(0,y-h) + O(h^3)$$

we obtain

$$u(h,y) = 2u(0,y) - \tfrac{1}{2}u(0,y+h) - \tfrac{1}{2}u(0,y-h)$$

$$+ hu_x(0,y) - \tfrac{1}{2}h^2q(0,y) + O(h^3)$$

$$u_x(0,y) = \frac{1}{2h}[2u(h,y)+u(0,y+h)+u(0,y-h)-4u(0,y)]$$

$$+ \tfrac{1}{2}hq(0,y) + O(h^2).$$

This leads to the difference equation

$$\frac{1}{2h}\{(2h+4\alpha)u(0,y) - \alpha[2u(h,y)+u(0,y+h)+u(0,y-h)]\}$$

$$= \phi(y) + \tfrac{\alpha}{2}hq(0,y).$$

Since $\alpha > 0$, the corresponding matrix is an M-matrix. A theorem similar to Theorem 13.16 holds true. The method converges like $O(h^2)$. If one multiplies the difference equation by $1/\alpha$, the passage to the limit $\alpha \to \infty$ is immediately possible. □

14. Variational methods

We consider the variational problem

$$I[u] = \min\{I[w] \mid w \in W\}, \tag{14.1}$$

where

$$I[w] = \iint_{\overline{G}} [a_1 w_x^2 + a_2 w_y^2 + 2Q(x,y,w)]\,dxdy.$$

Here G is to be a bounded region in $\mathbb{R}^2$ to which the Gauss integral theorem is applicable, and $a_1, a_2 \in C^1(\overline{G}, \mathbb{R})$, and $Q \in C^2(\overline{G} \times \mathbb{R}, \mathbb{R})$ where

$$a_1(x,y) \geq \alpha > 0, \quad a_2(x,y) \geq \alpha > 0,$$

$$0 \leq Q_{zz}(x,y,z) \leq \delta, \quad (x,y) \in \overline{G}, \quad z \in \mathbb{R}.$$

The function space W will be characterized more closely below. The connection with boundary value problems is established by the following theorem (cf., e.g., Gilbarg-Trudinger 1977, Ch. 10.5).

Theorem 14.2: A function $u \in C^2(G, \mathbb{R}) \cap C^0(\overline{G}, \mathbb{R})$ is a solution of the boundary value problem

$$\begin{aligned} -[a_1 u_x]_x - [a_2 u_y]_y + Q_z(x,y,u) &= 0, \quad (x,y) \in G \\ u(x,y) &= 0, \quad (x,y) \in \partial G \end{aligned} \tag{14.3}$$

if and only if it satisfies condition (14.1) with

$$W = \{w \in C^2(G, \mathbb{R}) \cap C^0(\overline{G}, \mathbb{R}) \mid w(x,y) = 0 \text{ for all } (x,y) \in \partial G\}.$$

In searching for the minimum of the functional $I[w]$, it has turned out to be useful to admit functions which are not everywhere twice continuously differentiable. In practice one approximates the twice continuously differentiable solutions of the boundary value problem (14.3) with piecewise once continuously differentiable functions, e.g. piecewise polynomials. Then one only has to make sure that the functions are continuous across the boundary points.

We will now focus on the space in which the functional $I[w]$ will be considered.

<u>Definition 14.4</u>: Let $K(G, \mathbb{R})$ be the vector space of all functions $w \in C^0(\bar{G}, \mathbb{R})$ such that:

(1) $w(x,y) = 0, \quad (x,y) \in \partial G$.

(2) w is absolutely continuous, both as a function of x with y held fixed, and as a function of y with x held fixed.

(3) $w_x, \; w_y \in L^2(G, \mathbb{R})$.

We define the following norm (the *Sobolev norm*) on $K(G, \mathbb{R})$:

$$\|w\|_H = [\iint_G (w^2 + w_x^2 + w_y^2)\,dxdy]^{1/2}.$$

We denote the closure of the space $K(G, \mathbb{R})$ with respect to this norm by $H(G, \mathbb{R})$. □

We can extend w continuously over all of $\mathbb{R}^2$ by setting $w(x,y) \equiv 0$ outside of $\bar{G}$. Then condition (2) implies that w is almost everywhere partially differentiable, and that for arbitrary $(a,b) \in \mathbb{R}^2$ (cf. Natanson 1961, Ch. IX):

$$\left.\begin{aligned} w(x,y) &= \int_a^x w_x(t,y)\,dt \\ &= \int_b^y w_y(x,t)\,dt. \end{aligned}\right\} \quad (x,y) \in \mathbb{R}^2$$

The following remark shows that variational problem (14.1) can also be considered in the space $H(G, \mathbb{R})$.

<u>Remark 14.5</u>: Let $u \in C^2(G, \mathbb{R}) \cap C^0(\overline{G}, \mathbb{R})$ be a solution of problem (14.3). Then we have

$$I[u] = \min\{I[w] \mid w \in H(G, \mathbb{R})\}.$$

When the boundary ∂G is sufficiently smooth, the converse also holds. For example, it is enough that ∂G be piecewise continuously differentiable and all the internal angles of the corners of the region be less than 2π. □

The natural numerical method for a successive approximation of the minimum of the functional $I[w]$ is the *Ritz method:*

Choose n linearly independent functions f_ν, $\nu = 1(1)n$, from the space $K(G, \mathbb{R})$. These will span an n-dimensional vector space V_n. Then determine $v \in V_n$, the minimum of the functionals $I[w]$ in V_n:

$$I[v] = \min\{I[w] \mid w \in V_n\}.$$

Each $w \in V_n$ can be represented as a linear combination of the f_ν:

$$w(x,y) = \sum_{\nu=1}^{n} \beta_\nu f_\nu(x,y).$$

In particular, we have

$$v(x,y) = \sum_{\nu=1}^{n} c_\nu f_\nu(x,y),$$

$$I[w] = \hat{I}(\beta_1,\ldots,\beta_n).$$

From the necessary conditions

$$\frac{\partial \hat{I}}{\partial c_\nu}(c_1,\ldots,c_n) = 0, \quad \nu = 1(1)n$$

one obtains a system of equations for the coefficients c_ν:

$$\iint_{\bar{G}}\Big[a_1(f_\nu)_x \sum_{\mu=1}^{n} c_\mu (f_\mu)_x + a_2(f_\nu)_y \sum_{\mu=1}^{n} c_\mu (f_\mu)_y + f_\nu Q_z\big(x,y,\sum_{\mu=1}^{n} c_\mu f_\mu\big)\Big]dxdy = 0, \quad \nu = 1(1)n. \tag{14.6}$$

Whenever the solution u of the boundary value problem (14.3) has a "good" approximation by functions in V_n, one can expect the error $u - v$ to be "small" also. Thus the effectiveness of the method depends very decidedly on a suitable choice for the space V_n. These relationships will be investigated carefully in a later part of the chapter. Now we will consider the practical problems which arise in solving the system of equations numerically. It will turn out that the choice of a special basis for V_n is also important.

In the following we will generally assume that $Q(x,y,z)$ is of the special form

$$Q(x,y,z) = \frac{1}{2}\sigma(x,y)z^2 - q(x,y)z,$$

where $\sigma(x,y) \geq 0$ for $(x,y) \in \bar{G}$. In this case, the system of equations (14.6) and the differential equation (14.3) are linear. The system of equations has the form

$$A\,c = d$$

where

$$A = (a_{\mu\nu}),\ c = (c_1,\ldots,c_n)^T, \text{ and } d = (d_1,\ldots,d_n)^T \text{ with}$$

$$a_{\mu\nu} = \iint_{\bar{G}} [a_1(f_\mu)_x(f_\nu)_x + a_2(f_\mu)_y(f_\nu)_y + \sigma f_\mu f_\nu]dxdy,$$

$$d_\mu = \iint_{\bar{G}} q f_\mu \, dxdy.$$

A is symmetric and positive semidefinite. Since the functions f_ν are linearly independent, A is even positive definite. Therefore, v is uniquely determined.

We begin with four classic choices of basis functions

f_ν, which are all of demonstrated utility for particular problems:

(1) $x^k y^\ell$: monomials

(2) $g_k(x)g_\ell(y)$: products of orthogonal polynomials

(3) $\begin{cases} \sin(kx)\sin(\ell y) \\ \sin(kx)\cos(\ell y) \\ \cos(kx)\cos(\ell y) \end{cases}$: trigonometric monomials

(4) $B_k(x)B_\ell(y)$: products of cardinal splines.

If the functions chosen above do not vanish on ∂G, they must be multiplied by a function which does vanish on ∂G and is never zero on G. It is preferable to choose basis functions at the onset which are zero on ∂G. For example, if $G = (0,1)^2$, one could choose

$$x(1-x)y(1-y),\quad x^2(1-x)y(1-y),\quad x(1-x)y^2(1-y),\quad x^2(1-x)y^2(1-y),$$

or

$$\sin(\pi x)\sin(\pi y),\quad \sin(2\pi x)\sin(\pi y),\quad \sin(\pi x)\sin(2\pi y),\quad \sin(2\pi x)\sin(2\pi y).$$

For $G = \{(r\cos\phi, r\sin\phi) \mid r \in [0,1), \phi \in [-\pi,\pi)$, a good choice is:

$$r^2-1,\quad (r^2-1)\sin\phi,\quad (r^2-1)\cos\phi,\quad (r^2-1)\sin 2\phi,\quad (r^2-1)\cos 2\phi.$$

Usually choice (2) is better than (1), since one obtains smaller numbers off of the main diagonal of A. The system of equations is then numerically more stable. For periodic solutions, however, one prefers choice (3). Choice (4) is particularly to be recommended when choices (1)-(3) give a poor approximation to the solution.

A shared disadvantage of choices (1)-(4) is that the matrix A is almost always dense. As a result, we have to compute $n(n+3)/2$ integrals in setting up the system of equations. The solution then requires tedious general methods such as the Gauss algorithm or the Cholesky method. The computational effort thus generally grows in direct proportion with n^3. One usually chooses $n \leq 100$.

The effort just described can be reduced by choosing initial functions with smaller support. The products

$$f_\mu f_\nu, \quad (f_\mu)_x (f_\nu)_x, \quad (f_\mu)_y (f_\nu)_y$$

will differ from zero only when the supports of f_μ and f_ν have nonempty intersection. In all other cases, the $a_{\mu\nu}$ are zero. A is *sparse*. In this case, specialized, faster methods are available to solve the system of equations. Estimates of this type are called *finite element* methods. The expression "finite element" refers to the support of the initial functions. In the sequel we present a few simple examples.

Example 14.7: *Linear polynomials on a triangulated region.* We assume that the boundary of our region is a polygonal line. Then we may represent $\overline{G}$ as the union of N closed triangles Δ_ρ, as in Figure 14.8. It is required that the intersection of two arbitrary distinct triangles be either empty or consist of exactly one vertex or exactly one side. Those vertices of the triangles Δ_ρ which do not belong to ∂G, will be denoted by ξ^ν. Let them be enumerated from 1 to n. We then define functions f_ν, $\nu = 1(1)n$, by the following rules:

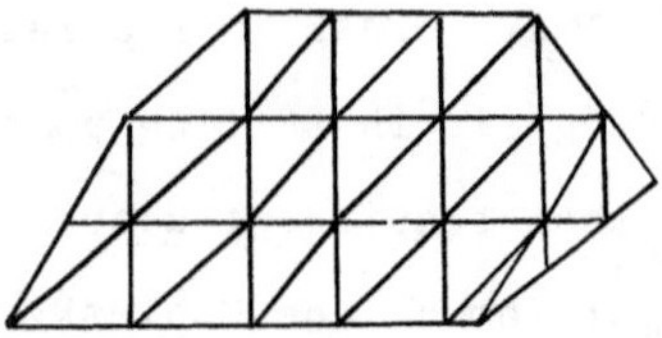

Figure 14.8. Triangulation of a region

(1) $f_\nu \in C^0(\bar{G}, \mathbb{R})$

(2) f_ν restricted to Δ_ρ is a first degree polynomial in $\mathbb{R}^2$, $\rho = 1(1)N$.

(3) $f_\nu(\xi^\mu) = \delta_{\nu\mu}$

(4) $f_\nu(x,y) = 0$ for $(x,y) \in \partial G$.

The functions f_ν are uniquely determined by properties (1)-(4). They belong to the space $K(G, \mathbb{R})$, and f_ν vanishes on every triangle Δ_ρ which does not contain vertex ξ^ν. If the triangulation is such that each vertex ξ^ν belongs to at most k triangles, then each row and column of A will contain at most $k + 1$ elements different from zero. In the special case

$$\xi^\nu = (r_\nu h, s_\nu h)^T, \quad r_\nu, s_\nu \in \mathbb{Z}$$

we can give formulas for the basis functions f_ν. The functions are given in the various triangles in Illustration 14.9. The coefficients for matrix A can be computed from this. We will demonstrate this for the special differential equation

$$-\Delta u(x,y) = q(x,y)$$

Thus we have $a_1 = a_2 \equiv 1$, $\sigma \equiv 0$, and

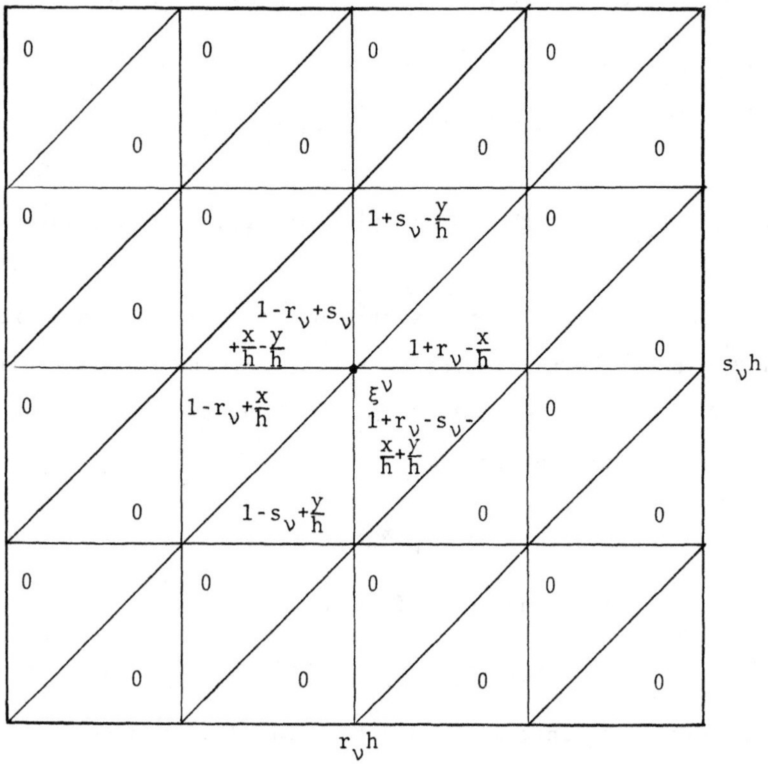

Figure 14.9. Initial functions f_ν for a special triangulation

$$a_{\mu\nu} = \iint_G [(f_\mu)_x(f_\nu)_x + (f_\mu)_y(f_\nu)_y]dxdy$$

Since $(f_\sigma)_x$ and $(f_\sigma)_y$ are $1/h$, $-1/h$, or 0, depending on the triangle, it follows that

$$a_{\mu\nu} = \begin{cases} 4 & \text{for } \mu = \nu \\ -1 & \text{for } s_\nu = s_\mu \text{ and } r_\nu = r_\mu+1 \text{ or } r_\nu = r_\mu-1 \\ -1 & \text{for } r_\nu = r_\mu \text{ and } s_\nu = s_\mu+1 \text{ or } s_\nu = s_\mu-1 \\ 0 & \text{otherwise.} \end{cases}$$

In this way we obtain the following "five point difference

operator" which is often also called a *difference star:*

$$\begin{bmatrix} 0 & -1 & 0 \\ -1 & 4 & -1 \\ 0 & -1 & 0 \end{bmatrix} = 4I - (T_{h,1} + T_{-h,1} + T_{h,2} + T_{-h,2}).$$

Here the $T_{k,i}$ are the translation operators from Chapter 10.

The left side of the system of equations is thus the same for this finite element method as for the simplest difference method (cf. Ch. 13). On the right side here, however, we have the integrals

$$d_\mu = \iint_{\bar{G}} qf_\mu dxdy$$

while in the difference method we had

$$h^2 q(r_\mu h,\ s_\mu h).$$

In practice, the integrals will be evaluated by a sufficiently accurate quadrature formula. In the case at hand the following formula, which is exact for first degree polynomials (cf., e.g. Witsch 1978, Theorem 5.2), is adequate:

$$\iint_\Delta g(x,y)dxdy \approx h^2\, [\tfrac{1}{6}g(0,0) + \tfrac{1}{6}g(h,0) + \tfrac{1}{6}g(0,h)],$$

where Δ is the triangle with vertices $(0,0)$, $(h,0)$, $(0,h)$. Since the f_μ will be zero on at least two of the three vertices, it follows that

$$d_\mu \approx \frac{h^2}{6}(+1+1+1+1+1+1)q(r_\mu h, s_\mu h) = h^2 q(r_\mu h, s_\mu h). \qquad \square$$

<u>Example 14.10</u>: *Linear product approach on a rectangular subdivision.* We assume that $\bar{G}$ is the union of N closed rectangles with sides parallel to the axes, so that $\bar{G}$ may

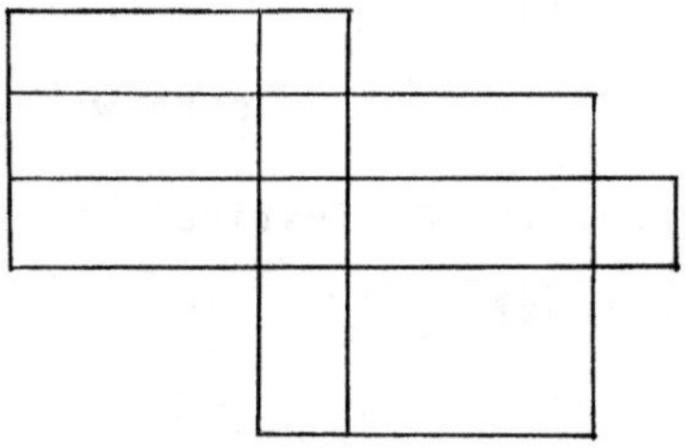

Figure 14.11. Subdivision into rectangles

be subdivided as in Figure 14.11. We require that the intersection of two arbitrary, distinct rectangles be either empty or consist of exactly one vertex or exactly one side. We denote by ξ^{ν} $(\nu = 1(1)n)$ those vertices of the rectangles $\square_{\rho}$ which do not belong to ∂G. Then we define functions f_{ν} by the following rule:

(1) $f_{\nu} \in C^{0}(\overline{G}, \mathbb{R})$

(2) f_{ν} restricted to $\square_{\rho}$ is the product of two first degree polynomials in the independent variables x and y.

(3) $f_{\nu}(\xi^{\mu}) = \delta_{\nu\mu}$

(4) $f_{\nu}(x,y) = 0$ for $(x,y) \in \partial G$.

As in Example 14.7, the functions f_{ν} are uniquely determined by properties (1)-(4), and belong to the space $K(G, \mathbb{R})$. Each f_{ν} vanishes except on the four rectangles with common vertex ξ^{ν}. Thus each row and column of A has at most nine elements which differ from zero.

In the special case

$$\xi^{\nu} = (r_{\nu}h,\ s_{\nu}h)^T, \quad r_{\nu}, s_{\nu} \in \mathbb{Z}$$

we can again provide formulas for the basis functions f_ν, namely:

$$f_\nu = \begin{cases} (1-|\frac{x}{h}-r_\nu|)(1-|\frac{y}{h}-s_\nu|) & \text{for } |\frac{x}{h}-r_\nu| \le 1,\ |\frac{y}{h}-s_\nu| \le 1 \\ 0 & \text{otherwise.} \end{cases}$$

We can compute the partial derivatives of the f_ν on the interiors of the rectangles:

$$(f_\nu)_x = \begin{cases} -\frac{1}{h}(1-|\frac{y}{h}-s_\nu|) & \text{for } 0 < \frac{x}{h}-r_\nu < 1,\ |\frac{y}{h}-s_\nu| < 1 \\ \frac{1}{h}(1-|\frac{y}{h}-s_\nu|) & \text{for } -1 < \frac{x}{h}-r_\nu < 0,\ |\frac{y}{h}-s_\nu| < 1 \\ 0 & \text{otherwise} \end{cases}$$

$$(f_\nu)_y = \begin{cases} -\frac{1}{h}(1-|\frac{x}{h}-r_\nu|) & \text{for } 0 < \frac{y}{h}-s_\nu < 1,\ |\frac{x}{h}-r_\nu| < 1 \\ \frac{1}{h}(1-|\frac{x}{h}-r_\nu|) & \text{for } -1 < \frac{y}{h}-s_\nu < 0,\ |\frac{x}{h}-r_\nu| < 1. \\ 0 & \text{otherwise.} \end{cases}$$

The coefficients of the matrix A can be derived from this. We restrict ourselves to the Poisson equation

$$-\Delta u(x,y) = q(x,y).$$

By exploiting symmetries, we need consider only four cases in computing the integrals:

(1) $|r_\mu - r_\nu| > 1$ or $|s_\mu - s_\nu| > 1$: $a_{\mu\nu} = 0$

(2) $\mu = \nu$: $a_{\mu\mu} = \frac{4}{h^2}\int_0^h\int_0^h [(1-\frac{x}{h})^2 + (1-\frac{y}{h})^2]\,dxdy = \frac{8}{3}$

(3) $r_\nu = r_\mu + 1$ and $s_\nu = s_\mu + 1$:

$$a_{\mu\nu} = \frac{1}{h^2}\int_0^h\int_0^h [-(1-\frac{y}{h})(1-(1-\frac{y}{h})) - (1-\frac{x}{h})(1-(1-\frac{x}{h}))]\,dxdy = -\frac{1}{3}$$

(4) $r_\nu = r_\mu + 1$ and $s_\nu = s_\mu$:

$$a_{\mu\nu} = \frac{2}{h^2}\int_0^h\int_0^h [-(1-\frac{y}{h})(1-\frac{y}{h}) + (1-\frac{x}{h})(1-(1-\frac{x}{h}))]\,dxdy = -\frac{1}{3}.$$

We obtain the difference star

$$\begin{bmatrix} -\frac{1}{3} & -\frac{1}{3} & -\frac{1}{3} \\ -\frac{1}{3} & +\frac{8}{3} & -\frac{1}{3} \\ -\frac{1}{3} & -\frac{1}{3} & -\frac{1}{3} \end{bmatrix} = \frac{8}{3}I - \frac{1}{3}[T_{h,1}+T_{-h,1}+T_{h,2}+T_{-h,2}$$

$$+ (T_{h,1}+T_{-h,1})(T_{h,2}+T_{-h,2})].$$

The integrals d_μ can be evaluated according to the formula

$$\iint_{\square} g(x,y)\,dxdy \approx \frac{h^2}{4}[g(0,0) + g(h,0) + g(0,h) + g(h,h)],$$

where $\square$ is the rectangle with vertices $(0,0)$, $(h,0)$, $(0,h)$, (h,h). Therefore,

$$d_\mu \approx h^2 q(r_\mu h,\ s_\mu h). \qquad \square$$

Example 14.12: *Quadratic polynomial approach on a triangulated region* (cf. Zlamal 1968). Let region G be triangulated as in Example 14.7. We will denote the vertices of the triangles Δ_ρ *and* the midpoints of those sides which do not belong to ∂G by ξ^ν. Let these be numbered from 1 to n. We define functions f_ν, $\nu = 1(1)n$ by the following rule:

(1) $f_\nu \in C^0(\overline{G}, \mathbb{R})$

(2) $f_\nu(x,y)$ restricted to Δ_ρ is a second degree polynomial, $\rho = 1(1)N$.

(3) $f_\nu(\xi^\mu) = \delta_{\nu\mu}$

(4) $f_\nu(x,y) = 0$ for $(x,y) \in \partial G$.

As in the previous examples, the functions f_ν are uniquely determined by properties (1)-(4). Restricted to one side of a triangle, f_ν is a second degree polynomial of one variable. Since three conditions are imposed on each side of a triangle, f_ν is continuous in $\overline{G}$, and hence belongs to $K(G, \mathbb{R})$. It vanishes on every triangle which does not contain ξ^ν. □

With a regular subdivision of the region, most finite element methods lead to difference formulas. For the programmer, the immediate application of difference equations is simpler. However, the real significance of finite element methods does not depend on a regular subdivision of the region. The method is so flexible that the region can be divided into arbitrary triangles, rectangles, or other geometric figures. In carrying out the division, one can let oneself be guided by the boundary and any singularities of the solution. Inside of the individual geometric figures it is most certainly possible to use higher order approximations (such as polynomials of high degree or functions with special singularities). In these cases, the reduction to difference formulas will be too demanding. The programming required by such flexible finite element methods is easily so extensive as to be beyond the capacities of an individual programmer. In such cases one usually relies on commercial software packages.

We now turn to the questions of convergence and error estimates for the Ritz method.

Definition 14.13: Special *inner products* and *norms*. The quantities

$$\langle u,v\rangle_2 = \iint_{\overline{G}} uv\,dxdy$$

$$\langle u,v\rangle_I = \iint_{\overline{G}} [a_1u_xv_x + a_2u_yv_y + \sigma uv]\,dxdy$$

are inner products on $K(G, \mathbb{R})$. They induce norms

$$\|u\|_2 = \sqrt{\langle u,u\rangle_2}, \quad \|u\|_I = \sqrt{\langle u,u\rangle_I}. \quad \square$$

The following theorem will show how the norms $\|\cdot\|_2$, $\|\cdot\|_I$, and $\|\cdot\|_H$ can be compared to each other on $K(G, \mathbb{R})$.

Theorem 14.14: There exist constants $\gamma_1, \gamma_2 > 0$ such that for all $u \in K(G, \mathbb{R})$:

(1) $\gamma_1\|u\|_I \le \|u\|_H \le \gamma_2\|u\|_I$

(2) $\|u\|_2 \le \|u\|_H$

(3) $\|u\|_2 \le \gamma_2\|u\|_I$.

Proof: The second inequality is trivial, and the third follows from the second and the first. The proof of the first is as follows. Analogously to Theorem 4.10(1) we can show that $C_0^1(G, \mathbb{R})$ is dense in $K(G, \mathbb{R})$ with respect to $\|\cdot\|_H$. Thus it suffices to establish the inequalities for $u \in C_0^1(G, \mathbb{R})$. We begin by showing that there exists a constant $\gamma_0 > 0$ such that

$$\iint_{\bar{G}} u^2 dxdy \le \gamma_0 \iint_{\bar{G}} (u_x^2 + u_y^2)dxdy, \quad u \in C_0^1(G, \mathbb{R}).$$

Let $[-a,a] \times [-a,a]$ be a square containing G. Let $\tilde{u}$ denote that continuous extension of $u \in C_0^1(G, \mathbb{R})$ to $[-a,a] \times [-a,a]$ which vanishes outside of G. It follows that

$$\tilde{u}(t,y) = \int_{-a}^{t} \tilde{u}_x(x,y)dx.$$

Applying the Schwartz inequality we obtain

$$\tilde{u}(t,y)^2 \le (t+a)\int_{-a}^{t} \tilde{u}_x(x,y)^2 dx \le 2a\int_{-a}^{a} \tilde{u}_x(x,y)^2 dx.$$

It follows from this that

$$\int_{-a}^{a}\tilde{u}(t,y)^2dy \leq 2a\int_{-a}^{a}\int_{-a}^{a}\tilde{u}_x(x,y)^2dxdy,$$

$$\iint_{\overline{G}}u^2dxdy = \int_{-a}^{a}\int_{-a}^{a}\tilde{u}^2dxdy \leq 4a^2\int_{-a}^{a}\int_{-a}^{a}\tilde{u}_x^2dxdy$$

$$\leq 4a^2\int_{\overline{G}}\int_{\overline{G}}(\tilde{u}_x^2+\tilde{u}_y^2)dxdy.$$

Setting $\gamma_0 = 4a^2$ establishes our claim.

We now set

$$\tilde{\alpha} = \min_{(x,y)\varepsilon\overline{G}}\{\min[a_1(x,y),\ a_2(x,y)]\}$$

$$\tilde{\gamma}_0 = \max_{(x,y)\varepsilon\overline{G}}\{\max[a_1(x,y),\ a_2(x,y),\ \sigma(x,y)]\}$$

and use the above result to obtain the estimates

$$\|u\|_H^2 \leq (1+\gamma_0)\iint_{\overline{G}}(u_x^2+u_y^2)dxdy \leq \frac{1+\gamma_0}{\tilde{\alpha}}\|u\|_I^2$$

$$\|u\|_I^2 \leq \tilde{\gamma}_0\|u\|_H^2.$$

Inequality (1) then follows by letting

$$\gamma_1 = \sqrt{1/\tilde{\gamma}_0} \quad \text{and} \quad \gamma_2 = \sqrt{(1+\gamma_0)/\tilde{\alpha}}. \qquad \square$$

Let $\{u_\nu \mid \nu = 1(1)\infty\}$ and $\{v_\nu \mid \nu = 1(1)\infty\}$ be Cauchy sequences in $K(G, \mathbb{R})$ which converge to elements u and v in $H(G, \mathbb{R})$ with respect to the norm $\|\cdot\|_H$. Then $\{<u_\nu,v_\nu>_I \mid \nu = 1(1)\infty\}$ is a Cauchy sequence in $\mathbb{R}$, for it follows from the Schwarz inequality and Theorem 14.14 that

$$|<u_\nu,v_\nu>_I - <u_\mu,v_\mu>_I| = |<u_\nu-u_\mu,v_\nu>_I - <u_\mu,v_\mu-v_\nu>_I|$$

$$\leq \|u_\nu-u_\mu\|_I\ \|v_\nu\|_I + \|v_\mu-v_\nu\|_I\ \|u_\mu\|_I$$

$$\leq \gamma_1^{-2}(\|u_\nu-u_\mu\|_H\ \|v_\nu\|_H + \|v_\mu-v_\nu\|_H\ \|u_\mu\|_H).$$

If we define $\langle u,v\rangle_I = \lim_{\nu\to\infty} \langle u_\nu,v_\nu\rangle$ and $\|u\|_I = \sqrt{\langle u,u\rangle_I}$, Theorem 14.14 holds trivially for all $u \in H(G, \mathbb{R})$.

The space $H(G, \mathbb{R})$ is closed with respect to the norms $\|\cdot\|_H$ and $\|\cdot\|_I$. However, this is not the case with respect to the norm $\|\cdot\|_2$, as rather simple counterexamples will show. There is no inequality of the form $\|u\|_H \leq \gamma_3\|u\|_2$. Convergence for the Ritz method is first established for the norm $\|\cdot\|_I$, and convergence with respect to $\|\cdot\|_H$ and $\|\cdot\|_2$ then follow from the theorem.

<u>Theorem 14.15</u>: Let $u \in H(G, \mathbb{R})$ be such that

$$I[u] = \min\{I[\tilde{w}] \mid \tilde{w} \in H(G, \mathbb{R})\}$$

and let $w \in H(G, \mathbb{R})$ be arbitrary. Then we have:

$$\langle u,w\rangle_I = \langle q,w\rangle_2 \qquad (14.16)$$

$$I[u+w] = I[u] + \|w\|_I^2 \qquad (14.17)$$

Proof: For $\lambda \in \mathbb{R}$ it follows that

$$I[u] = \langle u,u\rangle_I - 2\langle u,q\rangle_2$$

$$I[u+\lambda w] = \langle u+\lambda w,\ u+\lambda w\rangle_I - 2\langle u+\lambda w,\ q\rangle_2$$

$$= I[u] + 2\lambda(\langle u,w\rangle_I - \langle w,q\rangle_2) + \lambda^2\langle w,w\rangle_I.$$

Since u is the minimum of the variation integral, the expression in the parentheses in the last equality must be zero. Otherwise, the difference $I[u+\lambda w] - I[u]$ will change sign as λ changes sign. The second conclusion follows with $\lambda = 1$.

It is also possible to derive equation (14.16) directly from the differential equation (14.3). For

$$Q(x,y,z) = \frac{1}{2}\sigma(x,y)z^2 - q(x,y)z$$

we multiply (14.3) by an arbitrary function $w \in K(G, \mathbb{R})$ (*test function*) and integrate over $\overline{G}$:

$$\iint_{\overline{G}} [-(a_1u_x)_x - (a_2u_y)_y + \sigma u - q]w\,dxdy = 0.$$

It follows from the Gauss integral theorem that

$$\iint_{\overline{G}} [a_1u_xw_x + a_2u_yw_y + \sigma uw]dxdy = \iint_{\overline{G}} qw\,dxdy.$$

This is equation (14.16). It is called the *weak form* of differential equation (14.3). With the aid of the Gauss integral theorem, it can also be derived immediately from similar differential equations which are not Euler solutions of a variational problem.

The system of equations $Ac = d$ can also be obtained by discretizing (14.16). This process is called the *Galerkin method:*

Let f_ν, $\nu = 1(1)n$, be the basis of a finite dimensional subspace V_n of $K(G, \mathbb{R})$. We want to find an approximation

$$v(x,y) = \sum_{\nu=1}^{n} c_\nu f_\nu(x,y)$$

such that

$$\langle v,f_\mu\rangle_I = \langle q,f_\mu\rangle_2, \qquad \mu = 1(1)n.$$

As in the Ritz method it follows that

$$a_{\mu\nu} = \langle f_\nu,f_\mu\rangle_I \quad \text{and} \quad d_\mu = \langle q,f_\mu\rangle_2.$$

A derivation of this type has the advantage of being applicable to more general differential equations. We prefer to proceed via variational methods because the error estimates follow directly from (14.17).

Theorem 14.18: Let V_n be an n-dimensional subspace of $K(G, \mathbb{R})$. Let $u \in H(G, \mathbb{R})$ and $v \in V_n$ be such that

$$I[u] = \min\{I[w] \mid w \in H(G, \mathbb{R})\}, \quad I[v] = \min\{I[w] \mid w \in V_n\}.$$

Then it is true that

(1) $I[u] \leq I[v]$

(2) $\|u-v\|_2 \leq \gamma_2 \|u-v\|_I \leq \gamma_2 \min_{v^* \in V_n} \|u-v^*\|_I.$

Here γ_2 is the positive constant from Theorem 14.14.

Proof: Inequality (1) is trivial. It follows from this, with the help of Theorem 14.15, that for every $v^* \in V_n$,

$$\|u-v\|_I^2 = I[v] - I[u] \leq I[v^*] - I[u] = \|u-v^*\|_I^2$$

The conclusion follows from Theorem 14.14. □

Thus the error $\|u-v\|_2$ in the Ritz method is small if there is some approximation $v^* \in V_n$ of the solution u for which $\|u-v^*\|_I$ is small. This requires a good approximation in the mean to u and the first derivatives of u.

Nevertheless, Theorem 14.18 is not well suited to error estimates in practice, because the unknown quantity u continues to appear on the right sides of the inequalities in (2). However, the following theorem makes it possible to obtain an *a posteriori* error estimate from the computable *defect* of an approximate solution.

Theorem 14.19: Let $u \in C^2(G, \mathbb{R}) \cap C^0(\bar{G}, \mathbb{R})$ be a solution of boundary value problem (14.3). Let the boundary of G consist of finitely many segments of differentiable curves. Further let $v \in C^2(\bar{G}, \mathbb{R})$ be an arbitrary function with $v(x,y) = 0$ for all $(x,y) \in \partial G$ and let

$$L = -\frac{\partial}{\partial x}(a_1 \frac{\partial}{\partial x}) - \frac{\partial}{\partial y}(a_2 \frac{\partial}{\partial y}) + \sigma.$$

Then it is true that

$$||u-v||_2 \leq \gamma_2^2 ||Lv-q||_2.$$

Here γ_2 is the positive constant from Theorem 14.14.

Proof: Let $\varepsilon(x,y) = u(x,y) - v(x,y)$. Since $Lu(x,y) = q(x,y)$, $L\varepsilon(x,y)$ is square integrable on G. Since ε vanishes on ∂G, it follows from the Gauss integral theorem that

$$\iint_{\overline{G}} \varepsilon L\varepsilon dxdy = \iint_{\overline{G}} (a_1\varepsilon_x^2 + a_2\varepsilon_y^2 + \sigma\varepsilon^2) dxdy = ||\varepsilon||_I^2.$$

It follows from Theorem 14.14 and the Schwartz inequality that

$$||\varepsilon||_2^2 \leq \gamma_2^2 ||\varepsilon||_I^2 \leq \gamma_2^2 \, ||\varepsilon_2|| \, ||L\varepsilon||_2. \qquad \square$$

We see from the estimate in the theorem that the error will be small in the sense of norm $||\cdot||_2$ if v is a twice continuously differentiable approximation of solution u, for then $Lv \approx q$. Of course the quality of the estimate depends on γ_2. This shows how important it is to determine good constants γ_2 for a region G and functions a_1 and a_2.

One further difficulty arises from the fact that the Ritz method normally produces an approximation from $K(G, \mathbb{R})$ instead of from $C^2(\overline{G}, \mathbb{R})$. This difficulty can be circumvented as follows. First cover G with a lattice and compute the functional values of the approximation on this lattice with the Ritz method. Then obtain a smooth approximation by using a sufficiently smooth interpolation between the functional values $v(\xi_\rho)$ at the lattice points ξ_ρ.

Unfortunately, bilinear interpolation is out of the question because it does not yield a twice continuously differentiable function. A two dimensional generalization of spline interpolation is possible, but complicated. The so-called Hermite interpolation is simpler. We will consider it extensively in the next chapter.

Up to now we have assumed that Q has the special form $\frac{1}{2}\sigma z^2 - qz$. In the following, let Q be an arbitrary function in $C^2(\overline{G} \times \mathbb{R}, \mathbb{R})$ with

$$Q_z(x,y,z) \geq 0, \quad 0 \leq Q_{zz}(x,y,z) \leq \delta, \quad (x,y) \in \overline{G},\ z \in \mathbb{R}.$$

Then one has the following generalizations of Theorems 14.15 and 14.18.

<u>Theorem 14.20</u>: Let $u \in K(G, \mathbb{R})$ be such that

$$I[u] = \min\{I[w] \mid w \in K(G, \mathbb{R})\}$$

and let $v \in K(G, \mathbb{R})$ be arbitrary. Then it is the case that

$$\iint_{\overline{G}} [a_1 u_x v_x + a_2 u_y v_y + Q_z(x,y,u)v]dxdy = 0,$$

$$I[u+v] = I[u] + \iint_{\overline{G}} [a_1 v_x^2 + a_2 v_y^2 + Q_{zz}(x,y,u+\theta v)v^2]dxdy, \quad 0<\theta<1.$$

<u>Theorem 14.21</u>: Let V_n be an n-dimensional subspace of $K(G, \mathbb{R})$. Further let $u \in K(G, \mathbb{R})$ and $v \in V_n$ be such that

$$I[u] = \min\{I[w] \mid w \in K(G, \mathbb{R})\} \quad \text{and} \quad I[v] = \min\{I[w] \mid w \in V_n\}.$$

Then there exists a positive constant γ_2 such that

(1) $I[u] \leq I[v]$

(2) $$\|u-v\|_2 \leq \gamma_2 \min_{v^* \varepsilon V_n} \{\iint_{\overline{G}} [a_1(u_x-v_x^*)^2 + a_2(u_y-v_y^*)^2 + \delta(u-v^*)^2]dxdy\}^{1/2}.$$

The constant γ_2 does not depend on Q or on δ.

Theorems 14.20 and 14.21 are proven analogously to Theorems 14.15 and 14.18. Inequality (2) of Theorem 14.21 implies that convergence of the Ritz method for semilinear differential equations is hardly different from convergence for linear differential equations.

15. Hermite interpolation and its application to the Ritz method

We will present the foundations of global and piecewise Hermite interpolation in this section. This interpolation method will aid us in smoothing the approximation functions and also in obtaining a particularly effective Ritz method. In the interest of a simple presentation we will dispense with the broadest attainable generality, and instead endeavor to explain in detail the more typical approaches. We begin with global Hermite interpolation for one independent variable.

Theorem 15.1: Let $m \varepsilon \mathbb{N}$ and $f \varepsilon C^{m-1}([a,b], \mathbb{R})$. Then:

(1) There exists exactly one polynomial f_m such that $\deg f_m \leq 2m-1$ and

$$f_m^{(\mu)}(a) = f^{(\mu)}(a), \quad f_m^{(\mu)}(b) = f^{(\mu)}(b), \quad \mu = 0(1)m-1.$$

f_m is called the Hermite interpolation polynomial for f.

(2) If f is actually $2m$-times continuously differentiable on $[a,b]$, then the function $f^{(\mu)} - f_m^{(\mu)}$, for $\mu = 0(1)2m-1$, has at least $2m - \mu$ zeros $x_{\mu\nu}$ in $[a,b]$, $\nu = 1(1)2m - \mu$. Here each zero is counted according to multiplicity. For each $x \in [a,b]$ there exists a $\theta \in (a,b)$ such that the following representation holds:

$$f^{(\mu)}(x) = f_m^{(\mu)}(x) + \frac{f^{(2m)}(\theta)}{(2m-\mu)!} \prod_{\nu=1}^{2m-\mu} (x-x_{\mu\nu}), \quad \mu = 0(1)2m-1.$$

The $x_{\mu\nu}$ (μ fixed) ordered by size are given by

$$x_{\mu\nu} = \begin{cases} a & \text{for } \nu = 1(1)m - \mu \\ b & \text{for } \nu = m+1(1)2m - \mu. \end{cases}$$

We have the inequality

$$\|f^{(\mu)}-f_m^{(\mu)}\|_\infty \le \hat{c}_{m\mu}(b-a)^{2m-\mu}\|f^{(2m)}\|_\infty, \quad \mu = 0(1)2m-1.$$

where

$$\hat{c}_{m\mu} = \begin{cases} \dfrac{m^m(m-\mu)^{m-\mu}}{(2m-\mu)^{2m-\mu}} \cdot \dfrac{1}{(2m-\mu)!} & \text{for } \mu = 0(1)m-1 \\[2ex] \dfrac{1}{(2m-\mu)!} & \text{for } \mu = m(1)2m-1. \end{cases}$$

This theorem can be generalized when f is only ℓ-times continuously differentiable on $[a,b]$ with $0 \le \ell < 2m$. In that case, an estimate for $\|f^{(\mu)}-f_m^{(\mu)}\|_\infty$ can be found in Swartz-Varga 1972, Theorem 6.1. For $\ell < m-1$, we require in (1) that:

$$f_m^{(\mu)}(a) = f_m^{(\mu)}(b) = 0, \quad \mu = \ell+1(1)m-1.$$

The constants $\hat{c}_{m\mu}$ are not optimal. Through numerical computations, Lehmann 1975 obtained improved values of $\hat{c}^*_{m\mu}$ for small m (cf. Table 15.2).

TABLE 15.2: $\hat{c}_{m\mu}$ (upper entry) and $\hat{c}^*_{m\mu}$ (lower entry) for $m = 1,2,3,4$.

μ	m=1	m=2	m=3	m=4
0	1.25000000000E-1	2.60416666667E-3	2.17013888889E-5	9.68812003968E-8
	1.25000000000E-1	2.60416666667E-3	2.17013888889E-5	9.68812003968E-8
1	1.00000000000E 0	2.46913580247E-2	2.88000000000E-4	1.66527864535E-6
	5.00000000000E-1	8.01875373875E-3	7.45355992500E-5	3.68952200589E-7
2		5.00000000000E-1	4.39453125000E-3	3.04831580552E-5
		8.33333333333E-2	5.20833333333E-4	3.10019841270E-6
3		1.00000000000E 0	1.66666666667E-1	6.82666666667E-4
		5.00000000000E-1	8.33333333333E-3	2.45076190282E-5
4			5.00000000000E-1	4.16666666667E-2
			1.00000000000E-1	5.95238095238E-4
5			1.00000000000E 0	1.66666666667E-1
			5.00000000000E-1	1.19047619048E-2
6				5.00000000000E-1
				1.07142857143E-1
7				1.00000000000E 0
				5.00000000000E-1

Proof of (1): The conditions on $f_m^{(\mu)}$ create $2m$ linear equations for the $2m$ coefficients of polynomial f_m. If the determinant of the system of equations were zero, then for certain right sides there would be two different polynomials f_m and $\tilde{f}_m$. Then $f_m - \tilde{f}_m$ would be a nonvanishing polynomial with $2m$ zeros and degree $\leq 2m-1$. Since that is a contradiction, the system of equations must have a unique solution.

Proof of (2): The difference $f - f_m$ has a zero of multiplicity m at each of the points a and b, and therefore a total of $2m$ zeros. It then follows from the generalized Rolle's Theorem that the derivatives

$$(f-f_m)^{(\mu)}, \qquad \mu = 0(1)2m-1$$

have at least $2m-\mu$ zeros on $[a,b]$. We denote these by $x_{\mu\nu}$ and order them by size with respect to ν. Obviously the first $m-\mu$ zeros are equal to a, and the last $m-\mu$ are equal to b. Now we consider the function

$$\phi_q(x) = f^{(\mu)}(x) - f_m^{(\mu)}(x) - q\prod_{\nu=1}^{2m-\mu}(x-x_{\mu\nu})$$

for fixed $\mu \in \{0,1,\ldots,2m-1\}$ and $q \in \mathbb{R}$. For a fixed $x_0 \in [a,b]$ with

$$x_0 \neq x_{\mu\nu}, \qquad \nu = 1(1)2m-\mu$$

one can then choose a $q \in \mathbb{R}$ such that $\phi_q(x_0)$ is equal to zero. Then $\phi_q(x)$ has at least $2m-\mu+1$ zeros in $[a,b]$. We again appeal to the generalized Rolle's Theorem to conclude that $\phi_q^{(2m-\mu)}(x)$ has at least one zero θ in (a,b). Then it follows from

$$\phi_q(x_0) = 0, \qquad \phi_q^{(2m-\mu)}(\theta) = 0$$

that

$$f^{(2m)}(\theta) - q(2m-\mu)! = 0,$$

$$f^{(\mu)}(x_o) - f_m^{(\mu)}(x_o) - \frac{f^{(2m)}(\theta)}{(2m-\mu)!} \prod_{\nu=1}^{2m-\mu}(x_o - x_{\mu\nu}) = 0.$$

When x_o is one of the zeros $x_{\mu\nu}$, the last equation holds for arbitrary $\theta \in (a,b)$. Therefore it holds for all $x \in [a,b]$. The equation, together with

$$|x - x_{\mu\nu}| \leq b-a$$

immediately implies the inequality for $\|f^{(\mu)} - f_m^{(\mu)}\|_\infty$, where $\mu = m(1)2m-1$. When $\mu = 0(1)m-1$, we can split the product. We have

$$\prod_{\nu=1}^{2m-\mu} |x - x_{\mu\nu}| \leq \begin{cases} (x-a)^{m-\mu}(b-x)^m & \text{for} \quad x \in [a,(a+b)/2] \\ (x-a)^m (b-x)^{m-\mu} & \text{for} \quad x \in ((a+b)/2,b]. \end{cases}$$

We want to find the extrema of

$$y(x) = (x-a)^{m-\mu}(b-x)^m \qquad x \in [a,(a+b)/2].$$

We have

$$y'(\tilde{x}) = (m-\mu)(\tilde{x}-a)^{m-\mu-1}(b-\tilde{x})^m - m(\tilde{x}-a)^{m-\mu}(b-\tilde{x})^{m-1} = 0$$

exactly when $\tilde{x} = [ma + (m-\mu)b]/(2m-\mu)$. The function $y(x)$ assumes its maximum at $\tilde{x}$. Since

$$\tilde{x}-a = (m-\mu)(b-a)/(2m-\mu) \quad \text{and} \quad b-\tilde{x} = m(b-a)/(2m-\mu)$$

it follows that

$$y(\tilde{x}) = \frac{m^m(m-\mu)^{m-\mu}}{(2m-\mu)^{2m-\mu}}(b-a)^{2m-\mu}.$$

The considerations for $x \in ((a+b)/2,b]$ are similar. The inequality follows for $\mu = 0(1)m-1$. □

Suppose a fixed $x \in [a,b]$ has been chosen. Then the assignment

$$A_\mu: f \to f^{(\mu)}(x) - f_m^{(\mu)}(x), \qquad \mu = 0(1)2m-1$$

defines a linear functional on $C^{2m}([a,b], \mathbb{R})$. It vanishes on the set of all polynomials of degree less than $2m$. The functional can be represented explicitly with the aid of a *Peano kernel*.

Definition 15.3: Let $m \in \mathbb{N}$, $x,t \in [a,b]$ and

$$g(x,t) = (x-t)_+^{2m-1} = \begin{cases} (x-t)^{2m-1} & \text{for } x > t \\ 0 & \text{for } x \leq t. \end{cases}$$

For fixed t, we denote the Hermite interpolation polynomial of $g(x,t)$ by $g_m(x,t)$. We set

$$G_m(x,t) = g(x,t) - g_m(x,t).$$

Then $\partial^\mu G_m/\partial x^\mu$ is called the *Peano kernel* of A_μ. □

The coefficients of the Hermite interpolation polynomial $g_m(x,t)$ are functions of t which can be represented explicitly with the aid of Cramer's Rule. Therefore, $g_m \in C^{2m-2}([a,b] \times [a,b], \mathbb{R})$. Since $g(x,t)$ is also a function in $C^{2m-2}([a,b] \times [a,b], \mathbb{R})$, the same is true for $G_m(x,t)$.

Theorem 15.4: Let $f \in C^{2m}([a,b], \mathbb{R})$ and let f_m be the Hermite interpolation polynomial for f. Then for all $x \in [a,b]$ we have the representation:

$$f^{(\mu)}(x) - f_m^{(\mu)}(x) = \frac{1}{(2m-1)!}\int_a^b f^{(2m)}(t)\frac{\partial^\mu}{\partial x^\mu}G_m(x,t)\,dt,$$
$$\mu = 0(1)2m-1.$$

Proof: We begin by showing that

$$\phi(x) = \int_a^b f^{(2m)}(t) G_m(x,t) dt$$

is a solution of the following boundary value problem:

$$\begin{aligned} \phi^{(2m)}(x) &= (2m-1)!\, f^{(2m)}(x) \\ \phi^{(\mu)}(a) &= \phi^{(\mu)}(b) = 0, \qquad \mu = 0(1)m-1. \end{aligned} \tag{15.5}$$

$G_m(x,t)/(2m-1)!$ will then be the *Green's function* for the boundary value problem (cf., e.g. Coddington-Levinson 1955).

Since $G_m \in C^{2m-2}([a,b] \times [a,b], \mathbb{R})$, it follows that ϕ is $(2m-2)$-times continuously differentiable on $[a,b]$. We have

$$\begin{aligned} \phi^{(2m-2)}(x) &= \int_a^b f^{(2m)}(t) \frac{\partial^{2m-2}}{\partial x^{2m-2}} G_m(x,t) dt \\ &= \int_a^x f^{(2m)}(t) \frac{\partial^{2m-2}}{\partial x^{2m-2}} G_m(x,t) dt \\ &\qquad + \int_x^b f^{(2m)}(t) \frac{\partial^{2m-2}}{\partial x^{2m-2}} G_m(x,t) dt. \end{aligned}$$

For $x \neq t$, $g(x,t)$, $g_m(x,t)$, and hence $G_m(x,t)$ are all arbitrarily often differentiable. It follows that $\phi \in C^{2m-1}([a,b], \mathbb{R})$. Differentiation yields

$$\begin{aligned} \phi^{(2m-1)}(x) &= \int_a^x f^{(2m)}(t) \frac{\partial^{2m-1}}{\partial x^{2m-1}} G_m(x,t) dt + \\ &\qquad + f^{(2m)}(x) \frac{\partial^{2m-2}}{\partial x^{2m-2}} G_m(x,x-0) \\ &\quad + \int_x^b f^{(2m)}(t) \frac{\partial^{2m-1}}{\partial x^{2m-1}} G_m(x,t) dt \\ &\qquad - f^{(2m)}(x) \frac{\partial^{2m-2}}{\partial x^{2m-2}} G_m(x,x+0). \end{aligned}$$

Since the $(2m-2)$-th partial derivative of $G_m(x,t)$ with respect to x is continuous in x and t, only the two integral terms remain. As above, it follows that

$\phi \in C^{2m}([a,b], \mathbb{R})$ and

$$\phi^{(2m)}(x) = \int_a^x f^{(2m)}\frac{\partial^{2m}}{\partial x^{2m}}G_m(x,t)dt + f^{(2m)}(x)\ \frac{\partial^{2m-1}}{\partial x^{2m-1}}G_m(x,x-0)$$
$$+ \int_x^b f^{(2m)}(t)\frac{\partial^{2m}}{\partial x^{2m}}G_m(x,t)dt - f^{(2m)}(x)\frac{\partial^{2m-1}}{\partial x^{2m-1}}G_m(x,x+0).$$

We have

$$\frac{\partial^{2m-1}}{\partial x^{2m-1}}g(x,t) = \begin{cases} (2m-1)! & \text{for } x > t \\ 0 & \text{for } x \leq t \end{cases}$$
$$\frac{\partial^{2m}}{\partial x^{2m}}g(x,t) = 0 \quad \text{for} \quad x \neq t$$

and $\frac{\partial^{\mu}}{\partial x^{\mu}}g_m(x,t)$ is continuous in x and t for $\mu = 0(1)2m-1$, and $\frac{\partial^{2m}}{\partial x^{2m}}g_m(x,t) \equiv 0$. Combining all this, we obtain

$$\frac{\partial^{2m-1}}{\partial x^{2m-1}}G_m(x,x-0) - \frac{\partial^{2m-1}}{\partial x^{2m-1}}G_m(x,x+0) = (2m-1)!$$
$$\frac{\partial^{2m}}{\partial x^{2m}}G_m(x,t) = 0, \qquad x \neq t.$$

From this it follows that

$$\phi^{(2m)}(x) = (2m-1)!\ f^{(2m)}(x).$$

In addition, it follows for $\mu = 0(1)m-1$ from the construction of $G_m(x,t)$ that

$$\phi^{(\mu)}(a) = \phi^{(\mu)}(b) = 0.$$

Thus ϕ is a solution of boundary value problem (15.5). The function

$$(2m-1)![f(x) - f_m(x)]$$

is obviously also a solution of (15.5). Since the boundary value problem has a unique solution, it follows that

$$f(x) - f_m(x) = \frac{1}{(2m-1)!}\ \phi(x)$$
$$= \frac{1}{(2m-1)!}\int_a^b f^{(2m)}(t)G_m(x,t)dt.$$

Differentiating this and substituting the derivatives of $\phi(x)$ obtained farther above yields the conclusion. □

Example 15.6: g, g_m, and G_m for $m = 1,2,3$.

Case m = 1:

$$g(x,t) = (x-t)_+, \quad g_1(x,t) = \frac{(x-a)(b-t)}{b-a}$$

$$G_1(x,t) = \begin{cases} -\dfrac{(b-x)(t-a)}{b-a} & \text{for } x \geq t \\ -\dfrac{(b-t)(x-a)}{b-a} & \text{for } x \leq t \end{cases}$$

$$G_{1x}(x,t) = \begin{cases} \dfrac{t-a}{b-a} & \text{for } x \geq t \\ -\dfrac{b-t}{b-a} & \text{for } x \leq t \end{cases}$$

$$G_{1x}(x,x-0) - G_{1x}(x,x+0) = \frac{x-a}{b-a} + \frac{b-x}{b-a} = 1.$$

Case m = 2:

$$g(x,t) = (x-t)_+^3$$

$$g_2(x,t) = -\frac{(b-t)^2(x-a)^2}{(b-a)^3}[2(b-t)(x-a) + 3(b-a)(t-x)]$$

$$G_2(x,t) = \begin{cases} \dfrac{(b-x)^2(t-a)^2}{(b-a)^3}[2(b-x)(t-a)+3(b-a)(x-t)] & \text{for } x \geq t \\ \dfrac{(b-t)^2(x-a)^2}{(b-a)^3}[2(b-t)(x-a)+3(b-a)(t-x)] & \text{for } x \leq t \end{cases}$$

$$G_{2x}(x,t) = \begin{cases} \dfrac{(t-a)^2}{(b-a)^3}\{-2(b-x)[2(b-x)(t-a)+3(b-a)(x-t)] \\ \qquad + (b-x)^2(3b-a-2t)\} & \text{for } x \geq t \\ \dfrac{(b-t)^2}{(b-a)^3}\{2(x-a)[2(b-t)(x-a)+3(b-a)(t-x)] \\ \qquad + (x-a)^2(3a-b-2t)\} & \text{for } x \leq t \end{cases}$$

$$G_{2xx}(x,t) = \begin{cases} \dfrac{(t-a)^2}{(b-a)^3}\{2[2(b-x)(t-a)+3(b-a)(x-t)] \\ \qquad\qquad - 4(b-x)(3b-a-2t)\} & \text{for } x \geq t \\ \dfrac{(b-t)^2}{(b-a)^3}\{2[2(b-t)(x-a)+3(b-a)(t-x)] \\ \qquad\qquad + 4(x-a)(3a-b-2t)\} & \text{for } x \leq t \end{cases}$$

$$G_{2xxx}(x,t) = \begin{cases} \dfrac{(t-a)^2}{(b-a)^3}\, 6(3b-a-2t) & \text{for } x \geq t \\ \dfrac{(b-t)^2}{(b-a)^3}\, 6(3a-b-2t) & \text{for } x \leq t \end{cases}$$

$$G_{2xxx}(x,x-0) - G_{2xxx}(x,x+0) = 6.$$

Case $m = 3$:

$$g(x,t) = (x-t)_+^5$$

$$g_3(x,t) = \frac{(b-t)^3(x-a)^3}{(b-a)^5}\{5(b-a)(t-x)[2(b-a)(t-x)+3(x-a)(b-t)] + 6(x-a)^2(b-t)^2\}$$

$$G_3(x,t) = \begin{cases} (x-t)^5 - g_3(x,t) & \text{for } x \geq t \\ \quad\quad - g_3(x,t) & \text{for } x \leq t. \end{cases}$$

Theorem 15.4 immediately yields an estimate for the interpolation error with respect to the norm $\|\cdot\|_2$. □

<u>Theorem 15.7</u>: Let $f \in C^{2m}([a,b], \mathbb{R})$. Then the inequality

$$\|f^{(\mu)} - f_m^{(\mu)}\|_2 \leq c_{m\mu}(b-a)^{2m-\mu}\|f^{(2m)}\|_2, \quad \mu = 0(1)2m-1$$

where

$$c_{m\mu} = \frac{1}{(2m-1)!}\left\{\int_0^1\int_0^1 [\frac{\partial^\mu}{\partial x^\mu}\tilde{G}_m(x,t)]^2 dx\, dt\right\}^{1/2}$$

holds true. Here $\tilde{G}_m(x,t)$ is the function G_m from Definition 15.3 for the interval $[0,1]$. The constants $c_{m\mu}$ can be computed for small m. The values in Table 15.8 were obtained by Lehmann 1975.

TABLE 15.8. $c_{m\mu}$ for $m = 1,2,3,4$.

μ	m=1	m=2	m=3	m=4
0	1.05409255339E-1	2.01633313311E-3	1.63169843917E-5	7.17567956106E-8
1	4.08248290464E-1	7.27392967453E-3	6.56734371321E-5	3.19767674247E-7
2		4.87950036474E-2	4.45212852385E-4	2.38010180208E-6
3		4.14039335605E-1	4.24705992865E-3	2.24457822314E-5
4			5.37215309350E-2	2.77638992969E-4
5			4.21294644506E-1	5.08920680460E-3
6				5.92874650749E-2
7				4.27311575545E-1

Proof: From Theorem 15.4 and an application of the Cauchy-Schwarz Inequality we obtain

$$|f^{(\mu)}(x) - f_m^{(\mu)}(x)|^2$$
$$\leq \frac{1}{((2m-1)!)^2} \int_a^b [f^{(2m)}(t)]^2 dt \int_a^b [\frac{\partial^\mu}{\partial x} G_m(x,t)]^2 dt.$$

By integration, this becomes

$$\int_a^b |f^{(\mu)}(x) - f_m^{(\mu)}(x)|^2 dx \leq \tilde{c}_{m\mu}^2(b-a) \int_a^b [f^{(2m)}(t)]^2 dt$$

where

$$\tilde{c}_{m\mu}(b-a) = \frac{1}{(2m-1)!} \left\{\int_a^b \int_a^b [\frac{\partial^\mu}{\partial x^\mu} G_m(x,t)]^2 dt\, dx\right\}^{1/2}.$$

Every interval can be mapped onto $[0,1]$ by an affine transformation. With that substitution, we get

$$\tilde{c}_{m\mu}(b-a) = (b-a)^{2m-\mu}\, \tilde{c}_{m\mu}(1).$$

Letting

$$c_{m\mu} = \tilde{c}_{m\mu}(1) = \frac{1}{(2m-1)!} \left\{\int_0^1 \int_0^1 [\frac{\partial^\mu}{\partial x^\mu} \tilde{G}_m(x,t)]^2 dx\, dt\right\}^{1/2}$$

yields the desired conclusion. □

The polynomials of degree less than or equal to $2m-1$ form a 2m-dimensional vector space. The canonical basis $(1,x,\ldots,x^{2m-1})$ is very impractical for actual computations with Hermite interpolation polynomials. Therefore we will define a new basis which is better suited to our purposes.

Definition 15.9: *Basis of the 2m-dimensional polynomial space.* The conditions

$$S_{\alpha,\ell,m}^{(\mu)}(\beta) = \delta_{\mu\ell} \cdot \delta_{\alpha\beta} \quad (\alpha,\beta = 0,1; \quad \mu,\ell = 0(1)m-1)$$

define a basis

$$\{S_{\alpha,\ell,m}(x) \mid \alpha = 0,1; \quad \ell = 0(1)m-1\}$$

of the 2m-dimensional space of polynomials of degree less than or equal to $2m-1$. □

It is easily checked that the Hermite interpolation polynomial f_m for a function $f \in C^{m-1}([a,b], \mathbb{R})$ has the following representation:

$$f_m(x) = \sum_{\ell=0}^{m-1} (b-a)^{\ell}\Big[f^{(\ell)}(a)S_{0,\ell,m}(\tfrac{x-a}{b-a}) + f^{(\ell)}(b)S_{1,\ell,m}(\tfrac{x-a}{b-a})\Big]. \qquad (15.10)$$

This corresponds to the Lagrange interpolation formula for ordinary polynomial interpolation. Table 15.11 gives the $S_{\alpha,\ell,m}(x)$ explicitly for $m = 1,2,3$. In order to attain great precision it is necessary to use Hermite interpolation formulas of high degree. This can lead to the creation of numerical instabilities. To avoid this, we pass from global interpolation over the interval $[a,b]$ to *piecewise interpolation* with polynomials of lower degree. We do this by partitioning $[a,b]$ into n subintervals, introducing $n-1$ intermediate points. The interpolation function is pieced together from the Hermite interpolation polynomials for each subinterval.

Theorem 15.12: Let $m,n \in \mathbb{N}$, $f \in C^{m-1}([a,b], \mathbb{R})$ and let

$$a = x_0 < x_1 < \ldots\ldots < x_{n-1} < x_n = b$$

be a partition of the interval $[a,b]$. For $x \in [x_i,x_{i+1}]$ and $i = 0(1)n-1$ we define

TABLE 15.11: $S_{\alpha,\ell,m}(x)$ for $m = 1,2,3$.

α	ℓ	m	$S_{\alpha,\ell,m}(x)$
0	0	1	$1-x$
1	0	1	x
0	0	2	$1 - 3x^2 + 2x^3$
0	1	2	$x - 2x^2 + x^3$
1	0	2	$3x^2 - 2x^3$
1	1	2	$-x^2 + x^3$
0	0	3	$1 - 10x^3 + 15x^4 - 6x^5$
0	1	3	$x - 6x^3 + 8x^4 - 3x^5$
0	2	3	$\frac{1}{2}x^2 - \frac{3}{2}x^3 + \frac{3}{2}x^4 - \frac{1}{2}x^5$
1	0	3	$10x^3 - 15x^4 + 6x^5$
1	1	3	$-4x^3 + 7x^4 - 3x^5$
1	2	3	$\frac{1}{2}x^3 - x^4 + \frac{1}{2}x^5$

$$f_m(x) = \sum_{\ell=0}^{m-1}(x_{i+1}-x_i)^{\ell}\Big[f^{(\ell)}(x_i)S_{0,\ell,m}\Big(\frac{x-x_i}{x_{i+1}-x_i}\Big) + f^{(\ell)}(x_{i+1})S_{1,\ell,m}\Big(\frac{x-x_i}{x_{i+1}-x_i}\Big)\Big].$$

Then f_m is the Hermite interpolation polynomial for f on each subinterval $[x_i,x_{i+1}]$ (cf. Theorem 15.1(1)). f_m is $(m-1)$-times continuously differentiable on $[a,b]$. Whenever f is actually $2m$-times continuously differentiable on $[a,b]$, the following inequalities hold:

$$\|f^{(\mu)}-f_m^{(\mu)}\|_\infty^* \le \hat{c}_{m\mu} h^{2m-\mu} \|f^{(2m)}\|_\infty$$

$$\|f^{(\mu)}-f_m^{(\mu)}\|_2 \le c_{m\mu} h^{2m-\mu} \|f^{(2m)}\|_2$$

$$h = \max\{x_{i+1}-x_i \mid i = 0(1)n-1\}, \quad \mu = 0(1)2m-1.$$

Here $\hat{c}_{m\mu}$ and $c_{m\mu}$ are the constants from Theorems 15.1 and 15.7. We denote by $\|\cdot\|_\infty^*$ the norm obtained from $\|\cdot\|_\infty$ by considering only one-sided limits at the partitioning points x_j.

The proof follows immediately from (15.10) and Theorems 15.1 and 15.7. Our two-fold goal is to use global and piecewise Hermite Interpolation both for smoothing the approximation functions in two independent variables and for obtaining a special Ritz Method in two dimensions. Therefore, we will generalize global and piecewise Hermite Interpolation to two variables. We follow the approach of Simonsen 1959 and Stancu 1964 (cf. also Birkhoff-Schultz-Varga 1968). As our basic region, we choose $[0,1] \times [0,1]$ instead of an arbitrary rectangle, thereby avoiding unnecessary complications in our presentation.

Definition 15.13: Let $m \in \mathbb{N}$. We define $H^{(m)}$ to be the vector space generated by the set

$$\{p(x)\cdot q(y) \mid p,q \text{ polynomials of degree less than or equal to } 2m-1\}.$$

This space has dimension $4m^2$. The functions

$$S_{\alpha,k,m}(x)S_{\beta,\ell,m}(y), \qquad \alpha,\beta = 0,1; \quad k,\ell = 0(1)m-1$$

constitute a basis. □

<u>Remark 15.14</u>: $f \in H^{(m)}$ if and only if

$$f(x,y) = \sum_{i=0}^{2m-1} \sum_{j=0}^{2m-1} a_{ij} x^i y^j.$$

We can impose $4m^2$ conditions on the interpolation. The properties demanded of the $S_{\alpha,\ell,m}$ by Definition 15.9 require

$$\partial_x^\mu \partial_y^\nu (S_{\alpha,k,m}(\gamma) S_{\beta,\ell,m}(\delta)) = \delta_{\mu k} \delta_{\nu \ell} \delta_{\alpha\gamma} \delta_{\beta\delta}$$

$$\alpha,\beta,\gamma,\delta = 0,1; \quad \mu,\nu,k,\ell = 0(1)m-1. \quad \square$$

If, instead of $H^{(m)}$, we choose the set of polynomials in two variables of degree less than or equal to $2m-1$, the theory becomes substantially more difficult. Then we have a vector space of dimension $m(2m+1)$. At each of the four vertices of the unit square we will need to give $m(2m+1)$ conditions. Even in the case $m = 1$ we run into difficulties, since we cannot prescribe the four functional values at the four vertices. We avoid this difficulty by choosing the interpolation polynomials from $H^{(m)}$.

<u>Theorem 15.15</u>: For each sequence

$$\{c_{\alpha,\beta,k,\ell} \in \mathbb{R} \mid \alpha,\beta = 0,1; \quad k,\ell = 0(1)m-1\}$$

there exists exactly one polynomial $f \in H^{(m)}$ such that

$$\frac{\partial^k \partial^\ell}{\partial x^k \partial y^\ell} f(\alpha,\beta) = c_{\alpha,\beta,k,\ell} \qquad \alpha,\beta = 0,1; \quad k,\ell = 0(1)m-1.$$

This polynomial has the representation

$$f(x,y) = \sum_{\alpha,\beta=0}^{1} \sum_{k,\ell=0}^{m-1} c_{\alpha,\beta,k,\ell} S_{\alpha,k,m}(x) S_{\beta,\ell,m}(y).$$

See Remark 15.14 for the proof. Table 15.16 gives the special basis for $H^{(m)}$ explicitly for $m = 1,2$.

TABLE 15.16: Basis of $H^{(m)}$ for $m = 1,2$.

α	k	m	β	ℓ	m	$S_{\alpha,k,m}(x)$	$\cdot\ S_{\beta,\ell,m}(y)$
0	0	1	0	0	1	$(1-x)$	$\cdot\ (1-y)$
0	0	1	1	0	1	$(1-x)$	$\cdot\ y$
1	0	1	0	0	1	x	$\cdot\ (1-y)$
1	0	1	1	0	1	x	$\cdot\ y$
0	0	2	0	0	2	$(1-3x^2+2x^3)$	$\cdot\ (1-3y^2+2y^3)$
0	0	2	0	1	2	$(1-3x^2+2x^3)$	$\cdot\ (y-2y^2+y^3)$
0	1	2	0	0	2	$(x-2x^2+x^3)$	$\cdot\ (1-3y^2+2y^3)$
0	1	2	0	1	2	$(x-2x^2+x^3)$	$\cdot\ (y-2y^2+y^3)$
0	0	2	1	0	2	$(1-3x^2+2x^3)$	$\cdot\ (3y^2-2y^3)$
0	0	2	1	1	2	$(1-3x^2+2x^3)$	$\cdot\ (-y^2+y^3)$
0	1	2	1	0	2	$(x-2x^2+x^3)$	$\cdot\ (3y^2-2y^3)$
0	1	2	1	1	2	$(x-2x^2+x^3)$	$\cdot\ (-y^2+y^3)$
1	0	2	0	0	2	$(3x^2-2x^3)$	$\cdot\ (1-3y^2+2y^3)$
1	0	2	0	1	2	$(3x^2-2x^3)$	$\cdot\ (y-2y^2+y^3)$
1	1	2	0	0	2	$(-x^2+x^3)$	$\cdot\ (1-3y^2+2y^3)$
1	1	2	0	1	2	$(-x^2+x^3)$	$\cdot\ (y-2y^2+y^3)$
1	0	2	1	0	2	$(3x^2-2x^3)$	$\cdot\ (3y^2-2y^3)$
1	0	2	1	1	2	$(3x^2-2x^3)$	$\cdot\ (-y^2+y^3)$
1	1	2	1	0	2	$(-x^2+x^3)$	$\cdot\ (3y^2-2y^3)$
1	1	2	1	1	2	$(-x^2+x^3)$	$\cdot\ (-y^2+y^3)$

The following theorem uses the Peano kernel to obtain a representation of the error in a Hermite interpolation.

Theorem 15.17: Let $f \in C^{4m}([0,1] \times [0,1], \mathbb{R})$. We define $f_m \in H^{(m)}$ by the condition

$$\frac{\partial^{\mu+\nu}}{\partial x^\mu \partial y^\nu} f_m(\alpha,\beta) = \frac{\partial^{\mu+\nu}}{\partial x^\mu \partial y^\nu} f(\alpha,\beta), \quad \alpha,\beta = 0,1; \quad \mu,\nu = 0(1)m-1.$$

Then for $\mu,\nu = 0(1)2m-1$ we have

$$\left\|\frac{\partial^{\mu+\nu}}{\partial x^\mu \partial y^\nu}(f-f_m)\right\|_2 \le c_{m\mu}\left\|\frac{\partial^{2m+\nu}}{\partial x^{2m}\partial y^\nu} f\right\|_2 + c_{m\nu}\left\|\frac{\partial^{\mu+2m}}{\partial x^\mu \partial y^{2m}}\right.$$
$$\left. + c_{m\mu}c_{m\nu}\left\|\frac{\partial^{4m}}{\partial x^{2m}\partial y^{2m}} f\right\|_2 .$$

Here $c_{m\mu}$ are the constants from Theorem 15.7.

Proof: We first show for $\mu,\nu = 0(1)m-1$ that

$$\begin{aligned}
&\frac{\partial^{\mu+\nu}}{\partial x^\mu \partial y^\nu} f(\xi,\eta) - \frac{\partial^{\mu+\nu}}{\partial x^\mu \partial y^\nu} f_m(\xi,\eta) \\
&\quad = \frac{1}{(2m-1)!}\left\{\int_0^1 \frac{\partial^{2m+\nu}}{\partial x^{2m}\partial y^\nu} f(t,\eta)\frac{\partial}{\partial x}G_m(\xi,t)\,dt \right. \\
&\qquad \left. + \int_0^1 \frac{\partial^{\mu+2m}}{\partial x^\mu \partial y^{2m}} f(\xi,s)\frac{\partial^\nu}{\partial x^\nu}G_m(\eta,s)\,ds\right\} \qquad (15.18) \\
&\qquad - \frac{1}{[(2m-1)!]^2}\int_0^1\int_0^1 \frac{\partial^{4m}}{\partial x^{2m}\partial y^{2m}} f(t,s)\frac{\partial^\mu}{\partial x^\mu}G_m(\xi,t)\frac{\partial^\nu}{\partial x^\nu}G_m(\eta,s)\,dt\,ds.
\end{aligned}$$

G_m is the function from Definition 15.3 corresponding to the interval $[0,1]$. We begin by assuming that f can be represented as a product, $f(x,y) = p(x)q(y)$, where $p,q \in C^{2m}([0,1], \mathbb{R})$. Then p and q can be approximated individually by means of the one dimensional Hermite interpolation. By Theorem 15.4 it is true for all $\xi,\eta \in [0,1]$ that

$$p_m^{(\mu)}(\xi) = p^{(\mu)}(\xi) - \frac{1}{(2m-1)!}\int_0^1 p^{(2m)}(t)\,\frac{\partial^\mu}{\partial x^\mu}G_m(\xi,t)\,dt$$
$$q_m^{(\nu)}(\eta) = q^{(\nu)}(\eta) - \frac{1}{(2m-1)!}\int_0^1 q^{(2m)}(s)\,\frac{\partial^\nu}{\partial x^\nu}G_m(\eta,s)\,ds.$$

Multiplying the two equations together yields

$$p_m^{(\mu)}(\xi)q_m^{(\nu)}(\eta) = p^{(\mu)}(\xi)q^{(\nu)}(\eta)$$

$$- \frac{1}{(2m-1)!}\left\{\int_0^1 p^{(2m)}(t)q^{(\nu)}(\eta)\frac{\partial^\mu}{\partial x^\mu}G_m(\xi,t)dt + \int_0^1 p^{(\mu)}(\xi)q^{(2m)}(s)\frac{\partial^\nu}{\partial x^\nu}G_m(\eta,s)ds\right\}$$

$$+ \frac{1}{((2m-1)!)^2}\int_0^1\int_0^1 p^{(2m)}(t)q^{(2m)}(s)\frac{\partial^\mu}{\partial x^\mu}G_m(\xi,t)\frac{\partial^\nu}{\partial x^\nu}G_m(\eta,s)dtds.$$

By means of the identification

$$\frac{\partial^{\mu+\nu}}{\partial x^\mu \partial y^\nu}f(\xi,\eta) = p^{(\mu)}(\xi)q^{(\nu)}(\eta)$$

$$\frac{\partial^{\mu+\nu}}{\partial x^\mu \partial y^\nu}f_m(\xi,\eta) = p_m^{(\mu)}(\xi)q_m^{(\nu)}(\eta)$$

we obtain the conclusion for the special case of $f(x,y) = p(x)q(y)$. Since the formula is linear, it will also hold for all arbitrary linear combinations of such products. This includes all polynomials of arbitrarily high degree. There is a theorem in approximation theory which says that every function f satisfying the differentiability conditions of the hypotheses (including the 4m-th order partial derivatives), can be approximated uniformly by a sequence of polynomials. Therefore Equation (15.18) also holds for arbitrary $f \in C^{4m}([0,1] \times [0,1], \mathbb{R})$. The Schwarz integral inequality, applied to (15.18), immediately yields the conclusion. □

We now pass from global Hermite interpolation on the unit square to *piecewise Hermite interpolation*. For this we allow regions whose closure is the union of closed squares $\square_\rho$ with sides parallel to the axes and of length h. We require that the intersection $\square_\rho \cap \square_\sigma$ $(\rho \neq \sigma)$ be either empty or consist of exactly one vertex or exactly one side.

As in the one dimensional case, we construct the interpolation function by piecing together the Hermite interpolation polynomials for the various squares $\Box_\rho$. All of the concepts and formulas developed for the unit square carry over immediately to squares $\Box_\rho$ with sides of length h.

Theorem 15.19: Let G be a region in $\mathbb{R}^2$ with the above properties and let $f \in C^{4m}(\bar{G}, \mathbb{R})$. We define $f_m:\bar{G} \to \mathbb{R}$ by the two conditions:

(1) f_m restricted to $\Box_\rho$ is a polynomial in $H^{(m)}$.

(2) At the vertices of square $\Box_\rho$ it is true that

$$\frac{\partial^{\mu+\nu}}{\partial x^\mu \partial y^\nu} f_m(x,y) = \frac{\partial^{\mu+\nu}}{\partial x^\mu \partial y^\nu} f(x,y), \qquad \mu,\nu = 0(1)m-1.$$

Then f_m is $(m-1)$-times continuously differentiable in $\bar{G}$. In the interiors of the squares, f_m is arbitrarily often differentiable. For $\mu,\nu = 0(1)2m-1$ it is true that

$$\left\|\frac{\partial^{\mu+\nu}}{\partial x^\mu \partial y^\nu}(f-f_m)\right\|_2 \leq c_{m\mu} h^{2m-\mu}\left\|\frac{\partial^{2m+\nu}}{\partial x^{2m}\partial y^\nu} f\right\|_2 + c_{m\nu} h^{2m-\nu}\left\|\frac{\partial^{\mu+2m}}{\partial x^\mu \partial y^{2m}} f\right\|_2$$
$$+ c_{m\mu} c_{m\nu} h^{4m-\mu-\nu}\left\|\frac{\partial^{2m}}{\partial x^{2m}\partial y^{2m}} f\right\|_2.$$

Here the $c_{m\mu}$ are the constants from Theorem 15.7.

Proof: Along the lines joining two neighboring vertices, f_m and the partial derivatives of f_m through order $m-1$ are already determined by the values $\partial^{\mu+\nu} f(x,y)/\partial x^\mu \partial y^\nu$ at the two vertices of the given side. From this it follows that f_m is $(m-1)$-times continuously differentiable in $\bar{G}$.

With the aid of Theorem 15.7, the inequality of Theorem 15.17 can immediately be carried over to the case of a square with sides parallel to the axes and of length h:

$$\int_{\Box_\rho} [\partial_x^\mu \partial_y^\nu (f-f_m)]^2 dxdy \le \Big\{c_{m\mu} h^{2m-\mu}\Big[\int_{\Box_\rho} (\partial_x^{2m}\partial_y^\mu f)^2 dxdy\Big]^{1/2}$$

$$+ c_{m\nu} h^{2m-\nu}\Big[\int_{\Box_\rho} (\partial_x^\mu \partial_y^{2m} f)^2 dxdy\Big]^{1/2}$$

$$+ c_{m\mu} c_{m\nu} h^{4m-\mu-\nu}\Big[\int_{\Box_\rho} (\partial_x^{2m}\partial_y^{2m} f)^2 dxdy\Big]^{1/2}\Big\}^2.$$

Summing over ρ and applying the Minkowski inequality for sums (cf. e.g. Beckenbach-Bellmann 1971, Ch. 1.20) gives the inequality for the norms. □

In the sequel, we will explain how the global and piecewise Hermite interpolation polynomials can be used as initial functions for the Ritz method (cf. Chapter 14). We first consider the one-dimensional problem

$$I(u) = \min\{I(w) \mid w \in W\}$$

where

$$I(w) = \int_0^1 [a(x)w'(x)^2 + 2Q(x,w(x))]dx$$

$$W = \{w \in C^2([0,1], \mathbb{R}) \mid w(0) = w(1) = 0\}.$$

An actual execution of the Ritz method requires us to choose a finite dimensional subspace V_n of W, and a particular basis for this subspace.

<u>Example 15.20</u>: *Basis of a one-dimensional Ritz method with global Hermite interpolation.* Let $m \ge 2$. Choose

$$\{S_{\alpha,\ell,m}(x) \mid \alpha = 0,1; \quad \ell = 1(1)m-1\}$$

as a basis. The functions $S_{\alpha,0,m}(x)$ are discarded because they do not vanish at the points $x = 0$ and $x = 1$. The space generated has dimension $2m-2$. □

Example 15.21: *Basis of a one dimensional Ritz method with piecewise Hermite interpolation.* Let $n \in \mathbb{N}$ and $h = 1/n$. For $b \in \mathbb{R}$ and $\ell = 0(1)m-1$, define

$$T_{b,\ell,m}(x) = \begin{cases} h^{\ell} S_{0,\ell,m}(\frac{x-b}{h}) & \text{for } x \in [b,b+h] \\ h^{\ell} S_{1,\ell,m}(\frac{x-(b-h)}{h}) & \text{for } x \in [b-h,b] \\ 0 & \text{otherwise.} \end{cases}$$

The restrictions of the $T_{b,\ell,m}(x)$ to $[0,1]$ yield $m(n+1)-2$ basis functions, for the following combination of indices:

$$b = 0,1: \quad \ell = 1(1)m-1$$

$$b = h,2h,\ldots,(n-1)h: \quad \ell = 0(1)m-1.$$

On $[0,1]$, the basis functions are $(m-1)$-times continuously differentiable and only differ from zero on one or two of the subintervals of the partition

$$0 < h < 2h < \ldots < (n-1)h < nh = 1$$

of the interval $[0,1]$. They all vanish at $x = 0$ and $x = 1$.

The $T_{b,\ell,m}(x)$ are given for $m = 1,2,3$ in Table 15.22. They are graphed in Figures 15.23 and 15.24. □

We will now discuss the two-dimensional variational problem with

$$I[w] = \iint_{\overline{G}} [a_1 w_x^2 + a_2 w_y^2 + 2Q(x,y,w)]dxdy$$

$$W = \{w \in C^2(G, \mathbb{R}) \cap C^0(\overline{G}, \mathbb{R}) \mid w(x,y) = 0 \text{ for } (x,y) \in \partial G\}.$$

This is the same problem as in Chapter 14, with the same conditions on a_1, a_2, and Q. For an actual execution of

TABLE 15.22: $T_{b,\ell,m}(x)$ for $m = 1,2,3$.

ℓ	m	$T_{b,\ell,m}(x)$ for $\|x-b\| \leq h$, $z = \frac{\|x-b\|}{h}$
0	1	$1 - z$
0	2	$1 - 3z^2 + 2z^3$
1	2	$h\ \mathrm{sgn}(x-b)(z-2z^2+z^3)$
0	3	$1 - 10z^3 + 15z^4 - 6z^5$
1	3	$h\ \mathrm{sgn}(x-b)(z-6z^3+8z^4-3z^5)$
2	3	$\frac{h^2}{2}(z^2-3z^3+3z^4-z^5)$
For $\|x-b\| \geq h$, $T_{b,\ell,m}(x) = 0$.		

the Ritz method, we must choose a finite dimensional subspace V_n of W and a particular basis of this subspace.

Example 15.25: *Basis of a two-dimensional Ritz method with global Hermite interpolation.* Let $G = [0,1] \times [0,1]$ and $m \geq 2$. Choose

$$\{S_{\alpha,k,m}(x)S_{\beta,\ell,m}(y) \mid \alpha,\beta = 0,1;\ k,\ell = 1(1)m-1\}$$

as a basis. The space generated has dimension $4(m-1)^2$. □

Example 15.26: *Basis of a two-dimensional Ritz method with piecewise Hermite interpolation.* Let G be a region in $\mathbb{R}^2$ which satisfies the subdivision properties of Theorem 15.19.

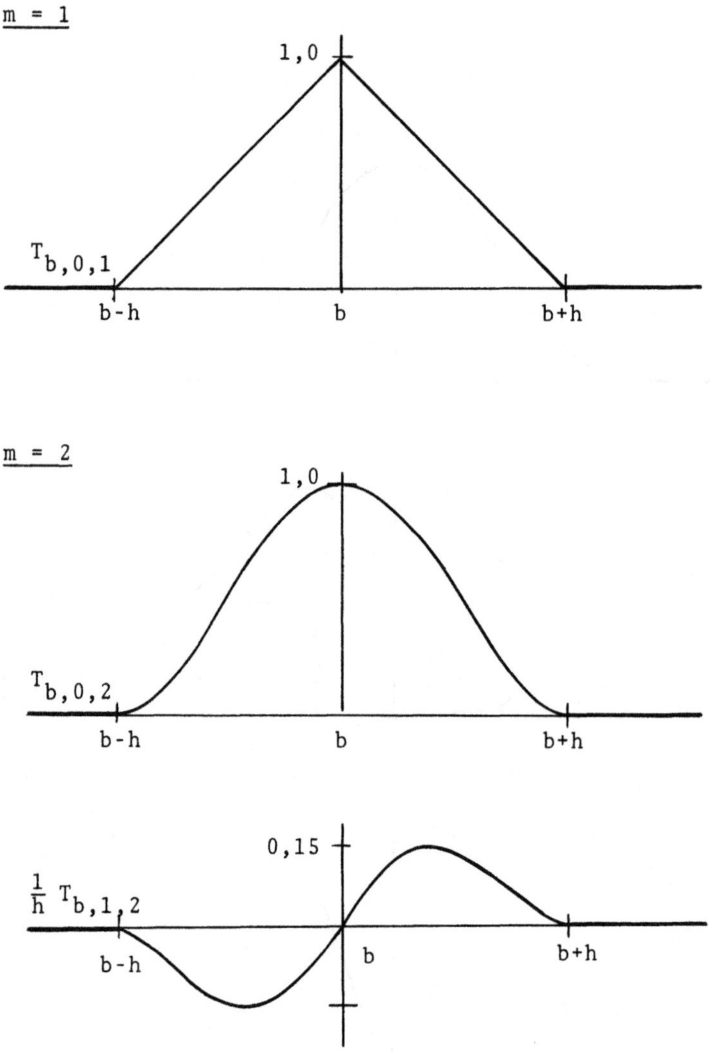

Figure 15.23: $T_{b,\ell,m}(x)/h^{\ell}$ for $m = 1,2$

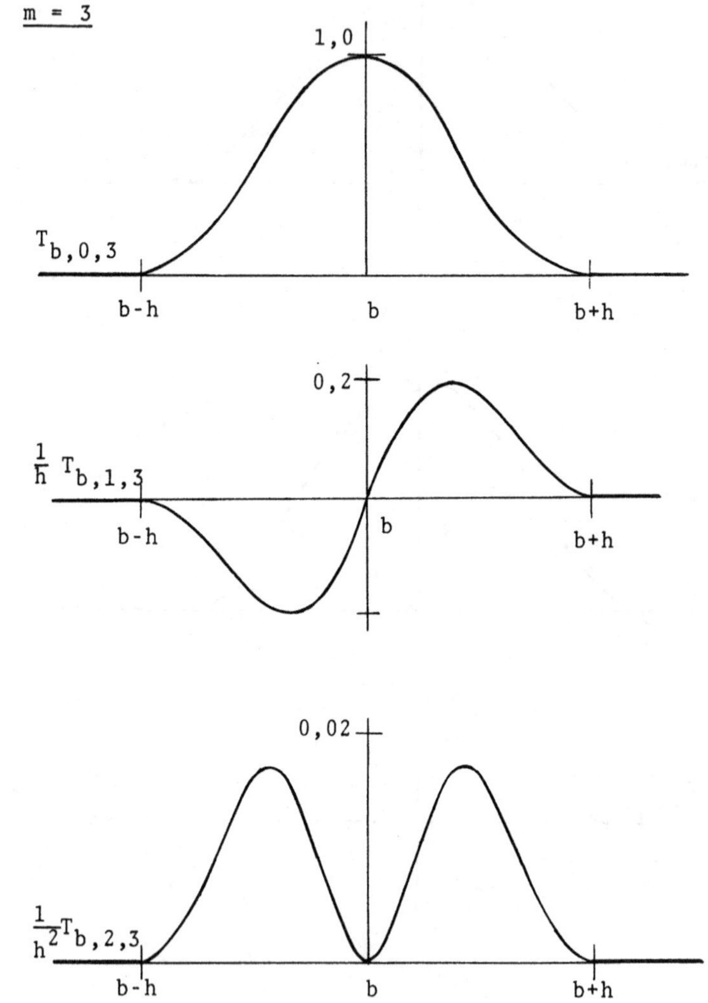

Figure 15.24: $T_{b,\ell,m}(x)/h^{\ell}$ for $m = 3$

Let E denote the set of all vertices of all of the squares $\square_\rho$ of the partition. Further, let $m,n \in \mathbb{N}$, and $h = 1/n$. We define the basis functions to be the restrictions of the functions

$$T_{b,k,m}(x)T_{\tilde{b},\ell,m}(y)$$

to $\overline{G}$ for the following combination of indices:

$$\begin{aligned}
&(b,\tilde{b}) \in E \cap G \quad \text{and} \quad k,\ell = 0(1)m-1; \\
&(b,\tilde{b}) \in E \cap \partial G \quad \text{and} \quad k,\ell = 0(1)m-1, \text{ but} \\
&\qquad k \geq 1 \quad \text{whenever} \quad (b,\tilde{b}+h) \in \partial G \quad \text{or} \quad (b,\tilde{b}-h) \in \partial G \\
&\qquad \ell \geq 1 \quad \text{whenever} \quad (b+h,\tilde{b}) \in \partial G \quad \text{or} \quad (b-h,\tilde{b}) \in \partial G.
\end{aligned}$$

The basis functions belong to $C^{m-1}(\overline{G}, \mathbb{R})$. They vanish on ∂G. For $m = 1$, we obtain the basis already discussed in Example 14.10, if the subdivision assumed there agrees with the one prescribed here. Thus piecewise Hermite interpolation supplies a generalization of this method to $m > 1$.

The matrix $A = (a_{\mu\nu})$ of the Ritz method for the basis chosen here can be given explicitly for the special case

$$Q(x,y,z) = q(x,y)z + \tfrac{1}{2}\sigma(x,y)z^2, \quad \sigma(x,y) \geq 0 \text{ for } (x,y) \in \overline{G}$$

by:

$$\begin{aligned}
a_{\mu\nu} = \iint_{\overline{G}} \Big[& a_1(x,y)T'_{b,k,m}(x)T_{\tilde{b},\ell,m}(y)T'_{b^*,k^*,m}(x)T_{\tilde{b}^*,\ell^*,m}(y) \\
&+ a_2(x,y)T_{b,k,m}(x)T'_{\tilde{b},\ell,m}(y)T_{b^*,k^*,m}(x)T'_{\tilde{b}^*,\ell^*,m}(y) \\
&+ \sigma(x,y)T_{b,k,m}(x)T_{\tilde{b},\ell,m}(y)T_{b^*,k^*,m}(x)T_{\tilde{b}^*,\ell^*,m}(y) \Big] dxdy.
\end{aligned}$$

In this case

$$T_{b,k,m}(x)T_{\tilde{b},\ell,m}(y)$$

is the μ-th basis function, and

$$T_{b^*,k^*,m}(x)\,T_{\tilde{b}^*,\ell^*,m}(y)$$

is the ν-th basis function. The integrals vanish whenever

$$\|(b,\tilde{b}) - (b^*,\tilde{b}^*)\|_2 > h\sqrt{2}$$

Therefore, each row and column of matrix A has at most $9m^2$ elements which differ from zero. At most four squares contribute to the integrals.

Theorem 14.18 supplies the inequalities

$$\|u-w\|_2 \le \gamma_2\|u-w\|_I \le \gamma_2\|u-w^*\|_I.$$

Here we have

u : solution of the variation problem in space W

w : Ritz approximation from space V_μ

w^* : arbitrary functions from V_μ.

We have the additional inequality:

$$\begin{aligned}\|u-w^*\|_I^2 \le & \max_{(x,y)\in\bar{G}} a_1(x,y)\|(u-w^*)_x\|_2^2 \\ & + \max_{(x,y)\in\bar{G}} a_2(x,y)\|(u-w^*)_y\|_2^2 \\ & + \max_{(x,y)\in\bar{G}} \sigma(x,y)\ \|u-w^*\|_2^2.\end{aligned}$$

In our problem we can apply Theorem 15.19 to choose w^* so that

$$\|u-w^*\|_2 \le M_0 h^{2m}$$

$$\|(u-w^*)_x\|_2 \le M_1 h^{2m-1}$$

$$\|(u-w^*)_y\|_2 \le M_2 h^{2m-1}.$$

The numbers M_0, M_1, and M_2 depend only on the derivatives of u and on m, but not on h. Altogether it follows that the Ritz method has convergence order $O(h^{2m-1})$:

$$\|u-w\|_2 \le Mh^{2m-1}.$$

In many practical cases it has been observed that the convergence order is actually $O(h^{2m})$. The explanation for this behavior can usually be found in the fact that the error must be a function of h^2 for reasons of symmetry. □

16. Collocation methods and boundary integral methods

Collocation methods are based on a very simple concept, which we will explain with the aid of the following *boundary value problem* (cf. Problem 13.2):

$$\begin{aligned} Lu(x,y) &= q(x,y), & (x,y) &\in G \\ u(x,y) &= \psi(x,y), & (x,y) &\in \Gamma. \end{aligned} \qquad (16.1)$$

Once again, G is a bounded region with boundary Γ and L is a linear, uniformly elliptical, second order differential operator of the form

$$\begin{aligned} Lu = & -a_{11}u_{xx} - 2a_{12}u_{xy} - a_{22}u_{yy} \\ & - b_1u_x - b_2u_y + gu \end{aligned}$$

where

$$a_{11},\ a_{12},\ a_{22},\ b_1,\ b_2,\ g \in C^\infty(\overline{G}, \mathbb{R})$$

and

$$q \in C^0(\overline{G}, \mathbb{R}), \quad \psi \in C^0(\Gamma, \mathbb{R}).$$

Further it is true for all $(x,y) \in \overline{G}$ that

$$a_{11}a_{22} - a_{12} > 0, \quad a_{11} > 0, \quad g \ge 0.$$

The execution of the method presupposes that we are given:

(1) n linearly independent *basis functions* v_j, $j = 1(1)n$, from $C^2(\overline{G}, \mathbb{R})$.

(2) n different *collocation points* $(x_k,y_k) \in \overline{G}$; of these, the first n_1 are to belong to G, and the remaining $n_2 = n - n_1$ are to lie in Γ.

The solution u of boundary value problem (16.1) will now be approximated by a linear combination w of the functions v_j, where we impose the following conditions on w:

$$\begin{aligned} Lw(x_k,y_k) &= q(x_k,y_k), \qquad k = 1(1)n_1 \\ w(x_k,y_k) &= \psi(x_k,y_k), \qquad k = n_1 + 1(1)n. \end{aligned} \tag{16.2}$$

In view of the fact that

$$w(x,y) = \sum_{j=1}^{n} c_j v_j(x,y), \quad c_j \in \mathbb{R}$$

the substitute problem (16.2) is concerned with the system of linear equations:

$$\sum_{j=1}^{n} \alpha_{kj} c_j = \beta_k, \qquad k = 1(1)n \tag{16.3}$$

$$\alpha_{kj} = \begin{cases} Lv_j(x_k,y_k) & \text{for } k \le n_1 \\ v_j(x_k,y_k) & \text{for } k > n_1 \end{cases}$$

$$\beta_k = \begin{cases} q(x_k,y_k) & \text{for } k \le n_1 \\ \psi(x_k,y_k) & \text{for } k > n_1. \end{cases}$$

In many actual applications, the system of equations can be simplified considerably by a judicious choice of the v_j. It is often possible to arrange matters so that either the differential equation or the boundary conditions are satisfied exactly by the functions v_j. We distinguish:

(A) *Boundary collocation:* We have $q \equiv 0$ and

$$Lv_j(x,y) = 0, \qquad j = 1(1)n, \ (x,y) \in G.$$

All (x_k,y_k) must lie in Γ, i.e. $n_1 = 0$, $n_2 = n$.

(B) *Interior collocation:* We have $\psi \equiv 0$ and

$$v_j(x,y) = 0, \qquad j = 1(1)n, \ (x,y) \in \Gamma.$$

All (x_k,y_k) must lie in G, i.e. $n_1 = n$, $n_2 = 0$.

The system of equations (16.3) does not always have a unique solution. When it does, the solution can be arbitrarily large. *A priori* conclusions about the error $u-w$ can only be drawn on the basis of very special hypotheses. This is the weakness of collocation methods. It is therefore essential to estimate the error *a posteriori*. Nevertheless, collocation methods with *a posteriori* error estimation frequently are superior to all other methods with respect to effort and accuracy.

Error estimates can be carried out in the norm $\|\cdot\|_2$, as explained in Section 14 (cf. Theorem 14.19). However, this seems unduly complicated in comparison with the simplicity of collocation methods. Therefore, one usually prefers to estimate errors with the aid of monotone principles. We wish to explain these estimates for the cases of boundary collocation and interior collocation. To this end, let

$$\varepsilon = u-w, \quad r = q-Lw, \quad \phi = \psi-w.$$

Then we have

$$\begin{aligned} L\varepsilon(x,y) &= r(x,y), \qquad (x,y) \in G \\ \varepsilon(x,y) &= \phi(x,y), \qquad (x,y) \in \Gamma. \end{aligned} \tag{16.4}$$

(A) For boundary collocation, we have $r(x,y) = 0$. It follows from the maximum-minimum principle (cf. Theorem 12.3) that for all $(x,y) \in \overline{G}$:

$$\min_{(x,y)\varepsilon\Gamma} \{\phi(x,y),0\} \le \varepsilon(x,y) \le \max_{(x,y)\varepsilon\Gamma} \{\phi(x,y),0\}.$$

Thus it suffices to derive estimates for ϕ. We assume that Γ consists of only finitely many twice continuously differentiable curves Γ_ℓ. Each arc then has a parametric representation

$$(x,y) = [\xi_1(t),\xi_2(t)], \quad t \varepsilon [0,1].$$

We set $\tilde{\phi}(t) = \phi(\xi_1(t),\xi_2(t))$ and compute $\tilde{\phi}$ for the finitely many points $t_j = jh$, $j = 0(1)m$, where $m \varepsilon \mathbb{N}$ and $h = 1/m$. Then it is obviously true for all $\hat{t} \varepsilon [0,1]$ that

$$\min_{j=0(1)m} |\tilde{\phi}(\hat{t})-\tilde{\phi}(t_j)| \le \frac{h}{2} \max_{t\varepsilon[0,1]} |\tilde{\phi}'(t)|.$$

$\tilde{\phi}$ can be interpolated linearly between the points t_j. The interpolation error will be at most

$$\frac{h^2}{4} \max_{t\varepsilon[0,1]} |\tilde{\phi}''(t)|.$$

Combining this and letting

$$d_1 = \tfrac{1}{2}h \max_{t\varepsilon[0,1]} |\tilde{\phi}'(t)|, \quad d_2 = \tfrac{1}{4}h^2 \max_{t\varepsilon[0,1]} |\tilde{\phi}''(t)|.$$

we have, either for $\nu = 1$ or for $\nu = 2$, that

$$\min_{(x,y)\varepsilon\Gamma_\ell} \phi(x,y) = \min_{t\varepsilon[0,1]} \tilde{\phi}(t) \ge \min_{j=0(1)m} \tilde{\phi}(t_j)-d_\nu$$

$$\max_{(x,y)\varepsilon\Gamma_\ell} \phi(x,y) = \max_{t\varepsilon[0,1]} \tilde{\phi}(t) \le \max_{j=0(1)m} \tilde{\phi}(t_j)+d_\nu.$$

For small h, coarse estimates for $\tilde{\phi}'$ or $\tilde{\phi}''$ suffice. Since

$$\tilde{\phi}^{(\nu)}(t) = \frac{d^\nu}{dt^\nu}\psi(\xi_1(t),\xi_2(t)) - \sum_{j=1}^{n} c_j \frac{d^\nu}{dt^\nu} v_j(\xi_1(t),\xi_2(t))$$

the quantities c_j, $j = 1(1)n$, are the deciding factors.

(B) For interior collocation, we have $\phi(x,y) = 0$. By Lemma 13.18, there exists a $\tilde{w} \in C^2(\overline{G}, \mathbb{R})$ such that

$$L\tilde{w}(x,y) \geq 1, \qquad (x,y) \in G$$
$$\tilde{w}(x,y) \geq 0, \qquad (x,y) \in \overline{G}.$$

We set

$$\beta = \max_{(x,y)\in\overline{G}} |r(x,y)| = ||r||_\infty$$

and obtain

$$L(\varepsilon+\beta\tilde{w})(x,y) \geq 0, \qquad (x,y) \in \overline{G}$$
$$(\varepsilon+\beta\tilde{w})(x,y) \geq 0, \qquad (x,y) \in \Gamma.$$

It follows from the monotone principle that

$$(\varepsilon+\beta\tilde{w})(x,y) \geq 0, \qquad (x,y) \in \overline{G}$$
$$\varepsilon(x,y) \geq -\beta\tilde{w}(x,y), \qquad (x,y) \in \overline{G}.$$

Analogously, one obtains

$$L(\varepsilon-\beta\tilde{w})(x,y) \leq 0, \qquad (x,y) \in \overline{G}$$
$$(\varepsilon-\beta\tilde{w})(x,y) \leq 0, \qquad (x,y) \in \Gamma$$
$$\varepsilon(x,y) \leq \beta\tilde{w}(x,y), \qquad (x,y) \in \overline{G}.$$

Combining this leads to

$$||\varepsilon||_\infty \leq \beta||\tilde{w}||_\infty .$$

Thus the computation of the error is reduced to a computation of

$$\beta = \max_{(x,y)\in\overline{G}} |q(x,y) - \sum_{j=1}^{n} c_j L v_j(x,y)|.$$

We next want to consider three examples of boundary collocation in detail. In all cases the differential equation will be

$$\Delta u(x,y) = 0.$$

Example 16.5: Let G be the unit circle. We use the polar coordinates

$$\begin{aligned} x &= r\cos t \\ y &= r\sin t \end{aligned} \qquad r \in [0,1],\ t \in [0,2\pi)$$

and let

$$v_j(x,y) = r^{j-1}\exp[i(j-1)t], \qquad j = 1(1)n$$

$$h = 2\pi/n$$

$$(x_k,y_k) = (\cos[(k-1)h],\ \sin[(k-1)h]), \quad k = 1(1)n.$$

Since the functions v_j are complex-valued, the coefficients c_j will also be complex. Naturally, one can split the entire system into real and imaginary parts. Then this approach fits in with the general considerations above. The system of equations (16.3) can be solved with the aid of a fast Fourier transform (cf. Stoer-Bulirsch 1980). □

Example 16.6: Let G be the annulus

$$G = \{(x,y) \in \mathbb{R}^2 \mid 0 < r_1^2 < x^2+y^2 < r_2^2\}.$$

For even $n = 2\hat{n}-2$ we set

$$v_1(x,y) = \log r$$

$$v_j(x,y) = r^{j-\hat{n}}\exp[i(j-\hat{n})t], \quad j = 2(1)n$$

$$h = 4\pi/n$$

$$(x_k,y_k) = (r_1\cos[(k-1)h],\ r_1\sin[(k-1)h]),\ k = 1(1)\hat{n}-1$$

$$(x_k, y_k) = (r_2\cos[(k-\hat{n})h],\ r_2\sin[(k-\hat{n})h]), \quad k = \hat{n}(1)n.$$

All functions v_j are bounded on the annulus. For $j = \hat{n}(1)n$ they correspond to the basis functions of the previous example. One cannot dispense with the functions $v_1, \ldots, v_{\hat{n}-1}$ because the region is not simply connected. v_1 through $v_{\hat{n}-1}$ cannot be approximated by $v_{\hat{n}}$ through v_n. One computes, e.g., that

$$\int_0^{2\pi} [v_1(r_2\cos t,\ r_2\sin t) - v_1(r_1\cos t,\ r_1\sin t)]dt = 2\pi\log(\frac{r_2}{r_1}) > 0$$

$$\int_0^{2\pi} [v_j(r_2\cos t,\ r_2\sin t) - v_j(r_1\cos t,\ r_1\sin t)]dt = 0, \quad j = 2(1)n.$$

This example shows that in each case a thorough theoretical examination of the problem is essential. □

Example 16.7: This example has its origins in nuclear technology. It has to do with flow through a porous body. Let the space coordinates be (x,y,s). We assume that the pressure, u, does not depend on s. Then a good approximation is given by $\Delta u(x,y) = 0$. Cylindrical channels are bored through the body, parallel to the s-axis:

$$\left.\begin{aligned} I_{\mu,\nu} &= \{(x,y) \in \mathbb{R}^2 \mid (x-2\mu)^2 + (y-2\nu)^2 < r_1^2\} \\ J_{\mu,\nu} &= \{(x,y) \in \mathbb{R}^2 \mid (x-2\mu-1)^2 + (y-2\nu-1)^2 < r_2^2\} \end{aligned}\right\} \quad \mu,\nu \in \mathbb{Z}.$$

Here r_1 and r_2 are fixed numbers with

$$0 < r_1 < 1, \quad 0 < r_2 < 1, \quad r_1+r_2 < \sqrt{2}.$$

Figure 16.8 depicts a section of this region for $r_1 = 1/4$ and $r_2 = 1/2$.

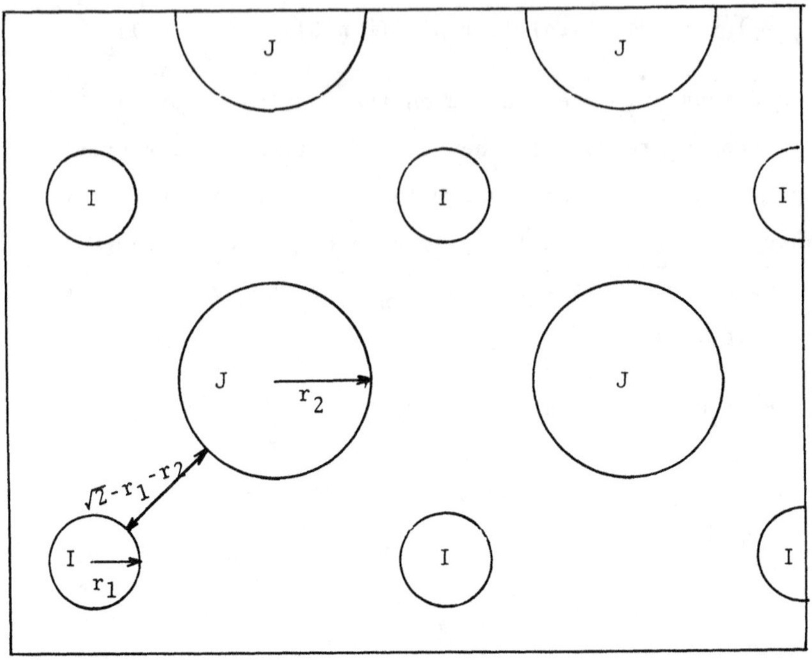

Figure 16.8: A region in Example 16.7

In each of the channels $I_{\mu,\nu}$ there is a pressure of 1, while in each of the $J_{\mu,\nu}$ there is a pressure of -1. The flow thus goes from channels I to channels J and increases monotonically with r_1 and r_2. Using the symmetries, one can reduce the problem to a restricted region G (cf. Figure 16.9). On the solid lines, $u(x,y) = 1$ or $u(x,y) = -1$; on the dashed lines, the normal derivative of u is zero. In this form, the problem can be solved with a difference method. The exact solution $u(x,y)$ of the boundary value problem is doubly periodic. For if (x,y) lies in the region between the channels, we have

$$u(x+2,y) = u(x,y+2) = u(x,y).$$

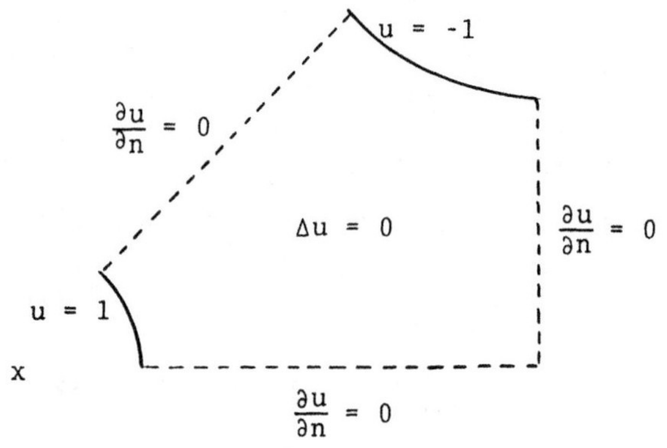

Figure 16.9: A region in Example 16.7.

Therefore it is natural to approximate u with the help of the simplest doubly periodic functions, the *Weierstrass P-functions* (cf. Magnus-Oberhettinger-Soni 1966, Ch. 10.5). Let $z = x+iy$. We denote the Weierstrass P-function with periods 2 and $2i$ by $p(z)$. The function is meromorphic, with a pole of second order at the points $2\mu + 2\nu i$ and with a zero of second order at $2\mu+1 + (2\nu+1)i$, where $\mu,\nu \in \mathbb{Z}$. The poles and zeros thus are at the centers of the channels. Therefore one can choose the basis functions v_j for the collocation method from the following set:

$$1, \ \log|p(z)|, \ \operatorname{Re}[p(z)^j], \ \operatorname{Im}[p(z)^j], \quad j \in \mathbb{Z} - \{0\}.$$

Because of all the symmetries, the set

$$1, \ \log|p(z)^2|, \ \operatorname{Re}[p(z)^{2j}]$$

suffices. We use the trial function

$$w(x,y) = \gamma \log|p(z)^2| + \sum_{j=-\ell}^{\ell} c_j \operatorname{Re}(p(z)^{2j})$$

and must determine the $n = 2\ell+2$ unknowns

$$\gamma,\ c_{-\ell},\ c_{-\ell+1},\dots,c_\ell.$$

To do this, we use the n collocation points; $\ell+1$ on the boundary of $I_{0,0}$ and $\ell+1$ on the boundary of $J_{0,0}$:

$$x_k+iy_k = \begin{cases} r_1\exp(\frac{\pi i}{4\ell}(k-1)) & \text{for } k = 1(1)\ell+1 \\ 1+i+r_2\exp(\frac{\pi i}{4\ell}(k-\ell-2)) & \text{for } k = \ell+2(1)n. \end{cases}$$

The linear system of equations is

$$w(x_k,y_k) = \begin{cases} 1 & \text{for } k = 1(1)\ell+1 \\ -1 & \text{for } k = \ell+2(1)n. \end{cases}$$

Table 16.10 gives the values of ℓ for several combinations of r_1 and r_2 necessary to obtain a precision of $10^{-3}/10^{-4}/10^{-5}$. When the difference $\sqrt{2}-r_1-r_2$ is not too small, one can obtain relatively high precision even with small ℓ. The flow depends only on γ. Table 16.11 contains the computed values γ for $\ell = 1,3,5,7$. The computing time to obtain the solution of the linear system is small in all of these cases compared to the effort required to estimate the error. The efficiency of a collocation method usually is highly dependent on the choice of collocation points. The situation is reminiscent of polynomial interpolation or numerical quadrature. Only there exists much less research into the optimal choice of support points for collocation methods. The least squares method is a modification of the collocation method in which the choice of collocation points is not quite as critical. In these procedures, one chooses m basis functions v_j and $n \geq m$ collocation points (x_k,y_k).

TABLE 16.10: $\ell_1/\ell_2/\ell_3$ for accuracies of $10^{-3}/10^{-4}/10^{-5}$

r_1 \ r_2	1/8	3/8	5/8	7/8
1/8	1/1/1	1/1/2	2/2/3	3/4/5
3/8	1/1/2	1/1/2	2/2/3	4/6/(9?)
5/8	2/2/3	2/2/3	3/4/(9?)	
7/8	3/4/5	4/6/(9?)		

TABLE 16.11: γ for $\ell = 1,3,5,7$

r_1 \ r_2	1/8	3/8	5/8	7/8
1/8	0.13823 0.13823 0.13823 0.13823	0.19853 0.19853 0.19853 0.19853	0.25011 0.25003 0.25003 0.25003	0.32128 0.31853 0.31852 0.31852
3/8	0.19853 0.19853 0.19853 0.19853	0.35218 0.35217 0.35217 0.35217	0.55542 0.55495 0.55495 0.55495	1.10252 1.06622 1.06599 1.06599
5/8	0.25011 0.25003 0.25003 0.25003	0.55542 0.55495 0.55495 0.55495	1.34168 1.32209 1.32207 1.32208	
7/8	0.32128 0.31853 0.31852 0.31852	1.10252 1.06622 1.06599 1.06599		

Again, n_1 of these are to lie in G, and n_2 in Γ. Condition (16.2) is replaced by:

$$\sum_{k=1}^{n_1} \delta_k [Lw(x_k,y_k) - q(x_k,y_k)]^2 + \sum_{k=n_1+1}^{n} \delta_k [w(x_k,y_k) - \psi(x_k,y_k)]^2 = \text{Min!} \tag{16.12}$$

Here the $\delta_k > 0$ are given weights and

$$w(x,y) = \sum_{j=1}^{m} c_j v_j(x,y).$$

Because of these conditions, the coefficients c_j, $j = 1(1)m$, can be computed as usual with balancing calculations (cf. Stoer-Bulirsch 1980, Chapter 4.8). Only with an explicit case at hand is it possible to decide if the additional effort (relative to simple collocation) is worthwhile. For $n = m$, one simply obtains the old procedure.

Occasionally there have been attempts to replace condition (16.12) with

$$\max\{\max_{k=1(1)n_1} \delta_k |Lw(x_k,y_k) - q(x_k,y_k)|,$$

$$\max_{k=n_1+1(1)n} \delta_k |w(x_k,y_k) - \psi(x_k,y_k)|\} = \text{Min!}$$

(minimization in the Chebyshev sense). Experience has demonstrated that this increases the computational effort tremendously. Consequently, any advantages with respect to the precision attainable become relatively minor.

We next discuss a boundary integral method for solving Problem (16.1), with $L = \Delta$, $q \equiv 0$, and $\psi \in C^1(\Gamma, \mathbb{R})$. The region G is to be a simply-connected subset of the closed unit disk $|z| \leq 1$, $z \in \mathbb{C}$, with a continuously differentiable boundary Γ. The procedure we are about to describe represents only one of several possibilities.

Let $\xi \in C^1([0,2\pi],\Gamma)$ be a parametrization of Γ without double points and with $\xi(0) = \xi(2\pi)$ and $\dot{\xi}_1^2 + \dot{\xi}_2^2 > 0$. Consider the trial function

$$u(z) = \int_0^{2\pi} \mu(t)\log|z-\xi(t)|dt, \qquad z \in G. \qquad (16.13)$$

If μ is continuous, $u \in C^0(\overline{G}, \mathbb{R})$ (cf. e.g. Kellog 1929). By differentiating, one shows in addition that u is harmonic in G. The boundary condition yields

$$\int_0^{2\pi} \mu(t)\log|z-\xi(t)|dt = \psi(z), \qquad z \in \Gamma. \qquad (16.14)$$

This is a linear Fredholm integral equation of the first kind with a weakly singular kernel. There exists a uniquely determined solution μ (cf. e.g. Jaswon 1963). The numerical method uses (16.14) to obtain first an approximation $\tilde{\mu}$ of μ at the discrete points $t_j = 2\pi(j-1)/n$, $j = 1(1)n$. Next (16.13) is used to obtain an approximation $\tilde{u}(z)$ of $u(z)$ for arbitrary $z \in G$.

The algorithm can be split into two parts, one dependent only on Γ and ξ, and the other only on ψ.

(A) *Boundary dependent part:*

(1) Computation of the weight matrix $W = (w_{jk})$ for n quadrature formulas

$$\int_0^{2\pi} f(t)\log|z_j-\xi(t)|dt = \sum_{k=1}^{n} w_{jk}f(t_k) + R(f)$$

$$z_j = \xi(t_j), \qquad j = 1(1)n$$

$$R(f_\nu) = 0 \quad \text{for} \quad f_\nu(t) = \begin{cases} 1 & \nu = 1 \\ \cos(\frac{\nu}{2}\, t) & \nu = 2(2)n \\ \sin(\frac{\nu-1}{2} t) & \nu = 3(2)n. \end{cases}$$

The matrix $(f_\nu(t_j))$ is regular. Therefore W is uniquely determined. Most of the computation is devoted to determining the n^2 integrals

$$\int_0^{2\pi} f_\nu(t)\log|z_j-\xi(t)|dt, \qquad \nu,j = 1(1)n.$$

(2) Triangulation of W into $W = LU$ using the Gauss algorithm or into $W = QR$ using the Householder transformations.

(B) *Boundary value dependent part:*

(1) Computation of $\tilde{\mu}(t_k)$ from the system of equations

$$\sum_{k=1}^{n} w_{jk}\tilde{\mu}(t_k) = \psi(z_j), \qquad j = 1(1)n.$$

Since $W = LU$ or $W = QR$, only $O(n^2)$ operations are required for this.

(2) Computation of $\tilde{u}(z)$ for $z \in G$ from (16.13). The integrand is a continuous 2π-periodic function. It seems natural to use a simple inscribed trapezoid rule with partition points t_j, $j = 1(1)n$:

$$\tilde{u}(z) = \frac{2\pi}{n} \sum_{k=1}^{n} \tilde{\mu}(t_k)\log|z-\xi(t_k)|. \tag{16.15}$$

If z does not lie in the vicinity of Γ, (16.15) actually yields good approximations for $u(z)$.

For boundary-close z, $-\log|z - \xi(t)|$ becomes extremely large on a small part of the interval $[0,2\pi]$. Then (16.15) is useless. The following procedure improves the results by several decimal places in many cases. But even this approach fails when the distances from the boundary are very small.

Let $\lambda(t)$ be that function $\tilde{\mu}(t)$ which results from boundary values $\psi \equiv 1$. Then, for $c \in \mathbb{R}$,

$$\tilde{u}_c(z) = c + \frac{2\pi}{n} \sum_{k=1}^{n} [\tilde{\mu}(t_k) - c\lambda(t_k)] \log|z - \xi(t_k)| \qquad (16.16)$$

are also approximations to $u(z)$. It is best to choose c so that

$$\tilde{\mu}(t_\ell) - c\lambda(t_\ell) = 0$$

whenever $|z-\xi(t_\ell)|$ is minimal. Since the computation of $\lambda(t)$ can proceed independently of the boundary values ψ, the effort in (16.15) is about the same as in (16.16). For each functional value $\tilde{u}(z)$ one needs $O(n)$ arithmetic operations. The method is thus economical when only a few functional values $\tilde{u}(z)$ are to be computed.

In the following example, we present some numerical results:

$$\psi(z) = \mathrm{Re}[\exp(z)] = \exp(x)\cos(y)$$

$$\xi_1(t) = 0.2\cos(t) + 0.3\cos(2t) - 0.3$$

$$\xi_2(t) = 0.7[0.5\sin(t-0.2)+0.2\sin(2t)-0.1\sin(4t)] + 0.1.$$

The region in question is the asymmetrically concave one shown in Figure 16.17. The approximation $\tilde{u}$ was computed on the rays 1, 2, and 3 leading from the origin to the points $\xi(0)$, $\xi(\pi)$, and $\xi(5\pi/3)$. R is the distance to the named points. Table 16.18 contains the absolute error resulting from the use of formula (16.15) (without boundary correction); Table 16.19 gives the corresponding values obtained from formula (16.16) (with boundary correction). We note that the method has no definitive convergence order.

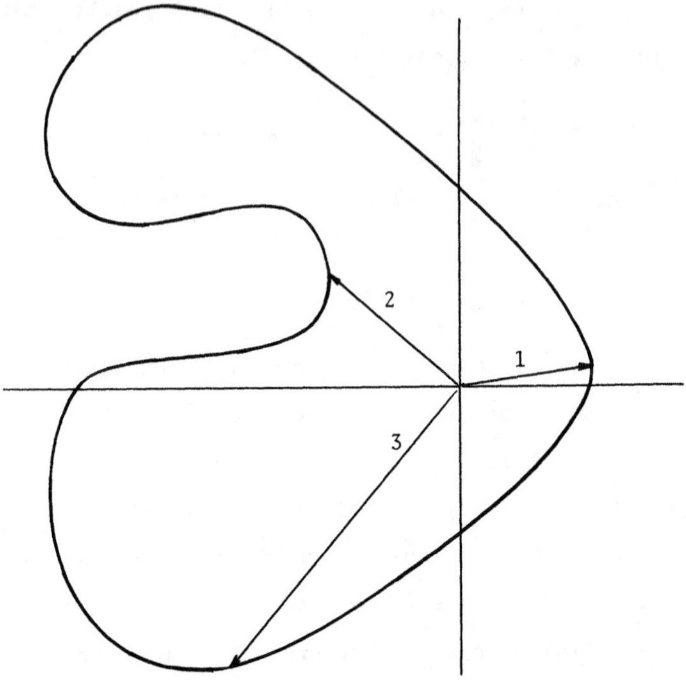

FIgure 16.17. Asymetrically concave region

	Ray 1			Ray 2			Ray 3		
n \ R	1/8	1/32	1/128	1/8	1/32	1/128	1/8	1/32	1/128
12	2.6E-2	5.5E-2	2.8E-1	1.3E-2	9.8E-3	7.0E-3	2.5E-2	6.9E-2	1.9E-1
24	7.5E-4	4.3E-3	8.1E-2	2.2E-4	2.2E-5	1.2E-4	5.5E-3	1.1E-2	6.7E-2
48	5.5E-7	3.0E-5	1.8E-2	1.9E-7	1.0E-6	7.0E-6	1.4E-4	2.2E-3	1.9E-2
96	1.1E-10	1.9E-7	2.5E-3	2.4E-12	5.0E-10	2.2E-7	3.3E-6	8.3E-4	3.3E-3

TABLE 16.18: Absolute error when computing without boundary correction

	Ray 1			Ray 2			Ray 3		
n \ R	1/8	1/32	1/128	1/8	1/32	1/128	1/8	1/32	1/128
12	3.1E-3	3.9E-3	4.7E-3	2.1E-2	1.4E-2	9.4E-3	1.3E-2	7.4E-3	4.6E-3
24	1.1E-4	1.6E-4	4.0E-4	1.5E-4	2.4E-5	7.5E-5	3.4E-4	2.0E-3	9.6E-4
48	5.4E-7	1.9E-6	4.3E-5	1.2E-7	2.6E-7	4.4E-8	4.0E-5	5.1E-4	2.8E-4
96	4.6E-12	2.4E-9	3.5E-6	1.3E-12	1.7E-12	4.7E-12	1.1E-6	2.4E-5	9.8E-5

TABLE 16.19: Absolute error when computing with boundary correction

PART III. SOLVING SYSTEMS OF EQUATIONS

17. Iterative methods for solving systems of linear and nonlinear equations

When we discretize boundary value problems for linear (nonlinear) elliptic differential equations, we usually obtain systems of linear (nonlinear) equations with a great many unknowns. The same holds true for the implicit discretization of initial boundary value problems for parabolic differential equations. For all practical purposes, the utility of such a discretization is highly dependent on the effectiveness of the methods for solving systems of equations.

In the case of systems of linear equations, one distinguishes between *direct* and *iterative* methods. Aside from rounding errors, the direct methods lead to an exact solution in finitely many steps (e.g. Gauss algorithm, Cholesky method, reduction method). Iterative methods construct a sequence of approximations, which converge to the exact solution (e.g. total step method, single step method, overrelaxation method). These are ordinarily much simpler to program than the direct methods. In addition, rounding errors play almost no role. However, in contrast to direct methods fitted to the problem (e.g. reduction methods), they require so much computing time that their use can only be defended when the demands for precision are quite modest. When using direct methods, one must remain alert to the fact that minimally different variants of a method can have entirely different susceptibilities to rounding errors.

We have only iterative methods for solving systems of non-linear equations. Newton's method (together with a few variants) occupies a special position. It usually requires only a few iterations. At each stage, we have to solve a system of linear equations. Experience shows that a quick direct method for solving the linear system is a necessary adjunct to Newton's method. An iterative method for solving the linear equations arising in a Newton's method is not to be recommended. It is preferable instead to apply an iterative method directly to the original non-linear system. The Newton's method/direct method combination stands to non-linear systems as direct methods to linear systems. However, the application is limited by the fact that frequently the linear systems arising at the steps of Newton's method are too complicated for the fast direct methods.

This section will serve as an introduction to the general theory of nonlinear iterative methods. A complete treatment may be found, e.g., in Ortega-Rheinboldt 1970.

In the following two sections, we examine overrelaxation methods (SOR) for systems of linear and nonlinear equations. After that, we consider direct methods.

Let $F: G \subset \mathbb{R}^n \to \mathbb{R}^n$ be a continuous function. We want to find a zero $x^* \in G$ of F, i.e. a solution of the equation

$$F(x) = 0 \tag{17.1}$$

lying in G. In functional analysis, one obtains a number of sufficient conditions for the existence of such a zero. Therefore, we will frequently assume that a zero $x^* \in G$ exists, and that there exists a neighborhood of x^* in which F has no other zeros. We further demand that G be an open set.

Iterative methods for determining a zero of F are based on a reformulation of (17.1) as an equivalent fixed point problem,

$$x = T(x), \tag{17.2}$$

so that x^* is a zero of F exactly when x^* is a fixed point of T. Then we set up the following iteration:

$$x^{(\nu)} = T(x^{(\nu-1)}), \quad \nu = 1(1)\infty. \tag{17.3}$$

One expects the sequence $\{x^{(\nu)} \mid \nu = 0(1)\infty\}$ to converge to x^* if the initial point $x^{(0)}$ is a sufficiently close approximation to x^*. But this is by no means true in every case. In addition to the question of convergence of the sequence, we naturally must give due consideration to the speed of the convergence, and to the simplicity, or lack thereof, of computing T. Before we begin a closer theoretical examination of these matters, we want to transform Equation (17.1) into the equivalent fixed point problem for a special case which frequently arises in practice.

<u>Example 17.4</u>: Suppose that the mapping F can be split into a sum, $F(x) = R(x) + S(x)$, in which S is only "weakly" dependent on x and R is constructively invertible. By the latter we mean that there exists an algorithm which is

realizable with respect to computing time, memory storage demand, and rounding error sensitivity, and for which the equation $R(y) = b$ can be solved for all b in a certain neighborhood of $-S(x^*)$. Such is the case, for example, when R is a linear map given by a nonsingular diagonal matrix or by a tridiagonal symmetric and positive definite matrix. If we set $T = R^{-1}\circ(-S)$ then equation $F(x) = 0$ is equivalent to the fixed point problem $x = T(x)$. When S, and therefore T also, depends only weakly on the point, one can expect the iterative method (17.3) to converge to x^* for sufficiently close approximations $x^{(0)}$ of x^*. □

<u>Definition 17.5</u>: Let $T: G \subset \mathbb{R}^n \to \mathbb{R}^n$ be a mapping and $x^* \in G$ a fixed point of T, i.e. $T(x^*) = x^*$. A point $y \in G$ belongs to the *attractive region* $I(T,x^*)$ of x^*, if the sequence (17.3) for $x^{(0)} = y$ remains in G and converges to x^*. The fixed point x^* is called *attractive* if it is an interior point of $I(T,x^*)$. The iteration (17.3) is called *locally convergent* if x^* is attractive. The mapping T is called *contracting* if there exists an $\alpha \in [0,1)$ and a norm $\|\cdot\|_T$ in $\mathbb{R}^n$ such that

$$\|T(x) - T(y)\|_T \le \alpha\|x-y\|_T \qquad x,y \in G. \quad □$$

Every contraction mapping is obviously continuous.

<u>Theorem 17.6</u>: Let $T: G \subset \mathbb{R}^n \to \mathbb{R}^n$ be a contraction mapping. Then it is true that:

(1) T has at most one fixed point $x^* \in G$. x^* is attractive.

(2) In case $G = \mathbb{R}^n$, there is exactly one fixed point x^*. Its attractive region is all of $\mathbb{R}^n$.

Proof of (1): Let x^* and y^* be two fixed points of T. Since T is contracting, there is an $\alpha \in [0,1)$ such that

$$\|x^*-y^*\|_T = \|T(x^*)-T(y^*)\|_T \le \alpha\|x^*-y^*\|_T.$$

It follows that $x^* = y^*$. We now choose $r \in \mathbb{R}_+$ so small that the closed ball

$$K_{T,r} = \{y \in \mathbb{R}^n \mid \|x^*-y\|_T \le r\}$$

lies entirely in G. It follows for all $z \in K_{T,r}$ that

$$\|T(z)-T(x^*)\|_T \le \alpha\|z-x^*\|_T \le r.$$

Therefore T maps the ball $K_{T,r}$ into itself. The sequence

$$x^{(\nu)} = T(x^{(\nu-1)}), \quad \nu = 1(1)\infty$$

is defined for $x^{(0)} \in K_{T,r}$ and satisfies the inequality

$$\|x^{(\nu)}-x^*\|_T \le \alpha^\nu\|x^{(0)}-x^*\|_T \le \alpha^\nu r, \qquad \nu = 1(1)\infty.$$

It follows that the sequence $\{x^{(\nu)} \mid \nu = 0(1)\infty\}$ converges to x^*.

Proof of (2): Let $x^{(0)} \in \mathbb{R}^n$ be chosen arbitrarily. Since T is contracting there is an $\alpha \in [0,1)$ so that for $\nu = 0(1)\infty$ it is true that

$$\begin{aligned}\|x^{(\nu+1)}-x^{(\nu)}\|_T &= \|T(x^{(\nu)})-T(x^{(\nu-1)})\|_T \\ &\le \alpha\|x^{(\nu)}-x^{(\nu-1)}\|_T \le \alpha^\nu\|x^{(1)}-x^{(0)}\|_T.\end{aligned}$$

From this it follows with the help of the triangle inequality that for all $\nu,\mu \in \mathbb{N}$,

$$\|x^{(\nu+\mu)}-x^{(\nu)}\|_T \le \sum_{\kappa=\nu}^{\nu+\mu-1}\|x^{(\kappa+1)}-x^{(\kappa)}\|_T \le \left(\sum_{\kappa=\nu}^{\nu+\mu-1}\alpha^\kappa\right)\|x^{(1)}-x^{(0)}\|_T$$

$$\le \frac{\alpha^\nu}{1-\alpha}\|x^{(1)}-x^{(0)}\|_T.$$

This says that $\{x^{(\nu)} \mid \nu = 0(1)\infty\}$ is a Cauchy sequence. Its limit value we denote by x^*. Since every contraction mapping is continuous, it follows that

$$x^* = \lim_{\nu\to\infty} x^{(\nu)} = \lim_{\nu\to\infty} T(x^{(\nu)}) = T(x^*).$$

Therefore x^* is a fixed point of T. The attractive region of x^* is all of $\mathbb{R}^n$. □

<u>Theorem 17.7</u>: Let A be a real $n \times n$ matrix, $b \in \mathbb{R}^n$ and let $T(x) = Ax + b$ be an affine mapping. Then T is contracting if and only if the spectral radius $\rho(A)$ is less than 1. In that case, T has exactly one fixed point. Its attractive region $I(T,x^*)$ is all of $\mathbb{R}^n$.

Proof: The conclusion follows at once from Lemma 9.17 and Theorem 17.6. □

<u>Theorem 17.8</u>: Let $T:G \subset \mathbb{R}^n \to \mathbb{R}^n$ be a map which has one fixed point $x^* \in G$. Let T be differentiable at x^*. Then

$$\rho(T'(x^*)) < 1,$$

implies that x^* is an attractive fixed point.

The proof is obtained by specializing the following theorem and making use of Lemma 9.17.

Theorem 17.8 says that the local convergence of an iterative method (17.3) for a differentiable map depends in a simple way on the derivative. Thus differentiable nonlinear maps behave like linear maps with respect to local convergence. This conclusion can be extended to nondifferentiable maps which are piecewise differentiable.

Theorem 17.9: Let $T: G \subset \mathbb{R}^n \to \mathbb{R}^n$ be a map which has one fixed point $x^* \in G$. Let T be continuous at x^*. Suppose there is an $m \in \mathbb{N}$ and maps

$$T_r: G \subset \mathbb{R}^n \to \mathbb{R}^n \qquad r = 1(1)m,$$

which are differentiable at x^*. Suppose that for each $x \in G$ there is an $s \in \{1,\ldots,m\}$ such that $T(x) = T_s(x)$. Suppose further that there is a vector norm $\|\cdot\|_T$ for which the corresponding matrix norm $\|\cdot\|_T$ satisfies

$$\|T_r'(x^*)\|_T < 1, \qquad r = 1(1)m.$$

Then x^* is an attractive fixed point.

Proof: Since T and T_r are continuous at x^*, we have, for each $r \in \{1,\ldots,m\}$ the alternative

(1) $T(x^*) = T_r(x^*)$, or

(2) There exists a neighborhood U_r of x^* in which T and T_r never agree.

Since we are only interested in the local behavior of T, we may disregard all r for which statement (2) is true. Therefore, without loss of generality, we may suppose that statement (1) is true for all r.

Since the maps T_r are differentiable at x^*, there exists a $\delta > 0$ for every $\varepsilon > 0$ such that, for all y with $\|y\|_T \leq \delta$ and all $r \in \{1,\ldots,m\}$ it is true that

$$\|T_r(x^*+y)-x^*-T_r'(x^*)y\|_T \leq \varepsilon\|y\|_T.$$

It follows for $r = 1(1)m$ that

$$||T_r(x^*+y)-x^*||_T \le (||T_r'(x^*)||_T+\varepsilon)\ ||y||_T.$$

Now we may choose ε so small that it is true for all r that

$$||T_r'(x^*)||_T + \varepsilon \le \gamma < 1.$$

For every initial vector $x^{(0)}$ satisfying

$$||x^{(0)}-x^*||_T \le \delta$$

it then follows that

$$||x^{(\nu)}-x^*||_T \le \gamma^\nu \cdot \delta, \qquad \nu = 1(1)\infty.$$

Therefore x^* is an attractive fixed point. □

In addition to the previously considered *single step method*

$$x^{(\nu)} = T(x^{(\nu-1)})$$

practical application also make use of *two step methods* (or *multistep methods*)

$$x^{(\nu)} = T(x^{(\nu-1)},x^{(\nu-2)}).$$

These do not lead to any new theory, since one can define a mapping $\tilde{T}\colon \mathbb{R}^{2n} \to \mathbb{R}^{2n}$ by setting

$$\tilde{T}\begin{pmatrix} x_1 \\ x_2 \end{pmatrix} = \begin{pmatrix} T(x_1,x_2) \\ x_1 \end{pmatrix}$$

which results in the single step method

$$\tilde{x}^{(\nu)} = \tilde{T}(\tilde{x}^{(\nu-1)}), \qquad \tilde{x}^{(\nu)} = \begin{pmatrix} x^{(\nu)} \\ x^{(\nu-1)} \end{pmatrix}.$$

$\tilde{T}$ is then significant for convergence questions. Of course this transformation is advisable only for theoretical considerations.

We are now ready to apply the theorems at hand to Newton's method. We start with a lemma to help us along.

Lemma 17.10: Let $F: G \subset \mathbb{R}^n \to \mathbb{R}^n$ be a mapping which has a zero at $x^* \in G$ and is differentiable at x^*. Let J be a mapping from G to $\mathrm{MAT}(n,n,\mathbb{R})$ which is continuous at x^*. Then the mapping

$$T(x) = x - J(x)F(x)$$

is differentiable at x^*, with Jacobian matrix

$$T'(x^*) = I - J(x^*)F'(x^*).$$

Proof: For every $\varepsilon > 0$ there exists a $\delta > 0$ so that for all $y,z \in \mathbb{R}^n$ satisfying $\|y\|_2 < \delta$, it is true that

$$\|F(x^*+y)-F(x^*)-F'(x^*)y\|_2 = \|F(x^*+y)-F'(x^*)y\|_2 \le \varepsilon\|y\|_2$$

$$\|[J(x^*+y) - J(x^*)]z\|_2 \le \varepsilon\|z\|_2.$$

This leads to the inequalities

$$\begin{aligned}
&\|T(x^*+y)-T(x^*)-[I-J(x^*)F'(x^*)]y\|_2 \\
&\quad = \|J(x^*+y)F(x^*+y)-J(x^*)F'(x^*)y\|_2 \\
&\quad \le \|[J(x^*+y)-J(x^*)]F(x^*+y)\|_2 + \|J(x^*)[F(x^*+y)-F'(x^*)y]\|_2 \\
&\quad \le \varepsilon\|F(x^*+y)\|_2 + \|J(x^*)\|_2\cdot\varepsilon\|y\|_2 \le \varepsilon\|y\|_2(\varepsilon+\|F'(x^*)\|_2 \\
&\qquad\qquad + \|J(x^*)\|_2)
\end{aligned}$$ □

Example 17.11: *Newton's method and variations.* The problem is to find a zero x^* of the mapping $F: G \subset \mathbb{R}^n \to \mathbb{R}^n$, where F is continuously differentiable in a neighborhood of x^*

and has a regular Jacobian matrix there. Then the basic fixed point problem underlying Newton's method is:

$$x = T(x) = x - J(x)F(x), \quad \text{where} \quad J(x) = F'(x)^{-1}.$$

By Lemma 17.10, T is differentiable at x^* and has Jacobian

$$T'(x^*) = I-J(x^*)F'(x^*) = I-F'(x^*)^{-1}F'(x^*) = 0.$$

This means that $\rho(T'(x^*)) = 0$. By Theorem 17.8, Newton's method converges for all initial values which lie sufficiently close to x^*.

Theorem 17.8 and Lemma 17.10 also establish that the fixed point x^* remains attractive when $J(x)$ is not the inverse of the Jacobian, but is merely an approximation thereto, since local convergence only demands

$$\rho(T'(x^*)) = \rho(I-J(x^*)F'(x^*)) < 1.$$

This is of considerable practical significance, since frequently considerable effort would be required to determine the Jacobian and its inverse exactly. It is also noteworthy that, by Lemma 17.10, it is not necessary for J itself to be differentiable. It suffices to have F differentiable at x^*. The following computation establishes how far $J(x)$ may deviate from the inverse of the Jacobian. Thus we let C be a perturbation matrix and let $J(x) = C[F'(x)]^{-1}$. Then by Lemma 17.10 we have

$$T'(x^*) = I-J(x^*)F'(x^*) = I-C[F'(x^*)]^{-1}F'(x^*) = I-C.$$

By Theorem 17.8, the iteration converges locally for $\rho(I-C) < 1$. For the special case $C = \lambda I$, we have convergence for $\lambda \in (0,2)$. □

The following two theorems will give a more precise concept of the attractive regions for Newton's method and for a simplified method. We suppose we are given the following situation:

$$G \subset \mathbb{R}^n \quad \text{convex}, \quad x^{(0)} \in G$$

$$K = \{x \in \mathbb{R}^n \mid \|x-x^{(0)}\| < r_0\}, \quad \overline{K} \subset G$$

$$F \in C^1(G, \mathbb{R}^n), \quad A = F'(x^{(0)}) \quad \text{regular}$$

$$\beta = \|A^{-1}\|, \quad \eta = \|A^{-1}F(x^{(0)})\|.$$

Theorem 17.12: *Newton-Kantorovich.*

Hypotheses:

(a) $\|F'(x)-F'(y)\| \le \gamma\|x-y\|, \qquad x,y \in G$

(b) $0 < \alpha = \beta\gamma\eta \le 1/2$

(c) $r_1 = 2\eta/(1+\sqrt{1-2\alpha}) \le r_0$.

Conclusions:

(1) The sequence

$$x^{(\nu+1)} = x^{(\nu)} - F'(x^{(\nu)})^{-1}F(x^{(\nu)}), \qquad \nu = 0(1)\infty$$

remains in K and converges to $x^* \in \overline{K}$.

(2) $\|x^{(\nu)}-x^*\| \le r_1 2^{-\nu}(\frac{\alpha}{1-\alpha})^{(2^\nu-1)}, \qquad \nu = 0(1)\infty.$

(3) x^* is the only zero of F in

$$K_2 = G \cap \{x \in \mathbb{R}^n \mid \|x-x^{(0)}\| < r_2 = (1+\sqrt{1-2\alpha})/(\beta\gamma)\}.$$

If $\alpha \ll 1/2$, the sequence converges very quickly, by (2). In a practical application of the method, after a few steps there will be only random changes per step. These arise because of inevitable rounding errors. The theorem permits an estimate on the error $\|x^* - x^{(\nu)}\|$. For this one takes

$x^{(\nu)}$ as the initial value $x^{(0)}$ for a new iteration, and computes upper bounds for γ, β, η and α. For $\alpha \leq 1/2$, the error $\|x^* - x^{(0)}\|$ is at most $r_1 = 2\eta/(1 + \sqrt{1-2\alpha})$.

For $\alpha = 1/2$, it is possible that convergence is only linear. The following example in $\mathbb{R}^1$ shows that this case can actually occur:

$$f(x) = -\frac{\eta}{\beta} + \frac{x}{\beta} - \frac{\gamma x^2}{2}, \quad \eta > 0,\ \beta > 0, \quad \gamma > 0$$

$$x^{(0)} = 0.$$

We have

$$|f'(x)-f'(y)| = \gamma|x-y|$$

$$1/|f'(0)| = \beta$$

$$|f(0)|/|f'(0)| = \eta.$$

For $\alpha < 1/2$, f has two different real zeros, $r_1 = 2\eta/(1+\sqrt{1-2\alpha})$ and $r_2 = 2\eta/(1-\sqrt{1-2\alpha})$. When $\alpha = 1/2$, they become the same. When $\alpha > 1/2$, f has no real zeros (see Figure 17.13). The example is so chosen that convergence of Newton's method is worse for no other f. The proof of Theorem 17.12 is grounded on this idea.

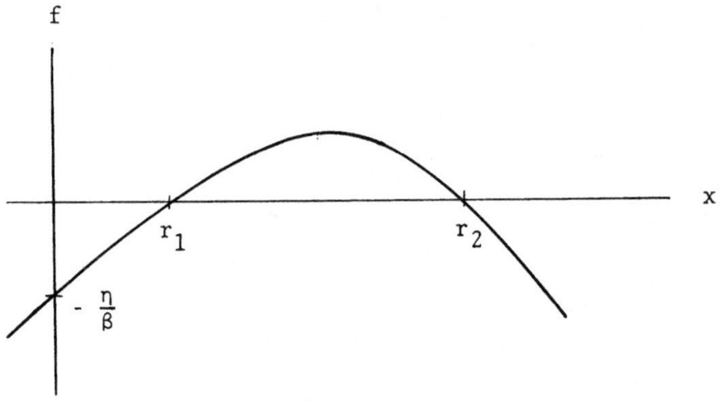

Figure 17.13. Typical graph for $\alpha < 1/2$.

$F'(x^{(\nu)})$ is always regular. However, for $\alpha = 1/2$, $F'(x^*)$ can be singular (as in our example).

We will use the following three lemmas in the proof of Theorem 17.12. In addition to assuming all the hypotheses of Theorem 17.12, we also make the definitions:

$$A_\nu = F'(x^{(\nu)}), \quad \beta_\nu = \|A_\nu^{-1}\|, \quad \eta_\nu = \|A_\nu^{-1}F(x^{(\nu)})\|,$$

$$\alpha_\nu = \beta_\nu\eta_\nu\gamma, \quad \rho_\nu = 2\eta_\nu/(1+\sqrt{1-2\alpha_\nu}).$$

Naturally these definitions only make sense if we also assume $x^{(\nu)} \in G$, A_ν is regular, and $\alpha_\nu \leq 1/2$. Therefore we will restrict ourselves temporarily to the set M of integers $\nu \geq 0$ for which these hypotheses are true. At this point it is not at all clear that there are any integers besides 0 which belong to M. However, it will later turn out that M contains all the positive integers.

<u>Lemma 17.14</u>: If $x^{(\nu+1)} \in G$, then $A_{\nu+1}$ is regular and $\beta_{\nu+1} \leq \beta_\nu/(1-\alpha_\nu)$.

Proof: Since

$$\|A_\nu - A_{\nu+1}\| \leq \gamma\|x^{(\nu)} - x^{(\nu+1)}\| = \gamma\eta_\nu$$

we have

$$\|A_\nu^{-1}(A_\nu - A_{\nu+1})\| \leq \alpha_\nu \leq 1/2.$$

Therefore, we have convergence for the series

$$S = \sum_{\mu=0}^{\infty} [A_\nu^{-1}(A_\nu - A_{\nu+1})]^\mu.$$

We have

$$[I - A_\nu^{-1}(A_\nu - A_{\nu+1})]S = I, \quad S[I - A_\nu^{-1}(A_\nu - A_{\nu+1})] = I.$$

The matrix inside the square brackets is therefore regular, and its inverse is S. But then

$$A_{\nu+1} = A_\nu[I-A_\nu^{-1}(A_\nu-A_{\nu+1})]$$

is also regular. For the norm of the inverses we have the inequalities

$$\beta_{\nu+1} \leq ||S|| \cdot ||A_\nu^{-1}|| \leq \sum_{\mu=0}^{\infty} \alpha_\nu^\mu \beta_\nu = \beta_\nu/(1-\alpha_\nu). \qquad \square$$

<u>Lemma 17.15</u>: If $x^{(\nu+1)} \in G$, then $\eta_{\nu+1} \leq \frac{1}{2}\eta_\nu\alpha_\nu/(1-\alpha_\nu)$.

Proof: We have shown above that $A_{\nu+1}^{-1} = SA_\nu^{-1}$ and

$$\eta_{\nu+1} = ||A_{\nu+1}^{-1}F(x^{(\nu+1)})|| \leq ||S|| \cdot ||A_\nu^{-1}F(x^{(\nu+1)})||$$

$$\leq ||A_\nu^{-1}F(x^{(\nu+1)})||/(1-\alpha_\nu).$$

It remains to show that

$$||A_\nu^{-1}F(x^{(\nu+1)})|| \leq \frac{1}{2}\,\alpha_\nu\eta_\nu.$$

For $t \in [0,1]$, we set

$$\phi(t) = (1-t)x^{(\nu)}+tx^{(\nu+1)},$$

$$R(t) = A_\nu^{-1}F(\phi(t)) + (1-t)(x^{(\nu+1)}-x^{(\nu)}).$$

Since G is convex, $\phi(t)$ remains in G. We clearly obtain the following:

$$\phi(0) = x^{(\nu)}, \qquad \phi(1) = x^{(\nu+1)},$$

$$R(0) = 0, \qquad R(1) = A_\nu^{-1}F(x^{(\nu+1)}),$$

$$\phi'(t) = x^{(\nu+1)}-x^{(\nu)},$$

$$R'(t) = [A_\nu^{-1}F'(\phi(t))-I](x^{(\nu+1)}-x^{(\nu)}), \qquad R'(0) = 0.$$

Since

$$\|F'(\phi(t))-A_\nu\| = \|F'(\phi(t))-F'(\phi(0))\|$$

$$\|F'(\phi(t))-A_\nu\| \leq \gamma\|\phi(t)-x^{(\nu)}\| = \gamma t\|x^{(\nu+1)}-x^{(\nu)}\|$$

we obtain the following estimate for $\|R'(t)\|$:

$$\|R'(t)\| \leq \beta_\nu \gamma t\|x^{(\nu+1)}-x^{(\nu)}\|^2 = \beta_\nu\gamma\eta_\nu^2 t = \alpha_\nu\eta_\nu t.$$

From this we obtain the desired inequality,

$$\|A_\nu^{-1}F(x^{(\nu+1)})\| = \|R(1)\| = \|\int_0^1 R'(t)dt\| \leq \frac{1}{2}\alpha_\nu\eta_\nu. \qquad \square$$

When $x^{(\nu+1)} \in G$, we set

$$\hat{\beta} = \beta_\nu/(1-\alpha_\nu) \geq \beta_{\nu+1} \qquad \text{(cf. Lemma 17.14)}$$

$$\hat{\alpha} = \hat{\beta}\gamma\eta_{\nu+1} \geq \alpha_{\nu+1}.$$

Then the last lemma implies that

$$\alpha_{\nu+1} \leq \hat{\alpha} \leq \frac{1}{2}\frac{\alpha_\nu^2}{(1-\alpha_\nu)^2} \leq \alpha_\nu \leq 1/2.$$

It follows that if $\nu \in M$ and $x^{(\nu+1)} \in G$, then $\nu+1 \in M$.

<u>Lemma 17.16</u>: If $x^{(\nu+1)} \in G$, then $\eta_\nu + \rho_{\nu+1} \leq \rho_\nu$.

Proof: Let

$$\hat{\rho} = 2\eta_{\nu+1}/(1+\sqrt{1-2\hat{\alpha}}) = \frac{1-\sqrt{1-2\hat{\alpha}}}{\hat{\beta}\gamma} \geq \rho_{\nu+1}.$$

From the inequality

$$\hat{\alpha} \leq \frac{1}{2}\frac{\alpha_\nu^2}{(1-\alpha_\nu)^2}$$

it follows that

$$1-\sqrt{1-2\hat{\alpha}} \leq \frac{1-\sqrt{1-2\alpha_\nu}}{1-\alpha_\nu} - \frac{\alpha_\nu}{1-\alpha_\nu}.$$

Since $\hat{\beta}(1-\alpha_\nu) = \beta_\nu$, it further follows that

$$\frac{1-\sqrt{1-2\hat{\alpha}}}{\hat{\beta}\gamma} \le \frac{1-\sqrt{1-2\alpha_\nu}}{\beta_\nu\gamma} - \frac{\alpha_\nu}{\beta_\nu\gamma}$$

$$\rho_{\nu+1} \le \hat{\rho} \le \rho_\nu - \eta_\nu. \quad \square$$

Proof of Theorem 17.12: If $\mu \le \nu$ and $\nu \in M$, then by Lemma 17.16,

$$\sum_{\tau=\mu}^{\nu} \eta_\tau + \rho_{\nu+1} \le \rho_\mu.$$

Since $\rho_0 = r_1 \le r_0$, we also have

$$\sum_{\tau=0}^{\nu} \eta_\tau + \rho_{\nu+1} \le r_1 \le r_0$$

$$\|x^{(\nu+1)}-x^{(0)}\| \le \sum_{\tau=0}^{\nu} \|x^{(\tau+1)}-x^{(\tau)}\| \le r_1-\rho_{\nu+1}.$$

It follows that $x^{(\nu+1)} \in \overline{K} \subset G$ and, from Lemmas 17.14 and 17.15, that $A_{\nu+1}$ is regular and $\alpha_{\nu+1} \le \alpha_\nu$. Altogether, this implies that $\nu+1 \in M$. Thus the set M contains all integers $\nu \ge 0$.

Next we show that $x^{(\nu+1)} \in K$ for $\nu = 0(1)\infty$. Since

$$\|x^{(\nu+1)}-x^{(0)}\| \le r_1 - \rho_{\nu+1}$$

it follows that either all $x^{(\nu+1)} \in K$, or there is a first ν such that $\rho_{\nu+1} = 0$, i.e. $\eta_{\nu+1} = 0$. This implies

$$x^{(\nu+1)} = x^{(\nu+2)} = x^{(\nu+3)} = \dots$$

Since $\eta_\nu \ne 0$ and $\alpha_\nu \ne 0$, we also have

$$\eta_\nu + \rho_{\nu+1} = \eta_\nu < 2\eta_\nu/(1+\sqrt{1-2\alpha_\nu})$$

$$\eta_\nu + \rho_{\nu+1} < \rho_\nu$$

$$\sum_{\tau=0}^{\nu} \eta_\tau + \rho_{\nu+1} < \rho_0 = r_1 \le r_0$$

$$\|x^{(\nu+1)} - x^{(0)}\| < r_1$$

$$x^{(\nu+1)} \in K.$$

Hence in this case, too, all $x^{(\nu)} \in K$.

Next we establish the error estimate (2). We begin with the inequality

$$\alpha_{\nu+1} \le \frac{1}{2} \frac{\alpha_\nu^2}{(1-\alpha_\nu)^2}$$

It implies that

$$1 - \alpha_{\nu+1} \ge \frac{1}{2} \frac{2-4\alpha_\nu+\alpha_\nu^2}{(1-\alpha_\nu)^2}$$

$$\frac{\alpha_{\nu+1}}{1-\alpha_{\nu+1}} \le \frac{\alpha_\nu^2}{2-4\alpha_\nu+\alpha_\nu^2} = \frac{\alpha_\nu^2}{1-2\alpha_\nu+(1-\alpha_\nu)^2}$$

$$\frac{\alpha_{\nu+1}}{1-\alpha_{\nu+1}} \le \frac{\alpha_\nu^2}{(1-\alpha_\nu)^2} \le \frac{\alpha_{\nu-1}^4}{(1-\alpha_{\nu-1})^4} \le \cdots$$

$$\frac{\alpha_\nu}{1-\alpha_\nu} \le \left(\frac{\alpha_0}{1-\alpha_0}\right)^{(2^\nu)} = \left(\frac{\alpha}{1-\alpha}\right)^{(2^\nu)} .$$

From $\eta_{\nu+1} \le \frac{1}{2} \eta_\nu \alpha_\nu/(1-\alpha_\nu)$ (cf. Lemma 17.15) we finally obtain

$$\eta_{\nu+1} \le \frac{1}{2}\left(\frac{\alpha}{1-\alpha}\right)^{(2^\nu)} \eta_\nu \le \frac{1}{4}\left(\frac{\alpha}{1-\alpha}\right)^{(2^\nu+2^{\nu-1})} \eta_{\nu-1} \le \cdots$$

$$\eta_\nu \le 2^{-\nu}\left(\frac{\alpha}{1-\alpha}\right)^{(2^\nu-1)} \eta_o$$

$$\rho_\nu = 2\eta_\nu/(1+\sqrt{1-2\alpha_\nu}) \le 2\eta_\nu/(1+\sqrt{1-2\alpha})$$

$$\rho_\nu \le r_1 2^{-\nu}\left(\frac{\alpha}{1-\alpha}\right)^{(2^\nu-1)} .$$

The sequence $x^{(\nu)}$ therefore converges to $x^* \in \overline{K}$. $||x^{(\nu)}-x^*||$ can be estimated by conclusion (2). Since F and F' are continuous, it also follows that

$$F(x^*) = -F'(x^*)(x^* - x^*) = 0.$$

For the proof of conclusion (3) we refer to Ortega-Rheinboldt (1970). □

The Jacobian $F'(x^{(\nu)})$ must be computed anew at each step of Newton's method. Frequently this computation involves considerable effort. Therefore it may be advantageous to replace $F'(x^{(\nu)})$ by a fixed, nonsingular matrix B which does not differ too greatly from $F'(x^{(\nu)})$. This procedure is especially advantageous whenever B has a simpler algebraic structure than the Jacobians. It is conceivable that linear systems of equations involving B may be solved with the aid of special direct methods (e.g. the reduction method of Schröder/Trottenberg or the Buneman algorithm), while the corresponding systems involving the Jacobians $F'(x^{(\nu)})$ are too complicated for such an approach.

We describe this situation briefly:

$$G \subset \mathbb{R}^n \quad \text{convex}, \quad x^{(0)} \in G$$

$$K = \{x \in \mathbb{R}^n \mid \|x - x^{(0)}\| < r_0\}, \quad \bar{K} \subset G$$

$$F \in C^1(G, \mathbb{R}^n), \; B \in \mathrm{MAT}(n,n,\mathbb{R}) \quad \text{regular}$$

$$\beta = \|B^{-1}\|, \quad \eta = \|B^{-1}F(x^{(0)})\|$$

$$\delta = \|B - F'(x^{(0)})\| .$$

<u>Theorem 17.17</u>:

Hypotheses:

(a) $\|F'(x) - F'(y)\| \leq \gamma\|x-y\|, \quad x,y \in G$

(b) $\beta\delta < 1$

(c) $\alpha = \beta\eta\gamma/(1-\beta\delta)^2 \leq 1/2$

(d) $r_1 = 2\eta/[(1+\sqrt{1-2\alpha})(1-\beta\delta)] \leq r_0$

Conclusions:

(1) The sequence

$$x^{(\nu+1)} = x^{(\nu)} - B^{-1}F(x^{(\nu)}), \quad \nu = 0(1)\infty$$

remains in $\overline{K}$ and converges to $x^* \varepsilon K$.

(2) It is true that

$$||x^{(\nu+2)}-x^{(\nu+1)}|| \leq c||x^{(\nu+1)}-x^{(\nu)}||, \quad \nu = 0(1)\infty$$

where

$$c = \beta\delta + \frac{2}{1+\sqrt{1-2\alpha}}(1-\beta\delta) \leq 1.$$

The theorem contains two interesting special cases:

(A) F is an affine map, i.e. $\gamma = \alpha = 0$, $r_1 = \eta/(1-\beta\delta)$ and $c = \beta\delta$.

(B) $B = F'(x^{(0)})$, $\delta = 0$, $\gamma > 0$. The conditions (a), (c), and (d) are then precisely the hypotheses of the preceding theorem. Our convergence conditions for the simplified method thus are the same conditions as for Newton's method.

Conclusion (2) of the theorem is of interest first of all for large ν. For the first few iterations, there are better estimates. Thus, we have

$$||x^{(2)}-x^{(1)}|| \leq \tilde{c}||x^{(1)}-x^{(0)}||$$

where $\tilde{c} = \beta\delta + \frac{1}{2}\alpha(1-\beta\delta)^2$.

For $\alpha = 1/2$, we have $c = 1$, independently of δ. In fact in such cases the convergence of the method is almost arbitrarily bad. This can be seen with an example from $\mathbb{R}^1$. Let

$$f(x) = x^2, \quad x^{(0)} = 1, \quad B = f'(1) = 2.$$

The constants in this example are:

$$\beta = 1/2, \quad \eta = 1/2, \quad \gamma = 2, \quad \delta = 0, \quad \alpha = 1/2.$$

This leads to the sequence

$$x^{(\nu+1)} = x^{(\nu)} - \tfrac{1}{2}(x^{(\nu)})^2 = x^{(\nu)}(1 - \tfrac{1}{2}x^{(\nu)}).$$

which converges to zero very slowly.

In practice one can apply the method when $\beta\delta << 1$ and $\alpha << 1/2$. In these cases, $\tilde{c} \approx \beta\delta + \alpha/2$ and $c \approx \beta\delta + \alpha$. Table 17.18 shows the effect of larger α or larger $\beta\delta$.

TABLE 17.18

α	$\beta\delta$	$\tilde{c}$	$\beta\delta+\frac{1}{2}\alpha$	c	$\beta\delta+\alpha$
1/4	1/2	0.531	0.625	0.646	0.750
1/4	1/4	0.320	0.375	0.470	0.500
1/4	0	0.125	0.125	0.293	0.250
1/8	1/2	0.516	0.563	0.567	0.625
1/8	1/4	0.285	0.313	0.350	0.375
1/8	0	0.063	0.063	0.134	0.125
1/16	1/2	0.508	0.531	0.532	0.563
1/16	1/4	0.268	0.281	0.298	0.313
1/16	0	0.031	0.031	0.065	0.063

The proof of Theorem 17.17 runs on a course roughly parallel to that of the proof of Theorem 17.12. Once again, it is based on three lemmas. We make the following definitions:

$$\eta_\nu = ||B^{-1}F(x^{(\nu)})||, \quad \delta_\nu = ||B - F'(x^{(\nu)})||,$$

$$\alpha_\nu = \beta\gamma\eta_\nu/(1-\beta\delta_\nu)^2, \quad \rho_\nu = 2\eta_\nu/[(1+\sqrt{1-2\alpha_\nu})(1-\beta\delta_\nu)].$$

ν runs through the set M of all nonnegative integers for which it is true that $x^{(\nu)} \in G$, $\beta\delta_\nu < 1$, and $\alpha_\nu \leq 1/2$.

<u>Lemma 17.19</u>: Let $x^{(\nu+1)} \in G$ and $\hat{\delta} = [\beta\delta_\nu + \alpha_\nu(1-\beta\delta_\nu)^2]/\beta$. Then

$$\beta\delta_{\nu+1} \le \beta\hat{\delta} < 1$$
$$1-\beta\delta_{\nu+1} \ge 1-\beta\hat{\delta} = [1-\alpha_\nu(1-\beta\delta_\nu)](1-\beta\delta_\nu) > 0.$$

Proof: We have

$$\beta\delta_{\nu+1} = \beta\|B-F'(x^{(\nu+1)})\| \le \beta\delta_\nu+\beta\|F'(x^{(\nu+1)})-F'(x^{(\nu)})\|$$

$$\beta\delta_{\nu+1} \le \beta\delta_\nu+\beta\gamma\|x^{(\nu+1)}-x^{(\nu)}\| = \beta\delta_\nu+\beta\gamma\eta_\nu$$

$$\beta\delta_{\nu+1} \le \beta\delta_\nu+\alpha_\nu(1-\beta\delta_\nu)^2 = \beta\hat{\delta}$$

$$= \tfrac{1}{2}(1+\beta^2\delta_\nu^2)-(\tfrac{1}{2}-\alpha_\nu)(1-\beta\delta_\nu)^2 \le \tfrac{1}{2}(1+\beta^2\delta_\nu^2) < 1.$$

From this it follows that

$$1-\beta\delta_{\nu+1} \ge 1-\beta\hat{\delta} = [1-\alpha_\nu(1-\beta\delta_\nu)](1-\beta\delta_\nu). \qquad \square$$

<u>Lemma 17.20</u>: If $x^{(\nu+1)} \in G$, then
$\eta_{\nu+1} \le [\beta\delta_\nu + \frac{1}{2}\alpha_\nu(1-\beta\delta_\nu)^2]\eta_\nu$.

Proof: As in the proof of Lemma 17.15, for $t \in [0,1]$ we set

$$\phi(t) = (1-t)x^{(\nu)} + tx^{(\nu+1)}$$

$$R(t) = B^{-1}F(\phi(t)) + (1-t)(x^{(\nu+1)}-x^{(\nu)}).$$

It follows that

$$\phi(0) = x^{(\nu)}, \quad \phi(1) = x^{(\nu+1)},$$

$$R(0) = 0, \qquad R(1) = B^{-1}F(x^{(\nu+1)}),$$

$$\phi'(t) = x^{(\nu+1)} - x^{(\nu)}$$

$$R'(t) = B^{-1}[F'(\phi(t))-B](x^{(\nu+1)}-x^{(\nu)}).$$

By hypothesis (a) of Theorem 17.17 it follows that

$$\|F'(\phi(t))-B\| \leq \|F'(x^{(\nu)})-B\| + \gamma\|\phi(t)-x^{(\nu)}\|$$

$$\|F'(\phi(t))-B\| \leq \delta_\nu + \gamma\eta_\nu t.$$

Therefore we have

$$\|R'(t)\| \leq \beta(\delta_\nu + \gamma\eta_\nu t)\eta_\nu$$

and finally

$$\eta_{\nu+1} = \|R(1)\| = \left\|\int_0^1 R'(t)dt\right\|$$

$$\eta_{\nu+1} \leq (\beta\delta_\nu + \tfrac{1}{2}\beta\gamma\eta_\nu)\eta_\nu = [\beta\delta_\nu + \tfrac{1}{2}\alpha_\nu(1-\beta\delta_\nu)^2]\eta_\nu. \qquad \square$$

With

$$\hat{\alpha} = \beta\gamma\eta_{\nu+1}/(1-\beta\hat{\delta})^2$$

$$\hat{\alpha} \geq \alpha_{\nu+1}$$

the last two lemmas yield

$$\hat{\alpha} \leq \frac{\beta\gamma\eta_\nu[\beta\delta_\nu + \frac{1}{2}\alpha_\nu(1-\beta\delta_\nu)^2]}{[1-\alpha_\nu(1-\beta\delta_\nu)]^2(1-\beta\delta_\nu)^2}.$$

Since

$$\begin{aligned}[1-\alpha_\nu(1-\beta\delta_\nu)]^2 &= 1-2\alpha_\nu(1-\beta\delta_\nu)[1-\tfrac{1}{2}\alpha_\nu(1-\beta\delta_\nu)] \\ &\geq 1-(1-\beta\delta_\nu)[1-\tfrac{1}{2}\alpha_\nu(1-\beta\delta_\nu)] \\ &= \beta\delta_\nu + \tfrac{1}{2}\alpha_\nu(1-\beta\delta_\nu)^2\end{aligned}$$

we have

$$\alpha_{\nu+1} \leq \hat{\alpha} \leq \alpha_\nu \leq 1/2.$$

It follows that when $\nu \in M$ and $x^{(\nu+1)} \in G$, then also $\nu+1 \in M$.

<u>Lemma 17.21</u>: If $x^{(\nu+1)} \in G$, then $\eta_\nu + \rho_{\nu+1} \leq \rho_\nu$.

Proof:

Case 1: $\eta_\nu = 0$. Then $x^{(\nu+1)} = x^{(\nu)}$ and $\rho_{\nu+1} = \rho_\nu = 0$.

Case 2: $\gamma = 0$. F is an affine map. Therefore:

$$\alpha_\nu = \alpha_{\nu+1} = 0, \quad \delta_\nu = \delta_{\nu+1}, \quad \eta_{\nu+1} \le \beta\delta_\nu\eta_\nu,$$

$$\rho_\nu = \eta_\nu/(1-\beta\delta_\nu), \quad \rho_{\nu+1} = \eta_{\nu+1}/(1-\beta\delta_{\nu+1}) \le \nu\delta_\nu\rho_\nu$$

$$\eta_\nu + \rho_{\nu+1} \le \eta_\nu + \beta\delta_\nu\rho_\nu = \rho_\nu.$$

Case 3: $\alpha_\nu = \beta\gamma\eta_\nu/(1-\beta\delta_\nu)^2 > 0$. Let

$$\hat\rho = 2\eta_{\nu+1}/[(1+\sqrt{1-2\hat\alpha})(1-\beta\hat\delta)] = \frac{(1-\sqrt{1-2\hat\alpha})(1-\beta\hat\delta)}{\beta\gamma}$$

Then $\hat\rho \ge \rho_{\nu+1}$. Lemma 17.20 implies that

$$\eta_{\nu+1} \le [\beta\delta_\nu + \tfrac{1}{2}\alpha_\nu(1-\beta\delta_\nu)^2]\eta_\nu.$$

Multiplying by $\beta\gamma$ yields

$$\hat\alpha(1-\beta\hat\delta)^2 \le [\beta\delta_\nu + \tfrac{1}{2}\alpha_\nu(1-\beta\delta_\nu)^2]\alpha_\nu(1-\beta\delta_\nu)^2.$$

We have

$$1-\beta\hat\delta = [1-\alpha_\nu(1-\beta\delta_\nu)](1-\beta\delta_\nu)$$

and therefore,

$$\hat\alpha[1-\alpha_\nu(1-\beta\delta_\nu)]^2 \le [\beta\delta_\nu + \tfrac{1}{2}\alpha_\nu(1-\beta\delta_\nu)^2]\alpha_\nu$$

$$2\hat\alpha \le \frac{2\beta\delta_\nu+\alpha_\nu(1-\beta\delta_\nu)^2}{[1-\alpha_\nu(1-\beta\delta_\nu)]^2}\,\alpha_\nu$$

$$1-\sqrt{1-2\hat\alpha} \le \frac{1-\alpha_\nu(1-\beta\delta_\nu)-\sqrt{1-2\alpha_\nu}}{1-\alpha_\nu(1-\beta\delta_\nu)}.$$

Multiplying by

$$\frac{1-\beta\hat\delta}{\beta\gamma} = \frac{1-\alpha_\nu(1-\beta\delta_\nu)}{\beta\gamma}(1-\beta\delta_\nu)$$

yields

$$\frac{(1-\sqrt{1-2\hat\alpha})(1-\beta\hat\delta)}{\beta\gamma} \le \frac{1-\alpha_\nu(1-\beta\delta_\nu)-\sqrt{1-2\alpha_\nu}}{\beta\gamma}(1-\beta\delta_\nu).$$

The left side is a different representation of $\hat{\rho}$. Therefore, we have shown that

$$\rho_{\nu+1} \leq \hat{\rho} \leq \frac{1-\alpha_\nu(1-\beta\delta_\nu)-\sqrt{1-2\alpha_\nu}}{\beta\gamma}(1-\beta\delta_\nu).$$

Now the right side of the inequality is equal to $\rho_\nu - \eta_\mu$. □

Proof of Theorem 17.17: If $\nu \in M$, then $\eta_\nu \leq \rho_\nu$ and by Lemma 17.21,

$$\sum_{\mu=0}^{\nu-1} \eta_\mu + \rho_\nu \leq \rho_0 = r_1 \leq r_0.$$

This implies that

$$\|x^{(\nu+1)}-x^{(0)}\| \leq r_0, \quad x^{(\nu+1)} \in \overline{K}.$$

The lemmas also imply that $\beta\delta_{\nu+1} < 1$, $\alpha_{\nu+1} \leq 1/2$. M therefore contains all $\nu \geq 0$. $x^{(\nu)}$ remains in $\overline{K}$. The sequence $\{x^{(\nu)} \mid \nu = 0(1)\infty\}$ has at most one limit point $x^* \in \overline{K}$, since

$$\sum_{\mu=0}^{\nu} \|x^{(\mu+1)}-x^{(\mu)}\| = \sum_{\mu=0}^{\nu} \eta_\mu, \quad \nu = 0(1)\infty$$

is bounded above by r_0. It remains to prove the error estimate (2). By Lemma 17.20, we have

$$\eta_{\nu+1} \leq [\beta\delta_\nu + \tfrac{1}{2}\alpha_\nu(1-\beta\delta_\nu)^2]\eta_\nu$$

or

$$\|x^{(\nu+2)}-x^{(\nu+1)}\| \leq [\beta\delta_\nu + \tfrac{1}{2}\alpha_\nu(1-\beta\delta_\nu)^2]\|x^{(\nu+1)}-x^{(\nu)}\|.$$

We can get a bound on $\beta\delta_\nu$ with the aid of the Lipschitz condition for F':

$$\beta\delta_\nu \leq \beta\delta_0+\beta\|F'(x^{(\nu)})-F'(x^{(0)})\| \leq \beta\delta_0+\beta\gamma\|x^{(\nu)}-x^{(0)}\|.$$

It follows from

that

$$\sum_{\mu=0}^{\nu-1} \eta_\mu+\rho_\nu = \sum_{\mu=0}^{\nu-1} \|x^{(\mu+1)}-x^{(\mu)}\| + \rho_\nu \le r_1$$

$$\beta\delta_\nu \le \beta\delta_o + \beta\gamma r_1-\beta\gamma\rho_\nu \le \beta\delta_o + \beta\gamma r_1-\beta\gamma\eta_\nu$$

$$\beta\delta_\nu \le \beta\delta + 2\beta\gamma\eta/[(1+\sqrt{1-2\alpha})(1-\beta\delta)]-\alpha_\nu(1-\beta\delta_\nu)^2$$

$$\beta\delta_\nu + \frac{1}{2}\alpha_\nu(1-\beta\delta_\nu)^2 \le \beta\delta + \frac{2\alpha}{1+\sqrt{1-2\alpha}}(1-\beta\delta) = c. \qquad \square$$

We follow this discussion of Newton's method with a definition of generalized single step methods. The starting point is an arbitrary iterative method

$$x^{(\nu)} = T(x^{(\nu-1)}).$$

In an actual computation, the components $x_i^{(\nu)}$, $i = 1(1)n$, of $x^{(\nu)}$ are computed sequentially from the components of $x^{(\nu-1)}$. Therefore it is advantageous, in looking at the right side of the equation, to use those components of $x^{(\nu)}$ which are already known, instead of the corresponding components of $x^{(\nu-1)}$. In practice this means that the components of $x^{(\nu)}$ immediately replace those of $x^{(\nu-1)}$ in memory. This not only saves memory, but simplifies the program. In many important cases (see Varga, 1962, Ch. 3), the new method converges better than the original method. The new method is called a *single step method* in contrast to the original *total step method*. By defining a modified operator $\hat{T}$, one can again regard a single step method as a total step method.

Before defining the operator $\hat{T}$, we will look at an important special case.

Example 17.22: Let

$$T(x) = (L+U)x + b$$

where $b \in \mathbb{R}^n$, L is a strictly lower $n \times n$ triangular matrix (diagonal identically zero), and U is a strictly upper $n \times n$ triangular matrix (diagonal identically zero). The corresponding total step method

$$x^{(\nu)} = T(x^{(\nu-1)}) = (L+U)x^{(\nu-1)} + b$$

is also known as the *Jacobi* method, and by Theorem 17.7, converges for $\rho(L+U) < 1$. In view of our discussion above, the single step method may be characterized by the rule

$$x^{(\nu)} = Lx^{(\nu)} + Ux^{(\nu-1)} + b.$$

This is also called the *Gauss-Seidel* method. We can again rewrite it formally as a total step method:

$$x^{(\nu)} = (I-L)^{-1}(Ux^{(\nu-1)}+b).$$

By Theorem 17.7, the method converges for $\rho((I-L)^{-1}U) < 1$.

In the following formulation the similarity between the two methods will become clearer.

Jacobi method:

$$x^{(\nu)} = x^{(\nu-1)} - [(I-L-U)x^{(\nu-1)}-b].$$

Gauss-Seidel method:

$$x^{(\nu)} = x^{(\nu-1)} - (I-L)^{-1}[(I-L-U)x^{(\nu-1)}-b]. \qquad \square$$

Definition 17.23: Let $T\colon G \subset \mathbb{R}^n \to \mathbb{R}^n$ be the fixed point operator of some total step method, with component mappings

$$t_i(y_1, y_2, \ldots, y_n), \qquad i = 1(1)n.$$

We define the components $\hat{t}_i$ of a mapping $\hat{T}: G \subset \mathbb{R}^n \to \mathbb{R}^n$ recursively by the rule

$$\hat{t}_i(y_1, y_2, \ldots, y_n) = t_i(w_1, w_2, \ldots, w_n)$$

$$w_j = \begin{cases} \hat{t}_j(y_1, y_2, \ldots, y_n) & \text{for } j < i \\ y_j & \text{otherwise.} \end{cases}$$

Then $x^{(\nu)} = \hat{T}(x^{(\nu-1)})$, $\nu = 1(1)\infty$ defines the *single step method* corresponding to T. We frequently use the notation

$$x^{(\nu)} = T(x^{(\nu-1)}/x^{(\nu)}).$$

This is to be interpreted as saying that $x^{(\nu)}$ is to be computed with the aid of the mapping T from $x^{(\nu-1)}$; however, insofar as they are already known, the components of $x^{(\nu)}$ are to be used in the computation. □

The following theorem focuses on some significant connections between total and single step methods.

<u>Theorem 17.24</u>: Let T and $\hat{T}$ be as in Definition 17.23. Let x^* be a fixed point of T. Then the following are true:

(1) x^* is a fixed point of $\hat{T}$.

(2) If T is continuous at x^*, then $\hat{T}$ is also continuous at x^*.

(3) If T is differentiable at x^*, then $\hat{T}$ is also differentiable at x^*. Let the Jacobian of T at x^* be partitioned as follows:

$$T'(x^*) = D - R - S$$

where $D = (d_{ij})$ is a diagonal matrix, $R = (r_{ij})$ is a strictly lower triangular matrix, and $S = (s_{ij})$ is a strictly upper triangular matrix. Then the Jacobian of $\hat{T}$

at x^* can be decomposed as follows:

$$\hat{T}'(x^*) = (I+R)^{-1}(D-S).$$

Proof: Conclusions (1) and (2) follow immediately from the definition of $\hat{T}$. By the recursive definition of $\hat{t}_i$, we have that at the fixed point,

$$\partial_j\hat{t}_i(x^*) = \sum_{\mu=1}^{i-1} \partial_\mu t_i(x^*)\partial_j\hat{t}_\mu(x^*), \qquad j < i$$

$$\partial_j\hat{t}_i(x^*) = \sum_{\mu=1}^{i-1} \partial_\mu t_i(x^*)\partial_j\hat{t}_\mu(x^*) + \partial_j t_i(x^*), \quad j \geq i.$$

In both cases, this means that

$$\partial_j\hat{t}_i(x^*) = -\sum_{\mu=1}^{i-1} r_{i\mu}\partial_j\hat{t}_\mu(x^*) + d_{ij}-s_{ij}, \qquad i,j = 1(1)n.$$

It follows from this that

$$\hat{T}'(x^*) = -R\hat{T}'(x^*) + D - S$$
$$\hat{T}'(x^*) = (I+R)^{-1}(D-S). \quad \square$$

The method considered in Example 17.11 had the form $T(x) = x - J(x)F(x)$. The corresponding single step method is

$$x^{(\nu)} = x^{(\nu-1)} - J(x^{(\nu-1)}/x^{(\nu)})F(x^{(\nu-1)}/x^{(\nu)}).$$

The Jacobian for this method can be determined with the aid of Theorem 17.24 for a special case. This will also be of significance in Sec. 19, when we develop the SOR method for nonlinear systems of equations.

Theorem 17.25: Let $F:G \subset \mathbb{R}^n \to \mathbb{R}^n$ and J be as in Lemma 17.10. Suppose further that $J(x)$ is a diagonal matrix. Let the Jacobian of F at the point x^* be partitioned as

$$F'(x^*) = D^* - R^* - S^*$$

where D^* is a diagonal matrix, R^* is a strictly lower triangular matrix, and S^* is a strictly upper triangular matrix. Then the Jacobian of the single step method

$$x^{(\nu)} = x^{(\nu-1)} - J(x^{(\nu-1)}/x^{(\nu)})F(x^{(\nu-1)}/x^{(\nu)})$$

at the point x^* may be represented as follows:

$$\begin{aligned}\hat{T}'(x^*) &= [I-J(x^*)R^*]^{-1}[I-J(x^*)F'(x^*)-J(x^*)R^*] \\ &= I - [I-J(x^*)R^*]^{-1}J(x^*)F'(x^*).\end{aligned}$$

Proof: By Lemma 17.10, the Jacobian of the corresponding total step method can be represented as

$$\begin{aligned}T'(x^*) &= I-J(x^*)F'(x^*) = I-J(x^*)(D^*-R^*-S^*) \\ &= [I-J(x^*)D^*] + J(x^*)R^* + J(x^*)S^*.\end{aligned}$$

The last sum is a splitting of $T'(x^*)$ into diagonal, lower, and upper triangular matrices. Applying Theorem 17.24 yields:

$$\begin{aligned}\hat{T}'(x^*) &= [I-J(x^*)R^*]^{-1}[I-J(x^*)D^* + J(x^*)S^*] \\ &= [I-J(x^*)R^*]^{-1}[I-J(x^*)F'(x^*) - J(x^*)R^*] \\ &= I - [I-J(x^*)R^*]^{-1}J(x^*)F'(x^*). \qquad \square\end{aligned}$$

Remark 17.26: Instead of the iterative method

$$x^{(\nu)} = x^{(\nu-1)} - J(x^{(\nu-1)}/x^{(\nu)})F(x^{(\nu-1)}/x^{(\nu)})$$

one occasionally also uses the method

$$x^{(\nu)} = x^{(\nu-1)} - J(x^{(\nu-1)})F(x^{(\nu-1)}/x^{(\nu)}).$$

One can show that the two methods have the same derivative at the fixed point. Therefore one has local convergence for both methods or for neither. This is not to say that the attractive regions are the same. □

18. Overrelaxation methods for systems of linear equations

In this section we will discuss a specialized iterative method for the numerical solution of large systems of linear equations. This is the method of overrelaxation developed by Young (cf. Young 1950, 1971). It is very popular, for with the same programming effort as required by the Gauss-Seidel method (see below or Example 17.22), one obtains substantially better convergence in many important cases.

Definition 18.1: *Gauss-Seidel* method, *successive overrelaxation (SOR)* method. Let $A \in MAT(n,n,\mathbb{R})$ be regular and let $b \in \mathbb{R}^n$. The splitting

$$A = D - R - S$$

is called the *triangular splitting* of A if the following hold true:

R is a strictly lower triangular matrix

S is a strictly upper triangular matrix

D is a regular matrix

$L = D^{-1}R$ is a strictly lower triangular matrix

$U = D^{-1}S$ is a strictly upper triangular matrix.

To solve the equation $Ax = b$, we define the iterative methods;

(1) *Gauss-Seidel* method:

$$x^{(\nu)} = D^{-1}(Rx^{(\nu)}+Sx^{(\nu-1)}+b) = Lx^{(\nu)}+Ux^{(\nu-1)}+D^{-1}b, \quad \nu = 1(1)\infty$$

or

$$x^{(\nu)} = x^{(\nu-1)}-D^{-1}(Dx^{(\nu-1)}-Rx^{(\nu)}-Sx^{(\nu-1)}-b), \quad \nu = 1(1)\infty.$$

(2) *Successive overrelaxation* or *SOR* method:

$$\begin{aligned} x^{(\nu)} &= x^{(\nu-1)}-\omega D^{-1}(Dx^{(\nu-1)}-Rx^{(\nu)}-Sx^{(\nu-1)}-b) \\ &= x^{(\nu-1)}-\omega(x^{(\nu-1)}-Lx^{(\nu)}-Ux^{(\nu-1)}-D^{-1}b). \end{aligned} \qquad \nu = 1(1)\infty$$

where $\omega \in \mathbb{R}$ is called the *relaxation parameter*. □

In the splitting $A = D - R - S$, D may possibly be a diagonal matrix. In that case, L and U are strictly triangular matrices regardless. Our definition, however, also encompasses the possibility that the D in the method contains more than simply the diagonal of A.

When D is a diagonal matrix, then the methods can also be described by:

$$x^{(\nu)} = x^{(\nu-1)} - D^{-1}(Ax^{(\nu-1)}/x^{(\nu)}-b)$$

$$x^{(\nu)} = x^{(\nu-1)} - \omega D^{-1}(Ax^{(\nu-1)}/x^{(\nu)}-b).$$

When $\omega = 1$, the successive overrelaxation method is the same as the Gauss-Seidel method. When $\omega > 1$, the changes in each iterative step are greater than in the Gauss-Seidel method. This explains the description as overrelaxation. However, it is also used for $\omega < 1$. For $\omega > 1$, convergence in many important cases is substantially better than for $\omega = 1$ (cf. Theorem 18.11).

From a theoretical viewpoint it is useful to rewrite these methods as equivalent total step methods. The following lemma is useful to this end. The transformation is without significance for practical computations.

Lemma 18.2: Let $A \in \mathrm{MAT}(n,n,\mathbb{R})$ have a triangular splitting, let $b \in \mathbb{R}^n$, and let

$$\mathscr{L}_\omega = I-\omega(D-\omega R)^{-1}A = (I-\omega L)^{-1}[(1-\omega)I+\omega U].$$

Then the method

$$x^{(\nu)} = \mathscr{L}_\omega x^{(\nu-1)} + \omega(D-\omega R)^{-1}b, \qquad \nu = 1(1)\infty$$

yields the same sequence $x^{(\nu)}$ as the SOR method

$$x^{(\nu)} = x^{(\nu-1)}-\omega D^{-1}(Dx^{(\nu-1)}-Rx^{(\nu)}-Sx^{(\nu-1)}-b), \quad \nu = 1(1)\infty.$$

Proof: The SOR method is easily reformulated as

$$(I-\omega L)x^{(\nu)} = (1-\omega)Ix^{(\nu-1)}+\omega Ux^{(\nu-1)}+\omega D^{-1}b.$$

Since L is a strictly lower triangular matrix, the matrix $I - \omega L$ is invertible. It follows that

$$x^{(\nu)} = (I-\omega L)^{-1}[(1-\omega)I+\omega U]x^{(\nu-1)} + \omega(D-\omega R)^{-1}b.$$

Further reformulation yields:

$$(I-\omega L)^{-1}[(1-\omega)I+\omega U] = (I-\omega L)^{-1}D^{-1}D[(1-\omega)I+\omega U]$$

$$= (D-\omega R)^{-1}[(1-\omega)D+\omega(D-R-A)] = (D-\omega R)^{-1}[(D-\omega R)-\omega A]$$

$$= I - \omega(D-\omega R)^{-1}A = \mathscr{L}_\omega. \qquad \square$$

The following theorem restricts the relaxation parameter ω to the interval $(0,2)$, since the spectral radius $\rho(\mathscr{L}_\omega)$ is greater than or equal to 1 for all other values of ω, and for the method to converge, one needs $\rho(\mathscr{L}_\omega) < 1$, by Theorem 17.7.

<u>Theorem 18.3</u>: Under the hypotheses and definitions of Lemma 18.2 it follows that:

(1) $\det(\mathscr{L}_\omega) = (1-\omega)^n$

(2) $\rho(\mathscr{L}_\omega) \geq |1-\omega|$.

Proof: Lemma 18.2 provides the representation

$$\mathscr{L}_\omega = (I-\omega L)^{-1}[(1-\omega)I+\omega U].$$

$\mathscr{L}_\omega$ is thus the product of two triangular matrices. Since the determinant of a triangular matrix is the product of the diagonal elements, we have

$$\det(I-\omega L)^{-1} = 1/\det(I-\omega L) = 1$$
$$\det[(1-\omega)I+\omega U] = (1-\omega)^n.$$

Conclusion (1) follows from the determinant multiplication theorem.

For the proof of (2) we observe that the determinant of a matrix is the product of the eigenvalues. By (1) however, the size of at least one of the eigenvalues is greater than or equal to $|1-\omega|$. □

The next theorem yields a positive result on the convergence of the SOR method. However, there is the substantial condition that the matrix A be symmetric and positive definite. This is satisfied for many discretizations of differential equations (cf. Sections 13 and 14). The condition "D is symmetric and positive definite" is not necessary, when D is a diagonal matrix. The diagonal of a symmetric positive definite matrix is always positive definite. Even when D contains more than the true diagonal of A, it is usually true in most applications that D is still symmetric positive definite.

<u>Theorem 18.4</u>: *Ostrowski* (cf. Ostrowski 1954). Let $A \in \mathrm{MAT}(n,n,\mathbb{R})$ have a triangular splitting and let

$$\mathscr{L}_\omega = I - \omega(D-\omega R)^{-1}A.$$

We further require that: (a) A and D are symmetric and positive definite, and (b) $\omega \in (0,2)$. Then it is true that:

(1) $\|A^{1/2}\mathscr{L}_\omega A^{-1/2}\|_2 < 1$ $\quad(\|\cdot\|_2$ = spectral norm)

(2) $\rho(\mathscr{L}_\omega) < 1$.

(3) All mappings $T(x) = \mathscr{L}_\omega x+c$ $(c \in \mathbb{R}^n)$ are contracting in the vector norm $\|x\|_A = (x^TAx)^{1/2}$.

(4) For each sequence $\{\omega_i \mid i \in \mathbb{N}\}$ with $\omega_i \in (0,2)$ and

$$\limsup_{i\to\infty} \omega_i < 2, \qquad \liminf_{i\to\infty} \omega_i > 0,$$

we have

$$\lim_{i\to\infty} \|\mathscr{L}_{\omega_i}\mathscr{L}_{\omega_{i-1}} \cdots \mathscr{L}_{\omega_1}\|_2 = 0.$$

In (1), $A^{1/2}$ denotes the symmetric positive definite matrix whose square is A.

Proof of (1): Let

$$M = (D-\omega R)/\omega = \frac{1}{\omega} D-R.$$

Then

$$B = A^{1/2}\mathscr{L}_\omega A^{-1/2} = I - A^{1/2}M^{-1}A^{1/2}.$$

Therefore

$$\begin{aligned} BB^T &= (I-A^{1/2}M^{-1}A^{1/2})(I-A^{1/2}(M^T)^{-1}A^{1/2}) \\ &= I-A^{1/2}M^{-1}(M^T+M-A)(M^T)^{-1}A^{1/2}. \end{aligned}$$

The parenthetic expression on the right can further be rewritten as

$$M^T+M-A = \frac{1}{\omega}D-R^T + \frac{1}{\omega}D-R-D+R+R^T = (\frac{2}{\omega} -1)D.$$

This matrix is symmetric and positive definite. Therefore it has a symmetric and positive definite root, which we denote by C. It follows that $BB^T + UU^T = I$ where $U = A^{1/2}M^{-1}C$. The matrices BB^T and UU^T are symmetric, positive semi-definite, and also simultaneously diagonalizable. Thus for each eigenvalue λ of BB^T there is an eigenvalue μ of UU^T such that $\lambda + \mu = 1$. All the eigenvalues λ and μ are nonnegative. Since U, and therefore UU^T also, is regular, it further follows that $\mu > 0$. Thus all the eigenvalues λ of BB^T satisfy the inequality $0 \le \lambda < 1$. It follows that

$$\|A^{1/2}\mathscr{L}_\omega A^{-1/2}\|_2 = \|B\|_2 = [\rho(BB^T)]^{1/2} < 1.$$

Proof of (2): Since the matrices B and $\mathscr{L}_\omega$ are similar, they have the same eigenvalues. From (1) it follows that

$$\rho(\mathscr{L}_\omega) = \rho(B) \le \|B\|_2 < 1.$$

Proof of (3): We have

$$\|T(x)-T(y)\|_A^2 = \|\mathscr{L}_\omega(x-y)\|_A^2 = [\mathscr{L}_\omega(x-y)]^T A[\mathscr{L}_\omega(x-y)]$$

$$= (x-y)^T(\mathscr{L}_\omega)^T A\mathscr{L}_\omega(x-y) = (x-y)^T A^{1/2}(B^T B)A^{1/2}(x-y).$$

Here $B = A^{1/2}\mathscr{L}_\omega A^{-1/2}$ is the matrix already used in the proof of (1). BB^T and B^TB have the same eigenvalues. Thus, by the proof of (1), the largest eigenvalue of B^TB satisfies the inequality

$$\lambda_m = \max\{\lambda \mid \lambda \text{ eigenvalue of } B^TB\} < 1.$$

Altogether it follows that

$$\|T(x)-T(y)\|_A^2 \le \lambda_m(x-y)^T A(x-y) = \lambda_m\|x-y\|_A^2$$

$$\|T(x)-T(y)\|_A \le \sqrt{\lambda_m}\,\|x-y\|_A.$$

Proof of (4): From the proof of (3) we have for all $\omega \in (0,2)$ that

$$\|\mathcal{L}_\omega\|_A \le \sqrt{\lambda_m} < 1.$$

Here both $\|\mathcal{L}_\omega\|_A$ and $\sqrt{\lambda_m}$ depend in general on ω. The norm $\|\cdot\|_A$ however does not depend on ω. We regard $\|\mathcal{L}_\omega\|_A$ as a function of ω. This function is continuous and attains its maximum in every interval $[a,b]$ where $0 < a < b < 2$. Then

$$\alpha = \max_{\omega\in[a,b]} \|\mathcal{L}_\omega\|_A < 1.$$

The endpoints a,b are chosen so that all elements of the sequence $\{\omega_i \mid i \in \mathbb{N}\}$ lie in the interval $[a,b]$. This leads to the inequality

$$\|\mathcal{L}_{\omega_i}\mathcal{L}_{\omega_{i-1}} \cdots \mathcal{L}_{\omega_1}\|_A \le \alpha^i, \qquad i \in \mathbb{N}.$$

It follows that

$$\lim_{i\to\infty} \|\mathcal{L}_{\omega_i}\mathcal{L}_{\omega_{i-1}} \cdots \mathcal{L}_{\omega_1}\|_A = 0.$$

Because of the equivalence of norms for finite dimensional vector spaces, the conclusion also obtains for the norm $\|\cdot\|_2$. □

We follow the proof with some remarks intended to increase understanding of the theorem.

Remark 18.5: One knows from examples that in general the spectral norm

$$\|\mathcal{L}_\omega\|_2 = [\rho(\mathcal{L}_\omega\mathcal{L}_\omega^T)]^{1/2}$$

of $\mathcal{L}_\omega$ is not less than 1. Thus the convergence of the SOR method is not necessarily monotonic in the norm $\|\cdot\|_2$. However, convergence is always monotonic in the vector norm $\|\cdot\|_A$, by (3). □

Remark 18.6: The significance of conclusion (4) is that the SOR method will also converge when ω changes from one step to the next. Such *nonstationary* methods are by no means unusual in practice. In a manner similar to the proof of (4) one can also prove convergence for the method when the matrix D is changed from one step to the next. It is only necessary to remain within the fixed set $D_U \leq D \leq D_0$, where the inequality signs are to be understood as applying componentwise. Finally, convergence is even assured when the sequence of equations and unknowns is permuted from step to step (see Theorem 19.13). □

Remark 18.7: The SOR method, like every single step method, is by definition substantially dependent on the sequential ordering of the equations. But it is worth noting that hypothesis (a) of Theorem 18.4 remains true when the ordering of the equations is changed. Let P be an arbitrary permutation matrix. When A and D are symmetric and positive definite, then so are PAP^{-1} and PDP^{-1}. Thus convergence of the method is assured, independent of the ordering of the equations, for symmetric and positive definite matrices. The speed of convergence, however, is dependent on the ordering of the equations. In certain special cases, it is possible to characterize particularly favorable orderings (see Young's Theorem 18.11). In the worst cases, with the least favorable orderings, one needs twice as many iterations. □

Definition 18.8: A matrix $B \in \mathrm{MAT}(n,n,\mathbb{R})$ is called *weakly cyclic* of *index 2*, if there exists a permutation matrix P, a matrix $B_1 \in \mathrm{MAT}(q,n-q,\mathbb{R})$, and a matrix $B_2 \in \mathrm{MAT}(n-q,q,\mathbb{R})$ such that

$$P\,B\,P^{-1} = \left(\begin{array}{c|c} 0 & B_1 \\ \hline B_2 & 0 \end{array}\right).$$

B is called *consistently ordered* if the eigenvalues of the matrix $\alpha L + \frac{1}{\alpha}U$ $(\alpha \in ¢,\ \alpha \neq 0)$ do not depend on α, where B is to be split into

$$B = L + U$$

with L a strictly lower triangular matrix and U a strictly upper triangular matrix. □

Example 18.9: If B already has the form

$$B = \left(\begin{array}{c|c} 0 & B_1 \\ \hline B_2 & 0 \end{array}\right)$$

then B is weakly cyclic of index 2 and consistently ordered. In the proof, we use block notation, beginning with

$$B\,x = \left(\begin{array}{c|c} 0 & B_1 \\ \hline B_2 & 0 \end{array}\right)\left(\begin{array}{c} x_1 \\ \hline x_2 \end{array}\right) = \lambda\left(\begin{array}{c} x_1 \\ \hline x_2 \end{array}\right).$$

It follows that

$$B_1x_2 = \lambda x_1, \quad B_2x_1 = \lambda x_2.$$

We let

$$L = \left(\begin{array}{c|c} 0 & 0 \\ \hline B_2 & 0 \end{array}\right), \quad U = \left(\begin{array}{c|c} 0 & B_1 \\ \hline 0 & 0 \end{array}\right)$$

and obtain

$$(\alpha L + \tfrac{1}{\alpha}U)\left(\begin{array}{c} x_1 \\ \hline \alpha x_2 \end{array}\right) = \left(\begin{array}{c|c} 0 & \frac{1}{\alpha}B_1 \\ \hline \alpha B_2 & 0 \end{array}\right)\left(\begin{array}{c} x_1 \\ \hline \alpha x_2 \end{array}\right) = \lambda\left(\begin{array}{c} x_1 \\ \hline \alpha x_2 \end{array}\right). \qquad \square$$

Example 18.10: When B is block tridiagonal (cf. Sec. 13) with vanishing diagonal blocks, then B is weakly cyclic of index 2 and consistently ordered.

In the following proof, we restrict ourselves to tridiagonal matrices. We choose that permutation which places all odd numbered rows and columns at the beginning, and has all even numbered rows and columns following. For example, for $n = 5$, the permutation matrix P is

$$P = \begin{pmatrix} 1 & 0 & 0 & 0 & 0 \\ 0 & 0 & 1 & 0 & 0 \\ 0 & 0 & 0 & 0 & 1 \\ 0 & 1 & 0 & 0 & 0 \\ 0 & 0 & 0 & 1 & 0 \end{pmatrix}.$$

From this we get that

$$PBP^{-1} = \left(\begin{array}{c|c} 0 & B_1 \\ \hline B_2 & 0 \end{array}\right).$$

It only remains to show that B is consistently ordered in the original ordering. Let $x \in \mathbb{R}^n$ be the eigenvector of $B = L + U$ for the eigenvalue λ:

$$b_{i,i-1}x_{i-1} + b_{i,i+1}x_{i+1} = \lambda x_i.$$

It follows that

$$\alpha^{i-1}b_{i,i-1}x_{i-1} + \alpha^{i-1}b_{i,i+1}x_{i+1} = \lambda\alpha^{i-1}x_i$$

$$(\alpha b_{i,i-1})(\alpha^{i-2}x_{i-1}) + (\tfrac{1}{\alpha}b_{i,i+1})(\alpha^{i}x_{i+1}) = \lambda(\alpha^{i-1}x_i).$$

This means that the vector

$$\tilde{x} = (x_1, \alpha x_2, \ldots, \alpha^{n-1}x_n)^T$$

is an eigenvector of $\alpha L + \frac{1}{\alpha}U$ for the same eigenvalue λ. □

We now come to the theorem of Young. Its significance lies in the fact that it accurately describes the behavior of the function $\rho(\mathcal{L}_\omega)$ in the interval $(0,2)$. Such information is important for the determination of an ω for which the convergence of the SOR method is more rapid.

Theorem 18.11: *Young*. Let $A \in \mathrm{MAT}(n,n,\mathbb{R})$ have a triangular splitting and let $\mathcal{L}_\omega = I-\omega(D-\omega R)^{-1}A$. Let the matrix $B = D^{-1}(R+S) = L+U$, where $\beta = \rho(B)$ and $\omega_b = 2/[1+(1-\beta^2)^{1/2}]$ be weakly cyclic of order 2 and consistently ordered. We further hypothesize that: (a) All eigenvalues of B are real and $\beta < 1$, or (b) A and D are symmetric and positive definite. Then it follows that:

(1) $\beta^2 = \rho(\mathcal{L}_1) < 1$.

(2) $\rho(\mathcal{L}_\omega) \geq \rho(\mathcal{L}_{\omega_b}) = \omega_b - 1, \qquad \omega \in (0,2)$

(3) $$\rho(\mathcal{L}_\omega) = \begin{cases} 1-\omega+\frac{1}{2}\omega^2\beta^2+\omega\beta\sqrt{1-\omega+\frac{1}{4}\omega^2\beta^2} & \text{for } \omega \in (0,\omega_b) \\ \omega-1 & \text{for } \omega \in [\omega_b,2). \end{cases}$$

(4) For $\omega < \omega_b$, $\rho(\mathcal{L}_\omega)$ is an eigenvalue of $\mathcal{L}_\omega$. It is simple, if β is a simple eigenvalue of B. All other eigenvalues of $\mathcal{L}_\omega$, for $\omega < \omega_b$, are less in absolute value. For $\omega \geq \omega_b$, all eigenvalues of $\mathcal{L}_\omega$ have magnitude $\omega - 1$.

Proof: We derive the proof from a series of intermediate conclusions:

(i) *All eigenvalues of* B *are real.* If condition (a) does not hold, then by (b) all the matrices A and D are symmetric and positive definite. Then

$$\tilde{B} = D^{-1/2}(D-A)D^{-1/2} = D^{-1/2}(R+S)D^{-1/2}$$

is also symmetric and hence has only real eigenvalues. Since B and $\tilde{B}$ are similar, B too has only real eigenvalues.

(ii) *If* μ *is an eigenvalue of* B, *then so is* $-\mu$. Since B is consistently ordered,

$$-B = -L + (-1)^{-1}U$$

has the same eigenvalues as B.

(iii) *For arbitrary* $z,w \in \mathbb{C}$ *the matrices* $zL + wU$ *and* $\pm\sqrt{zw}(L+U)$ *have the same eigenvalues.* The assertion is clear for $z = 0$ or $w = 0$, for then $zL + wU$ is a strictly upper or lower triangular matrix. Its eigenvalues are all zero. So now let $z \neq 0$ and $w \neq 0$. Then we can rearrange

$$zL + wU = \sqrt{zw}[(z/w)^{1/2}L + (z/w)^{-1/2}U].$$

Since B is consistently ordered, the square-bracketed expression has the same eigenvalues as $L + U$. In view of (ii), the conclusion follows.

(iv) *It is true for arbitrary* $z,w,\gamma \in \mathbb{C}$ *that:*

$$\det(\gamma I - zL - wU) = \det(\gamma I \pm \sqrt{zw}(L+U)).$$

The determinant of a matrix is equal to the product of its eigenvalues.

(v) *For* $\omega \in (0,2)$ *and* $\lambda \in \mathbb{C}$ *it is true that:*

$$\det(\lambda I - \mathscr{L}_\omega) = \det((\lambda+\omega-1)I \pm \omega\sqrt{\lambda}B).$$

It follows from the representation

$$\mathscr{L}_\omega = (I-\omega L)^{-1}[(1-\omega)I+\omega U]$$

that

$$\begin{aligned}\det(\lambda I - \mathscr{L}_\omega) &= \det(\lambda I - (I-\omega L)^{-1}[(1-\omega)I+\omega U]) \\ &= \det((I-\omega L)^{-1}(\lambda I - \lambda\omega L - (1-\omega)I - \omega U)).\end{aligned}$$

Since $\det(I-\omega L) = 1$ it further follows that

$$\det(\lambda I-\mathcal{L}_\omega) = \det(\lambda I-\lambda\omega L-(1-\omega)I-\omega U)$$
$$= \det((\lambda+\omega-1)I-\lambda\omega L-\omega U).$$

This, together with (iv) yields the conclusion.

(vi) $\beta = \rho(B) = 0$ *implies that for all* $\omega \in (0,2)$,

$$\rho(\mathcal{L}_\omega) = |1-\omega|.$$

Since the determinant of a matrix is the product of its eigenvalues, it follows from (v) that for $\rho(B) = 0$,

$$\prod_{r=1}^{n} (\lambda-\lambda_r) = (\lambda+\omega-1)^n.$$

Here the λ_i, $i = 1(1)n$, are the eigenvalues of $\mathcal{L}_\omega$. The conclusion follows immediately.

(vii) *Let* $\omega \in (0,2)$, $\mu \in \mathbb{R}$ *and* $\lambda \in \mathbb{C}$, $\lambda \neq 0$. *Further let*

$$(\lambda+\omega-1)^2 = \lambda\omega^2\mu^2.$$

Then μ *is an eigenvalue of* B *exactly when* λ *is an eigenvalue of* $\mathcal{L}_\omega$. The assertion follows with the aid of (v):

$$\det(\lambda I-\mathcal{L}_\omega) = \det(\pm\omega\mu\sqrt{\lambda}I\pm\omega\sqrt{\lambda}B) = (\omega\sqrt{\lambda})^n\det(\pm\mu I\pm B).$$

We are now ready to establish conclusions (1) - (4):

Proof of (1): By (vii), $\mu \neq 0$ is an eigenvalue of B if and only if μ^2 is an eigenvalue of $\mathcal{L}_1$. Thus $\beta^2 = \rho(\mathcal{L}_1)$. (a) implies $\beta^2 < 1$, and (b), by Theorem 18.4(2), implies $\rho(\mathcal{L}_1) < 1$.

Proof of (2): The conclusion $\rho(\mathcal{L}_\omega) \geq \rho(\mathcal{L}_{\omega_b})$ follows from considering the graph of the real valued function $f(\omega) = \rho(\mathcal{L}_\omega)$, defined in (3), over the interval $(0,2)$ (cf. also Remark 18.13).

Proof of (3) and (4): We solve the equation $(\lambda+\omega-1)^2 - \lambda\omega^2\mu^2 = 0$ given in (vii) for λ:

$$\lambda^2 - 2\lambda(1-\omega+\tfrac{1}{2}\omega^2\mu^2) + (\omega-1)^2 = 0$$

$$\lambda = 1-\omega + \tfrac{1}{2}\omega^2\mu^2 \pm \omega\mu(1-\omega + \tfrac{1}{4}\omega^2\mu^2)^{1/2}.$$

For $\omega \in [\omega_b, 2)$, the element under the radical is non-positive for all eigenvalues μ of B:

$$1-\omega + \tfrac{1}{4}\omega^2\mu^2 \leq 1-\omega + \tfrac{1}{4}\omega^2\beta^2 \leq 0.$$

Therefore it is true for all eigenvalues λ of $\mathscr{L}_\omega$ that

$$|\lambda|^2 = (1-\omega + \tfrac{1}{2}\omega^2\mu^2)^2 - \omega^2\mu^2(1-\omega + \tfrac{1}{4}\omega^2\mu^2) = (\omega-1)^2$$

$$|\lambda| = \omega-1.$$

It follows that $\rho(\mathscr{L}_\omega) = \omega-1$.

We now consider the case $\omega \in (0,\omega_b)$. In this case too there can exist eigenvalues μ of B for which the expression inside the above radical is non-positive. For the corresponding eigenvalues λ of $\mathscr{L}_\omega$ we again have $|\lambda| = \omega-1$. However, there is at least one eigenvalue of B (namely $\mu = \beta$) for which the expression under the radical is positive. We consider the set of all of these eigenvalues of B. The corresponding eigenvalues λ of $\mathscr{L}_\omega$ are real. The positive root gives the greater eigenvalue. For

$$\mu \geq 2(|\omega-1|)^{1/2}/\omega$$

the function

$$1-\omega + \tfrac{1}{2}\omega^2\mu^2 + \omega\mu(1-\omega + \tfrac{1}{4}\omega^2\mu^2)^{1/2}$$

grows monotonically with μ. The maximum is thus obtained for $\mu = \beta$. It follows that

$$\rho(\mathcal{L}_\omega) = 1-\omega + \frac{1}{2}\omega^2\beta^2 + \omega\beta[1-\omega + \frac{1}{4}\omega^2\beta^2]^{1/2}.$$

$\rho(\mathcal{L}_\omega)$ is an eigenvalue of $\mathcal{L}_\omega$ by (vii). The monotonicity also implies that $\rho(\mathcal{L}_\omega)$ is a simple eigenvalue of $\mathcal{L}_\omega$ whenever $\beta = \rho(B)$ is a simple eigenvalue of B. All of the other eigenvalues of $\mathcal{L}_\omega$ are smaller. □

Remark 18.12: In the literature, the matrix A is called 2-cyclic whenever B is weakly cyclic of index 2. If one allows matrices other than the true diagonal of A for the matrix D, then B depends not only on A, but also on the particular choice of D. Therefore it seemed preferable to us to impose the hypotheses directly on matrix B. □

Remark 18.13: Conclusion (1) of Young's Theorem means that the Gauss-Seidel method converges asymptotically twice as fast as the Jacobi method. For the SOR method, the speed of convergence for $\omega = \omega_b$ in many important cases is substantially greater than for $\omega = 1$ (cf. Table 18.20 in Example 18.15).

In (3) the course of the function $\rho(\mathcal{L}_\omega)$ is described exactly for $\omega \in (0,2)$. A consideration of the graph shows that the function decreases as the variable increases from 0 to ω_b. The limit of the derivative as $\omega \to \omega_b - 0$ is $-\infty$. On the interval $(\omega_b,2)$, the function increases linearly (see Figure 18.14). ω_b is easily computed when $\beta = \rho(B)$ is known. However that situation arises only in exceptional cases at the beginning of the iteration. As a rule, ω_b will be determined approximately in the course of the iteration. We start the iteration with an initial value $\omega_0 \in [1,\omega_b)$:

$$x^{(\nu)} = \mathcal{L}_{\omega_0} x^{(\nu-1)} + \omega_0(D-\omega_0 R)^{-1}b.$$

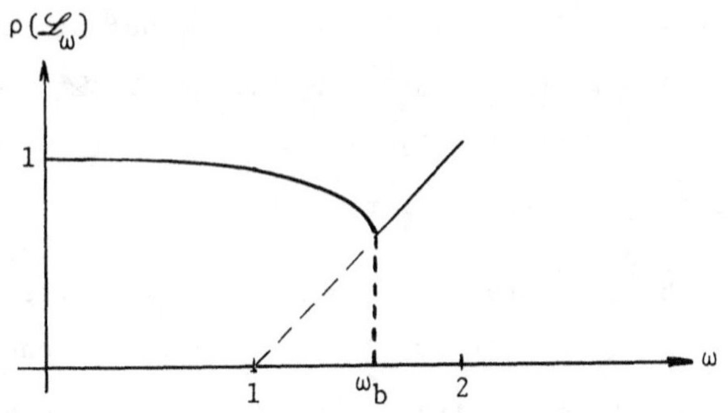

Figure 18.14. Typical behavior of $\rho(\mathcal{L}_\omega)$

For a solution x^* of the system of equations we have

$$x^* = \mathcal{L}_{\omega_0} x^* + \omega_0 (D-\omega_0 R)^{-1} b.$$

It follows that

$$x^{(\nu)} - x^* = \mathcal{L}_{\omega_0}(x^{(\nu-1)} - x^*) = (\mathcal{L}_{\omega_0})^{\nu-1}(x^{(1)} - x^*).$$

By (4), $\lambda_0 = \rho(\mathcal{L}_{\omega_0})$ is an eigenvalue of $\mathcal{L}_{\omega_0}$, and a simple one if $\beta = \rho(B)$ is a simple eigenvalue of B. This occurs, by a theorem of *Perron-Frobenius* (cf. Varga 1962), whenever, e.g., the elements of B are non-negative and B is irreducible. We now assume that λ_0 is a simple eigenvalue of $\mathcal{L}_{\omega_0}$ with eigenvector e. Then the *power method* can be used to compute an approximation of $\beta = \rho(B)$.

For sufficiently large ν it holds that:

$$x^{(\nu)} - x^* \approx a_\nu e, \quad x^{(\nu+1)} - x^* \approx \lambda_0 a_\nu e, \quad x^{(\nu+2)} - x^* \approx \lambda_0^2 a_\nu e.$$

It follows that

$$x^{(\nu+2)} - x^{(\nu+1)} \approx (\lambda_o - 1)\lambda_o a_\nu e$$

$$x^{(\nu+1)} - x^{(\nu)} \approx (\lambda_o - 1) a_\nu e$$

$$\frac{\|x^{(\nu+2)} - x^{(\nu+1)}\|_2^2}{\|x^{(\nu+1)} - x^{(\nu)}\|_2^2} \approx \lambda_o^2.$$

The equation

$$(\lambda_o + \omega_o - 1)^2 = \lambda_o \omega_o^2 \beta^2$$

makes it possible to determine an approximate value $\tilde{\beta}^2$ for β^2. Next compute $\tilde{\omega}_b$ from the formula

$$\tilde{\omega}_b = 2/[1+(1-\tilde{\beta}^2)^{1/2}]$$

and then continue the iteration with $\tilde{\omega}_b$.

The initial value ω_o must be distinctly less than ω_b, for otherwise the values of the eigenvalues of $\mathscr{L}_{\omega_o}$ will be too close together (cf. the formula in 18.11(3)) and the power method described here will converge only very slowly. But it is preferable to round up $\tilde{\omega}_b$, since the function $\rho(\mathscr{L}_\omega)$ grows more slowly for $\omega > \omega_b$ than for $\omega < \omega_b$ (cf. Figure 18.14). It is worthwhile to reduce the difference $2-\tilde{\omega}_b$ by about ten percent. □

In the following example we compare in an important special case the speed of convergence of the Jacobi, Gauss-Seidel, and SOR methods for $\omega = \omega_b$.

Example 18.15: *Sample Problem.* The five-point discretization of the problem

$$\Delta u(x,y) = q(x,y), \qquad (x,y) \in G = (0,1)^2$$

$$u(x,y) = \psi(x,y), \qquad (x,y) \in \partial G.$$

leads to a linear system of equations with coefficient matrix

$$A = \begin{pmatrix} \hat{A} & I & & \\ I & \hat{A} & \ddots & \\ & \ddots & \ddots & I \\ & & I & \hat{A} \end{pmatrix} \in \mathrm{MAT}(N^2, N^2, \mathbb{R}).$$

Here we have

$$\hat{A} = \begin{pmatrix} -4 & 1 & & \\ 1 & -4 & \ddots & \\ & \ddots & \ddots & 1 \\ & & 1 & -4 \end{pmatrix} \in \mathrm{MAT}(N, N, \mathbb{R})$$

$$N+1 = 1/h.$$

The eigenvalues of A, as we know from Section 13, are

$$\lambda_{\nu\mu} = -2(2-\cos \nu h\pi - \cos \mu h\pi), \qquad \nu,\mu = 1(1)N.$$

Let A be partitioned triangularly into $A = D - R - S$, where $D = -4I$. The iteration matrix

$$D^{-1}(R+S) = D^{-1}(D-A) = I-D^{-1}A$$

of the Jacobi method has eigenvalues $1 + \frac{1}{4}\lambda_{\nu\mu}$. Therefore

$$\beta = \rho(D^{-1}(R+S)) = \cos h\pi = 1- \frac{1}{2} h^2\pi^2 + O(h^4).$$

By Theorem 18.11 (Young) we further obtain

$$\rho(\mathscr{L}_1) = \beta^2 = \cos^2 h\pi = 1-h^2\pi^2 + O(h^4)$$

$$\omega_b = 2/(1+\sqrt{1-\beta^2}) = 2/(1+\sin h\pi)$$

$$\rho(\mathscr{L}_{\omega_b}) = \omega_b - 1 = 1-2h\pi + O(h^2).$$

Table 18.16 contains β, $\rho(\mathscr{L}_1)$, ω_b, and $\rho(\mathscr{L}_{\omega_b})$ for different step sizes.

TABLE 18.16: Spectral radii and ω_b.

h	β	$\rho(\mathcal{L}_1)$	ω_b	$\rho(\mathcal{L}_{\omega_b})$
1/8	0.92388	0.85355	1.4465	0.44646
1/16	0.98079	0.96194	1.6735	0.67351
1/32	0.99519	0.99039	1.8215	0.82147
1/64	0.99880	0.99759	1.9065	0.90650
1/128	0.99970	0.99940	1.9521	0.95209
1/256	0.99993	0.99985	1.9758	0.97575

Now let $\varepsilon^{(\nu)} = x^{(\nu)} - x^*$ be the absolute error of the ν-th approximation of an iterative method

$$x^{(\nu+1)} = Mx^{(\nu)} + c.$$

Here let M be an arbitrary matrix and let $x^* = Mx^* + c$. Since $\varepsilon^{(\nu)} = M^\nu \varepsilon^{(0)}$ and

$$\lim_{\nu\to\infty} \|M^\nu\|^{1/\nu} = \rho(M),$$

there is for each $\eta > 0$ a $\nu_0 \in \mathbb{N}$, such that

$$\left.\begin{aligned} &\|M^\nu\| \le (\rho(M)+\eta)^\nu, \\ &\|\varepsilon^{(\nu)}\| \le (\rho(M)+\eta)^\nu \|\varepsilon^{(0)}\|. \end{aligned}\right\} \quad \nu \ge \nu_0$$

The condition

$$\|\varepsilon^{(m)}\| \le \hat{\varepsilon}\|\varepsilon^{(0)}\|$$

thus leads to the approximation formula

$$m = \frac{\log \hat{\varepsilon}}{\log \rho(M)} . \qquad (18.17)$$

which is sufficiently accurate for practical purposes.

In summary we obtain the following relations for the iteration numbers of the methods considered above:

$$\begin{aligned}
&\text{Jacobi:} && m_J \approx \frac{\log \hat{\varepsilon}}{-h^2\pi^2/2} \\
&\text{Gauss-Seidel } (\omega=1): && m_1 \approx \frac{\log \hat{\varepsilon}}{-h^2\pi^2} \qquad (18.18)\\
&\text{SOR } (\omega=\omega_b): && m_{\omega_b} \approx \frac{\log \hat{\varepsilon}}{-2h\pi}\,.
\end{aligned}$$

Here the exact formulas for the spectral radii were replaced by the approximations given above. The Jacobi method thus requires twice as many iterations as the Gauss-Seidel method in order to obtain the same degree of accuracy. In practice, one frequently requires that $\hat{\varepsilon} = 1/1000$. Since $\log 1/1000 = -6.91$, we get

$$\begin{aligned}
m_1 &\approx \frac{6.91}{\pi^2}\, h^{-2} = 0.7/h^2 \\
m_{\omega_b} &\approx \frac{6.91}{2\pi}\, h^{-1} = 1.1/h \qquad (18.19)\\
m_1/m_{\omega_b} &\approx 0.64/h.
\end{aligned}$$

Table 18.20 contains m_1, h^2m_1, m_{ω_b}, hm_{ω_b}, and m_1/m_{ω_b} for various step sizes. These values were computed using Formula (18.17) and exactly computed spectral radii. One sees that the approximate formulas (18.18) and (18.19) are also accurate enough.

TABLE 18.20: Step sizes for reducing the error to 1/1000.

h	m_1	h^2m_1	m_{ω_b}	hm_{ω_b}	m_1/m_{ω_b}
1/8	43	0.682	8	1.071	5
1/16	178	0.695	17	1.092	10
1/32	715	0.699	35	1.098	20
1/64	2865	0.700	70	1.099	40
1/128	11466	0.700	140	1.099	81
1/256	45867	0.700	281	1.099	162

For each iterative step, the Jacobi and Gauss-Seidel methods require $4N^2$ floating point operations. The SOR method, in contrast, requires $7N^2$ operations. From (18.19) we get as the total number of operations involved ($\hat{\varepsilon} = 1/1000$):

Jacobi:	$1.4 \cdot 4N^4 \approx 6N^4$
Gauss-Seidel ($\omega=1$):	$0.7 \cdot 4N^4 \approx 3N^4$
SOR ($\omega=\omega_b$):	$1.1 \cdot 7N^3 \approx 8N^3$.

The sample problem is particularly suited to a theoretical comparison of the three iterative methods. Practical experience demonstrates that these relations do not change significantly in more complex situations. However, there exist substantially faster direct methods for solving the sample problem (cf. Sections 21, 22). SOR is primarily recommended, therefore, for non-rectangular regions, for differential equations with variable coefficients, and for certain nonlinear differential equations. □

19. Overrelaxation methods for systems of nonlinear equations

In this chapter we extend SOR methods to systems of nonlinear equations. The main result is a generalization of Ostrowski's theorem, which assures the global convergence of SOR methods and some variants thereof.

In the following we let G denote an open subset of $\mathbb{R}^n$.

Definition 19.1: *SOR method for nonlinear systems of equations.* Let $F \in C^1(G, \mathbb{R}^n)$, and let F have a Jacobian with an invertible diagonal $D(x)$. Then we define the *SOR* method for solving the nonlinear equation $F(x) = 0$ by generalizing the method in Definition 18.1:

$$x^{(0)} \in G$$

$$x^{(\nu)} = x^{(\nu-1)} - \omega D^{-1}(x^{(\nu-1)}/x^{(\nu)})F(x^{(\nu-1)}/x^{(\nu)}) = \hat{T}(x^{(\nu-1)})$$

$$\omega \in (0,2), \quad \nu = 1(1)\infty. \tag{19.2}$$

Ortega-Rheinholdt 1970 calls this the *single-step SOR Newton* method.

If F has a zero $x^* \in G$, then x^* is a fixed point of $\hat{T}$. This immediately raises the following questions:

(1) When is x^* attractive?

(2) How should the relaxation parameter ω be chosen?

(3) Under which conditions is the convergence of the method global, i.e., when does it converge for all initial values $x^{(0)} \in G$?

(4) To what extent can the substantial task of computing the partial derivatives of F be avoided?

(5) Do there exist similar methods for cases where F is not differentiable?

The first and second questions can be answered immediately with the help of Theorems 17.8 and 17.25.

Theorem 19.3: Let the Jacobian of F at the point x^* be partitioned triangularly (cf. Definition 18.1) into $F'(x^*) = D^* - R^* - S^*$, where D^* is an (invertible) diagonal matrix. Then

$$\rho(I-\omega[D^*-\omega R^*]^{-1}F'(x^*)) < 1,$$

implies that x^* is attractive.

Proof: By Theorem 17.25 we have

$$\begin{aligned}\hat{T}'(x^*) &= I-[I-\omega(D^*)^{-1}R^*]^{-1}\omega(D^*)^{-1}F'(x^*)\\ &= I-\omega[D^*-\omega R^*]^{-1}F'(x^*).\end{aligned}$$

The conclusion then follows from Theorem 17.8. □

The SOR method for nonlinear equations has the same convergence properties locally as the SOR method for linear equations. The matrix

$$I - \omega[D^*-\omega R^*]^{-1}F'(x^*)$$

indeed corresponds to the matrix $\mathscr{L}_\omega$ of Lemma 18.2. Thus the theorems of Ostrowski and Young (Theorems 18.4, 18.11), with respect to local convergence at least, carry over to the nonlinear case. The speed of convergence corresponds asymptotically, i.e. for $\nu \to \infty$, to the rate of convergence for the linear case. Subject to the corresponding hypotheses, the optimal ω can be determined as in Remark 18.13. If a sufficiently accurate initial value $x^{(0)}$ is available for the iteration, the situation is practically the same as for linear systems. This also holds true for the easily modified method (cf. Remark 17.26)

$$x^{(\nu)} = x^{(\nu-1)} - \omega D^{-1}(x^{(\nu-1)})F(x^{(\nu-1)}/x^{(\nu)}).$$

The following considerations are aimed at a generalization of Ostrowski's theorem. Here convergence will be established independently of Theorem 17.8.

The method (19.2) will be generalized one more time, so that it will no longer be necessary to compute the diagonal of the Jacobian $F'(x)$. The hypothesis "F differentiable" can then be replaced by a Lipschitz condition. Then questions (4) and (5) will also have a positive answer. In an important special case, one even obtains global convergence.

Definition 19.4: A mapping $F \in C^0(G, \mathbb{R}^n)$ is called a *gradient mapping* if there exists a $\phi \in C^1(G, \mathbb{R}^1)$ such that $F(x)^T = \phi'(x)$, $x \in G$. We write $F = \text{grad}\ \phi$. □

In the special case of a simply connected region G, the gradient mappings may be characterized with the aid of a well-known theorem of Poincaré (cf. Loomis-Steinberg 1968, Ch. 11.5).

Theorem 19.5: *Poincaré.* Let G be a simply connected region and let $F \in C^1(G, \mathbb{R}^n)$. Then F is a gradient mapping if and only if $F'(x)$ is always symmetric.

Our interest here is only in open and convex subsets of $\mathbb{R}^n$, and these are always simply connected. In the sequel then, we always presuppose that G is an open, convex subset of $\mathbb{R}^n$.

Definition 19.6: Let $\phi: G \to \mathbb{R}^1$ and for all $x,y \in G$ and all $\alpha \in (0,1)$ let

$$r(x,y,\alpha) = \alpha\phi(x) + (1-\alpha)\phi(y) - \phi(\alpha x + (1-\alpha)y).$$

Then ϕ is called, respectively, a

convex function	if $r(x,y,\alpha) \geq 0$,
strictly convex function	if $r(x,y,\alpha) > 0$,
uniformly convex function	if $r(x,y,\alpha) \geq c\alpha(1-\alpha)\|x-y\|_2^2$,

for all $x,y \in G$ with $x \neq y$, and for all $\alpha \in (0,1)$. Here c is a positive constant which depends only on ϕ. □

The following theorem characterizes the convexity properties of ϕ with the aid of the second partial derivatives.

Theorem 19.7: A function $\phi \in C^2(G, \mathbb{R}^1)$ is convex, strictly convex, or uniformly convex, if and only if the matrix $A(x)$ of the second partial derivatives of ϕ satisfies the following inequalities for all $x \in G$ and all nonzero $z \in \mathbb{R}^n$, respectively,

$z^T A(x) z \geq 0$ *(positive semidefinite)*

$z^T A(x) z > 0$ *(positive definite)*

$z^T A(x) z \geq \tilde{c} z^T z$ *(uniformly positive definite in* x *and* z*).*

Here $\tilde{c} > 0$ depends only on A, not on x or z.

Proof: For $x,y \in G$, $x \neq y$, we define

$$\rho(t) = r(x, x+t(y-x), \alpha), \quad t \in [0,1].$$

Then we have

$$\rho(t) = \alpha\phi(x) + (1-\alpha)\phi(x+t(y-x)) - \phi(x+t(1-\alpha)(y-x))$$

and $\rho(0) = 0$, $\rho'(0) = 0$. It follows that

$$\rho(1) = \int_0^1 (1-s)\rho''(s)ds$$

$$\rho(1) = (1-\alpha)\int_0^1 (1-s)(y-x)^T A(x+s(y-x))(y-x)ds$$

$$- (1-\alpha)^2\int_0^1 (1-s)(y-x)^T A(x+s(1-\alpha)(y-x))(y-x)ds.$$

In the second integral, we can make the substitution $\tilde{s} = (1-\alpha)s$, and then call $\tilde{s}$ again s, and combine the integrals:

$$\rho(1) = \int_0^1 \tau(s)(y-x)^T A(x+s(y-x))(y-x)\,ds$$

where

$$\tau(s) = \begin{cases} \alpha s & \text{for } 0 \le s \le 1-\alpha \\ (1-\alpha)(1-s) & \text{for } 1-\alpha \le s \le 1. \end{cases}$$

The mean value theorem for integrals then provides a suitable $\theta \in (0,1)$ for which

$$r(x,y,\alpha) = \rho(1) = \frac{1}{2}\,\alpha(1-\alpha)(y-x)^T A(x+\theta(y-x))(y-x).$$

The conclusion of the theorem now follows easily from Definition 19.6. □

If ϕ is only once continuously differentiable, the convexity properties can be checked with the aid of the first derivative.

Theorem 19.8: Let $\phi \in C^1(G, \mathbb{R}^1)$, $F = \text{grad}\,\phi$, and

$$p(x,y) = [F(y)-F(x)]^T(y-x).$$

Then ϕ is convex, strictly convex, or uniformly convex, if and only if $p(x,y)$ satisfies the following inequalities, respectively,

$$p(x,y) \ge 0,$$
$$p(x,y) > 0,$$
$$p(x,y) \ge c^*\|y-x\|_2^2.$$

Here $c^* > 0$ depends only on F.

Proof: Again, let $\rho(t) = r(x,x+t(y-x),\alpha)$, $x,y \in G$, $x \ne y$, $t \in [0,1]$. Then we have

$$\rho(t) = \alpha\phi(x) + (1-\alpha)\phi(x+t(y-x)) - \phi(x+t(1-\alpha)(y-x))$$

and

$$\rho(1) = r(x,y,\alpha) = \int_0^1 \rho'(t)dt$$
$$= (1-\alpha)\int_0^1 [F(x+t(y-x))-F(x+t(1-\alpha)(y-x))]^T(y-x)dt.$$

It remains to prove that the inequalities in Theorem 19.8 and Definition 19.6 are equivalent. We content ourselves with a consideration of the inequalities related to uniform convexity. Suppose first that always

$$p(x,y) \geq c^*\|y-x\|_2^2.$$

Then it follows that

$$p(x+t(1-\alpha)(y-x),x+t(y-x)) \geq c^*\alpha^2t^2\|y-x\|_2^2$$

$$[F(x+t(y-x))-F(x+t(1-\alpha)(y-x))]^T(y-x) \geq c^*\alpha t\|y-x\|_2^2$$

$$r(x,y,\alpha) \geq \tfrac{1}{2} c^*\alpha(1-\alpha)\|y-x\|_2^2.$$

The quantity c in Definition 19.6 thus corresponds to $\frac{1}{2}c^*$ here. Now suppose that always

$$r(x,y,\alpha) \geq c\alpha(1-\alpha)\|y-x\|_2^2.$$

Then it follows that

$$\alpha\phi(x)+(1-\alpha)\phi(y) \geq \phi(x+(1-\alpha)(y-x))+c\alpha(1-\alpha)\|y-x\|_2^2$$

$$\phi(y)-\phi(x) \geq \frac{\phi(x+(1-\alpha)(y-x))-\phi(x)}{1-\alpha} + c\alpha\|y-x\|_2^2.$$

Since this inequality holds for all $\alpha \in (0,1)$, by passing to the limit $\alpha \to 1$, we obtain

$$\phi(y)-\phi(x) \geq F(x)^T(y-x) + c\|y-x\|_2^2.$$

Analogously, we naturally also obtain

$$\phi(x)-\phi(y) \geq F(y)^T(x-y) + c\|y-x\|_2^2.$$

Adding these two inequalities yields

$$0 \geq -[F(y)-F(x)]^T(y-x) + 2c\|y-x\|_2^2. \quad \square$$

The following theorem characterizes the solution set of the equation $F(x) = 0$ for the case where F is the gradient of a convex map.

<u>Theorem 19.9</u>: Let $\phi \in C^1(G, \mathbb{R}^1)$ be convex and let $F = \operatorname{grad} \phi$. Then:

(1) The level sets $N(\gamma,\phi) = \{x \in G \mid \phi(x) \leq \gamma\}$ are convex for all $\gamma \in \mathbb{R}$.

(2) x^* is a zero of F exactly when ϕ assumes its global minimum at x^*. The set of all zeros of F is convex.

(3) If ϕ is strictly convex, then F has at most one zero x^* in G.

(4) If ϕ is uniformly convex and $G = \mathbb{R}^n$, then F has exactly one zero x^*, and the inequality

$$c^*\|x^*\|_2 \leq \|F(0)\|_2.$$

is valid, where c^* is the constant from Theorem 19.8.

Proof of (1): Let $\alpha \in (0,1)$ and $y,x \in N(\gamma,\phi)$. Then it follows from the convexity of ϕ that

$$\phi(\alpha x+(1-\alpha)y) \leq \alpha\phi(x)+(1-\alpha)\phi(y) \leq \alpha\gamma+(1-\alpha)\gamma = \gamma.$$

Thus $\alpha x + (1-\alpha)y$ also belongs to $N(\gamma,\phi)$.

Proof of (2): Let x^* be a zero of F and let $x \in G$, $x \neq x^*$, be arbitrary. By the mean value theorem of differentiation, there is a $\lambda \in (0,1)$ such that

$$\phi(x) = \phi(x^*) + [F(x^*+\lambda(x-x^*))]^T(x-x^*).$$

It follows from Theorem 19.8 that

$$p(x^*,x^*+\lambda(x-x^*)) = [F(x^*+\lambda(x-x^*))]^T\lambda(x-x^*) \geq 0.$$

Thus we obtain

$$\phi(x) - \phi(x^*) \geq 0.$$

Therefore x^* is a global minimum of ϕ. The reverse conclusion is trivial, since G is open. It follows from (1) that the set of all zeros of F is convex, for if x_o^* is a particular zero of F, then

$$\{x^* \in G \mid F(x^*) = 0\} = N(\phi(x_o^*),\phi).$$

Proof of (3): If ϕ is a strictly convex function, then in the above proof we have the stronger inequality

$$p(x^*,x+\lambda(x-x^*)) = [F(x^*+\lambda(x-x^*))]^T\lambda(x-x^*) > 0.$$

It follows that

$$\phi(x) > \phi(x^*).$$

Therefore x^* is the only point at which ϕ can assume the global minimum.

Proof of (4): By (3), F has at most one zero. Thus we need to show that F has at least one zero. To this end we examine ϕ along the lines

$$x = bt, \quad t \in \mathbb{R}, \quad b \in \mathbb{R}^n \text{ fixed with } \|b\|_2 = 1.$$

There it is true that

$$\phi(bt) = \phi(0) + F(0)^T bt + \int_0^t [F(bs)-F(0)]^T b\,ds.$$

Since

$$[F(bs)-F(0)]^T bs \geq c^* s^2$$

it follows that

$$\phi(bt) \geq \phi(0) + F(0)^T bt + \frac{1}{2} c^* t^2.$$

For all $x = bt$ with

$$\|x\|_2 = |t| > 2\|F(0)\|_2/c^*$$

this inequality means that $\phi(x) > \phi(0)$. Therefore ϕ assumes its minimum in the ball

$$\{x \in \mathbb{R}^n \mid \|x\|_2 \leq 2\|F(0)\|_2/c^*\}.$$

This minimum is a global minimum in all of $\mathbb{R}^n$. F therefore has at least one zero x^* in the above ball. One can even establish the inequality $c^*\|x^*\|_2 \leq \|F(0)\|_2$, for we have

$$[F(x^*)-F(0)]^T(x^*-0) \geq c^*\|x^*-0\|_2^2$$

$$-F(0)^T x^* \geq c^*\|x^*\|_2^2$$

$$\|F(0)\|_2 \geq c^*\|x^*\|_2. \quad \square$$

<u>Definition 19.10</u>: Let $\phi \in C^1(G, \mathbb{R}^1)$ be strictly convex,

$$F = \text{grad}\,\phi = (f_1, f_2, \ldots, f_n),$$

and $G^* \subset G$ a convex set with at least one interior point. Then we define

$$\text{mid}_j(F,G^*) = \inf \frac{h}{f_j(x+he^{(j)})-f_j(x)}, \qquad j = 1(1)n.$$

Here the $e^{(j)}$ are the unit vectors parallel to the axes in $\mathbb{R}^n$, and $h \in \mathbb{R}$ and $x \in \mathbb{R}^n$ run through all possible combinations satisfying

$$x \in G^*, \quad x+he^{(j)} \in G^*, \quad h \neq 0.$$

Further let

$$MID(F,G^*) = \mathrm{diag}(\mathrm{mid}_j(F,G^*)).$$

The notation MID stands for "maximal inverse diagonal". □

Since G^* has at least one interior point, the infimum is always formed over a nonempty set of combinations of x and h. Further, we always have

$$[F(x+he^{(j)}) - F(x)]^T he^{(j)} > 0$$

$$[f_j(x+he^{(j)}) - f_j(x)]h > 0$$

and therefore, also

$$\frac{h}{f_j(x+he^{(j)})-f_j(x)} > 0.$$

Nevertheless, the infimum of these quantities can be zero.

When ϕ is uniformly convex, $\mathrm{mid}_j(F,G^*)$ has an upper bound which is independent of G^*. It follows from

$$[F(x+he^{(j)}) - F(x)]^T he^{(j)} \geq c^*\|he^{(j)}\|_2^2$$

that

$$f_j(x+he^{(j)}) - f_j(x) \geq c^*h$$

$$\mathrm{mid}_j(F,G^*) \leq 1/c^*.$$

In the sequel, a lower bound for $\mathrm{mid}_j(F,G^*)$ is more important than an upper bound. That requires Lipschitz conditions for f_j.

<u>Theorem 19.11</u>: If, for all $x \in G^*$ with $x+he^{(j)} \in G^*$, the inequalities

$$|f_j(x+he^{(j)})-f_j(x)| \le |h|L_j, \quad L_j > 0, \quad j = 1(1)n$$

hold, then MID(F,G*) is positive definite and

$$\mathrm{mid}_j(F,G^*) \ge 1/L_j, \qquad j = 1(1)n.$$

The proof is trivial. □

Theorem 19.12: Let ϕ be twice continuously differentiable and let G^* be compact. Then

$$\mathrm{mid}_j(F,G^*) = 1/\max_{x\in G^*} \partial_{jj}^2\phi(x), \qquad j = 1(1)n.$$

Proof: Since

$$f_j(x+he^{(j)})-f_j(x) = \partial_j\phi(x+he^{(j)})-\partial_j\phi(x) \le |h|\max_{x\in G^*} \partial_{jj}^2\phi(x)$$

it follows from the previous theorem that in any case

$$\mathrm{mid}_j(F,G^*) \ge 1/\max_{x\in G^*} \partial_{jj}^2\phi(x).$$

For the proof of the reverse inequality, we use the definition of the partial derivatives:

$$\lim_{h\to 0} \frac{\partial_j\phi(x+he^{(j)})-\partial_j\phi(x)}{h} = \partial_{jj}^2\phi(x)$$

$$\lim_{\substack{h\to 0\\ h\neq 0}} \frac{h}{\partial_j\phi(x+he^{(j)})-\partial_j\phi(x)} = 1/\partial_{jj}^2\phi(x)$$

$$\mathrm{mid}_j(F,G^*) = \inf \frac{h}{\partial_j\phi(y+he^{(j)})-\partial_j\phi(y)} \le 1/\partial_{jj}^2\phi(x). \qquad \square$$

In the previous section, we considered the iterative solution of linear systems of equations. We want to explain briefly how these fit in here in the case of real, symmetric, and positive definite matrices. Let $F(x) = Ax - b$ be an affine map with a real, symmetric, and positive definite matrix A. The map is the gradient of the function

$$\phi(x) = \frac{1}{2} x^T Ax - b^T x.$$

By Theorem 19.7, ϕ is uniformly convex. The constant c of that theorem is the smallest eigenvalue of A. The function $p(x,y)$ of Theorem 19.8 is

$$p(x,y) = (y-x)^T A(y-x).$$

Thus c^* is also the smallest eigenvalue of A. Finally by Theorem 19.12

$$\mathrm{mid}_j(F,G^*) = 1/a_{jj}, \quad j = 1(1)n.$$

These considerations hold for all G^*.

The next example shows that $\mathrm{MID}(F,G^*)$ can be positive definite even in the case of a nondifferentiable function F. Let $\phi \in C^1(\mathbb{R}, \mathbb{R})$ with

$$\phi(x) = \begin{cases} 2x^2 & \text{for} \quad x \geq 0 \\ x^2 & \text{for} \quad x \leq 0 \end{cases}$$

and

$$F(x) = \begin{cases} 4x & \text{for} \quad x \geq 0 \\ 2x & \text{for} \quad x \leq 0. \end{cases}$$

Then $\mathrm{MID}(F, \mathbb{R}^1) = 1/4$.

The next theorem presupposes that

$G \subset \mathbb{R}^n$ is open and convex,

$\phi \in C^1(G, \mathbb{R}^n)$,

$F = \mathrm{grad}\ \phi = (f_1, f_2, \ldots, f_n)$,

$Q \in \mathrm{MAT}(n,n, \mathbb{R})$ is a positive definite diagonal matrix,

$Q = \mathrm{diag}\ (q_j)$.

Theorem 19.13: *Hypotheses:*

(a) ϕ is strictly convex.

(b) There is a $\gamma \in \mathbb{R}$ with a compact level set $G^* = \{x \in G \mid \phi(x) \leq \gamma\}$ which consists of more than one point.

(c) $q_j < 2\,\mathrm{mid}_j(F,G^*)$, $j = 1(1)n$.

Conclusions:

(1) There is exactly one $x^* \in G$ with $F(x^*) = 0$. x^* is an interior point of G^*.

(2) Either $x = x^*$ or $\phi(y) < \phi(x)$ for $y = x\text{-}QF(x/y)$.

(3) Every sequence

$$x^{(0)} \in G^* \quad \text{arbitrary}$$
$$x^{(\nu+1)} = x^{(\nu)}\text{-}QF(x^{(\nu)}/x^{(\nu+1)}), \quad \nu = 0(1)\infty$$

converges to x^*.

Proof of (1): Since G^* is compact, ϕ assumes its minimum on G^*. $\phi(x)$ is larger outside of G^*. Therefore there is an $x^* \in G^*$ such that

$$\phi(x^*) = \min_{x \in G} \phi(x), \quad F(x^*) = 0.$$

ϕ is strictly convex. Therefore there is at most one point with these properties. Since G^* contains other points x besides x^*, $\phi(x^*) < \gamma$ and x^* is an interior point of G^*.

Proof of (2): The computation $y = x\text{-}QF(x/y)$ can be split naturally into n individual steps. With each of these individual steps, only one component is changed. We call these intermediate results $y^{(0)} = x, y^{(1)}, y^{(2)}, \ldots, y^{(n)} = y$. The individual computations are

$$y^{(j)} = y^{(j-1)} - \lambda_j e^{(j)}$$

where $\lambda_j = q_j f_j(y^{(j-1)})$. Since $q_j > 0$, either $\lambda_j \neq 0$ or $f_j(y^{(j-1)}) = 0$. In the first case we define

$$y(t) = y^{(j-1)} - t\lambda_j e^{(j)} \qquad t \in [0,1]$$
$$\rho(t) = \phi(y(t)).$$

This leads to

$$\rho'(t) = -\lambda_j f_j(y(t))$$
$$\rho'(0) = -\lambda_j f_j(y(0)) = -q_j f_j(y(0))^2 < 0$$
$$\rho'(t)-\rho'(0) = \lambda_j [f_j(y(0))-f_j(y(t))].$$

Here we need hypothesis (c). For $t > 0$ and $y(t) \in G^*$, this hypothesis implies:

$$q_j < \frac{2t\lambda_j}{f_j(y(0))-f_j(y(t))}$$

$$1 < \frac{2t|f_j(y(0))|}{|f_j(y(0))-f_j(y(t))|}$$

$$|\rho'(t)-\rho'(0)| \leq 2t\ q_j f_j(y(0))^2.$$

Since

$$\rho(t) = \rho(0) + \rho'(0)t + \int_0^t [\rho'(s)-\rho'(0)]ds$$

the last inequality leads to

$$\rho(t) < \rho(0) - q_j f_j(y(0))^2 t + q_j f_j(y(0))^2 t^2$$

$$\phi(y(t)) < \phi(y^{(j-1)}) - t(1-t)q_j f_j(y(0))^2$$

$$\phi(y(t)) < \phi(y^{(j-1)}) \leq \gamma.$$

Thus $y(t)$ cannot leave the set G^*. For all $t \in (0,1]$ we have

$$\phi(y(t)) < \phi(y^{(j-1)})$$

and in particular,

$$\phi(y^{(j)}) < \phi(y^{(j-1)}).$$

Thus, either $f_j(y^{(j-1)})$ is always zero or $\phi(y) < \phi(x)$. (2) is proven.

Proof of (3): The sequence $\{\phi(x^{(\nu)}) \mid \nu = 0(1)\infty\}$ converges, for by (2),

$$\phi(x^{(\nu-1)}) \geq \phi(x^{(\nu)}) \geq \phi(x^{(\nu+1)}) \geq \ldots \geq \phi(x^*).$$

Let x^{**} be an arbitrary limit point of the sequence $\{x^{(\nu)}\}$. It follows from continuity considerations that $\phi(y^{**}) = \phi(x^{**})$, where $y^{**} = x^{**} - QF(x^{**}/y^{**})$. By (2) however, this means that $F(x^{**}) = 0$ and hence by (1) that $x^{**} = x^*$. The only possible limit point of the sequence $\{x^{(\nu)} \mid \nu = 1(1)\infty\}$ in the compact set G^* is x^*. Thus the sequence is convergent. □

This last theorem requires a few clarifying remarks.

Remark 19.14: If $\phi \in C^1(\mathbb{R}^n, \mathbb{R})$ is *uniformly* convex, then every level set $\{x \in \mathbb{R}^n \mid \phi(x) \leq \gamma\}$ is compact (cf. proof of Theorem 19.9). The iteration therefore converges for all $x^{(0)} \in \mathbb{R}^n$. □

Remark 19.15: If $\mathrm{MID}(F,G^*)$ is not positive definite, then there is no matrix Q with the stated properties. However, for $\phi \in C^2(G, \mathbb{R}^1)$, $\mathrm{MID}(F,G^*)$ is always positive definite. The same is also true when the components of F satisfy Lipschitz conditions (cf. Theorem 19.11). □

Remark 19.16: In practice, the starting point is usually F, and not ϕ. Since the Jacobian $F'(x)$ of F is always symmetric, there exists a function ϕ with $F = \operatorname{grad} \phi$. If, in addition, $F'(x)$ is always positive definite, then ϕ is even strictly convex. It is most difficult to establish that hypothesis (b) of Theorem 19.13 is satisfied. $MID(F,G^*)$ can be determined from bounds on the diagonal elements of $F'(x)$. In the actual iteration one needs only F and not ϕ. □

Remark 19.17: The first convergence proof for an SOR method for convex maps was given by Schechter 1962. In Ortega-Rheinboldt 1970, Part V, several related theorems are proven. Practical advice based on actual executions can be found, among other places, in Meis 1971 and Meis-Törnig 1973. □

We present an application of Theorem 19.13 in the form of the following example.

Example 19.18: Let $\hat{G}$ be a rectangle in $\mathbb{R}^2$ with sides parallel to the axes, and let a boundary value problem of the first kind be given on this rectangle for the differential equation

$$-(a_1 u_x)_x - (a_2 u_y)_y + H(x,y,u) = 0.$$

The five point difference scheme (see Section 13) leads, for fixed h, to the system of equations

$$F(w) = Aw + \tilde{H}(w) = 0$$

where $A \in MAT(n,n,\mathbb{R})$ is symmetric and positive definite, $w = (w_1, w_2, \ldots, w_n)$, (x_j, y_j) = lattice points of the discretization, and $\tilde{H}(w) = (H(x_1,y_1,w_1), \ldots, H(x_n,y_n,w_n))$. A can be split into

$$A = D - R - R^T,$$

where D is a diagonal matrix, R is a strictly lower triangular matrix with nonnegative entries, and $D^{-1}(R+R^T)$ is weakly cyclic of index 2. Let

$$p_j(z) = \int_0^z H(x_j,y_j,\tilde{z})\,d\tilde{z}$$

$$P(w) = (p_1(w_1),\ldots,p_n(w_n))$$

$$\phi(w) = \frac{1}{2} w^T A w + P(w).$$

Then obviously $F(w) = \operatorname{grad}\phi$. Under the hypotheses

$$0 \le H_z(x,y,z) \le \delta, \qquad (x,y) \in \hat{G}, \quad z \in \mathbb{R}$$

ϕ is uniformly convex in $\mathbb{R}^n$ and $MID(F,G^*)$ is positive definite for every compact convex set $G^* \subset \mathbb{R}^n$. In particular, for $j = 1(1)n$:

$$\partial^2_{jj}\phi(w) = a_{jj} + H_z(x_j,y_j,w)$$

$$a_{jj} \le \partial^2_{jj}\phi(w) \le a_{jj} + \delta$$

$$1/a_{jj} \ge \operatorname{mid}_j(F,G^*) \ge 1/(a_{jj}+\delta).$$

a_{jj} contains the factor $1/h^2$. For small h, therefore, $a_{jj} \gg \delta$. Every sequence

$$w^{(0)} \quad \text{arbitrary}$$

$$w^{(\nu+1)} = w^{(\nu)} - Q(Aw^{(\nu-1)}/w^{(\nu)} + \tilde{H}(w^{(\nu-1)}))$$

converges, if

$$0 < q_j < 2/(a_{jj}+\delta).$$

The condition $H_z(x,y,u(x,y)) \le \delta$ appears to be very restrictive. In most cases, however, one has a priori knowledge of an estimate

$$\alpha \leq u(x,y) \leq \beta, \qquad (x,y) \in \hat{G}.$$

When the maximum principle applies, e.g. for $H(x,y,0) = 0$, this estimate follows at once from the boundary values. H is significant then only on $\hat{G} \times [\alpha,\beta]$. The function can be changed outside this set, without changing the solution of the differential equation. This change can be undertaken so that $H \in C^1(\hat{G} \times \mathbb{R}, \mathbb{R})$ and $H_z(x,y,z)$ are bounded. We will demonstrate this procedure for the case $H(x,y,z) = e^z$. If $\alpha \leq u(x,y) \leq \beta$, one defines

$$H^*(z) = \begin{cases} e^\alpha(z-\alpha+1) & \text{for} \quad z < \alpha \\ e^z & \text{for} \quad \alpha \leq z \leq \beta \\ e^\beta(z-\beta+1) & \text{for} \quad z > \beta. \end{cases}$$

It follows that

$$e^\alpha \leq H^{*\prime}(z) \leq e^\beta$$

and

$$\frac{1}{a_{jj}+e^\beta} \leq \mathrm{mid}_j(F,G^*) \leq \frac{1}{a_{jj}+e^\alpha}\ .$$

One begins the iteration with

$$Q = \omega\,\mathrm{diag}\left(\frac{1}{a_{jj}+e^\beta}\right), \qquad 0 < \omega < 2.$$

It may be possible, in the course of the computation, to replace β with a smaller number. Should one have erred in estimating α and β, one can correct the error on the basis of the results. The initial bounds on α and β therefore do not have to be precisely accurate. To speed convergence, one is best off in the final phase to chose for Q an approximation of

$$\omega\ \mathrm{diag}\left(\frac{1}{a_{jj}+\partial_j\tilde{H}_j(w)}\right). \qquad \square$$

20. Band width reduction for sparse matrices

When boundary value problems are solved with difference methods or finite element methods, one is led to systems of equations which may be characterized by *sparse matrices*. That is, on each row of these matrices there are only a few entries different from zero. The distribution of these entries, however, varies considerably from one problem to another.

In contrast, for the classical Ritz method and with many collocation methods, one usually finds matrices which are *full* or almost full; their elements are almost all different from zero.

Corresponding to the different types of matrices are different types of algorithms for solving linear systems of equations. We would like to differentiate *four groups* of *direct* methods.

The first group consists of the *standard* elimination methods of Gauss, Householder, and Cholesky, along with their numerous variants. At least for full matrices, there are as a rule no alternatives to these methods. Iterative methods too are seldom better. For sparse matrices, the computational effort of the standard methods is too great. For if the number of equations is n, then the required number of additions and multiplications for these methods is always proportional to n^3.

The second group of direct methods consists of the specializations of the standard methods for band matrices. In this section we treat Gaussian elimination for band matrices. Two corresponding FORTRAN programs may be found in

Appendix 5. The methods of Householder and Cholesky may be adapted in similar ways. The number of computations in this group is proportional to nw^2, where w is the band width of the matrix (cf. Definition 20.1 below).

The third group also consists of a modification of the standard methods. Its distinguishing characteristic is the manner of storing the matrix. Only those entries different from zero, along with their indices, are committed to memory. The corresponding programs are relatively complicated, since the number of nonzero elements increases during the computation. In many cases, the matrix becomes substantially filled. Therefore it is difficult to estimate the number of computations involved. We will not discuss these methods further. A survey of these methods can be found in Reid 1977.

The fourth group of direct methods is entirely independent of those discussed so far. Its basis lies in the special properties of the differential equations associated with certain boundary value problems. These methods thus can be used only to solve very special systems. Further, the methods differ greatly from each other. In the following two sections we consider two typical algorithms of this type. Appendix 6 contains the FORTRAN program for what is known as the Buneman algorithm. The computational effort for the most important methods in this group is proportional to merely $n \log n$ or n. These are known, therefore, as *fast direct methods*. Please note that n is defined differently in Sections 21 and 22.

We are now ready to investigate band matrices.

<u>Definition 20.1</u>: Let $A = (a_{ij}) \in \mathrm{MAT}(n,n,\mathbb{C})$ and $A \neq 0$. Then

$$w = 1 + 2\max\{d \mid d = |i-j| \text{ where } a_{ij} \neq 0 \text{ or } a_{ji} \neq 0\}$$

is called the *band width* of A. For a zero matrix, we define the band width to be 1. □

A matrix for which $w \ll n$ will be called a *band matrix*. Thus we do not have a precise mathematical concept in mind; the precise term is band width. In the following examples, an x represents a nonzero element, and a blank represents a zero.

A diagonal matrix has a band width $w = 1$, and a tridiagonal matrix has band width $w = 3$. A matrix of the following type has band width 5:

$$\begin{bmatrix} x & x & x & & & & \\ x & x & x & x & & & \\ x & x & x & x & x & & \\ & x & x & x & x & x & \\ & & x & x & x & x & x \\ & & & x & x & x & x \\ & & & & x & x & x \end{bmatrix}.$$

Every full matrix has the maximal band width of $2n-1$. However, many sparse matrices also have this bandwidth, e.g. the matrix:

$$\begin{bmatrix} x & & & & & & x \\ & x & & & & & \\ & & x & & & & \\ & & & x & & & \\ & & & & x & & \\ & & & & & x & \\ & & & & & & x \end{bmatrix}.$$

Now let $A \in MAT(n,n,\mathbb{C})$ be a matrix with band width less than or equal to w, w odd. Every linear system containing the matrix A,

$$\sum_{j=1}^{n} a_{ij}x_j = b_i, \qquad i = 1(1)n$$

is equivalent to the system

$$\sum_{j=1}^{w} \tilde{a}_{ji}x_{i+j-k-1} = \tilde{a}_{w+1,i}, \qquad i = 1(1)n$$

where $k = (w-1)/2$ and

$$\tilde{a}_{ji} = \begin{cases} a_{i,i+j-k-1} & \text{if } j \le w,\ 1 \le i+j-k-1 \le n \\ b_i & \text{if } j = w+1 \\ 0 & \text{otherwise.} \end{cases}$$

For $p < 1$ or $p > n$, set $x_p = 0$.

The quantities $\tilde{a}_{ji}$ form a matrix $\tilde{A} \in MAT(w+1,n,\mathbb{C})$ (cf. Figure 20.2). For $w << n$, $\tilde{A}$ occupies substantially less core memory than A and b. The ratio is $(w+1)/(n+1)$.

$$\begin{bmatrix} 0 & 0 & a_{31} & a_{42} & a_{53} & a_{64} & a_{75} & a_{86} & a_{97} \\ 0 & a_{21} & a_{32} & a_{43} & a_{54} & a_{65} & a_{76} & a_{87} & a_{98} \\ a_{11} & a_{22} & a_{33} & a_{44} & a_{55} & a_{66} & a_{77} & a_{88} & a_{99} \\ a_{12} & a_{23} & a_{34} & a_{45} & a_{56} & a_{67} & a_{78} & a_{89} & 0 \\ a_{13} & a_{24} & a_{35} & a_{46} & a_{57} & a_{68} & a_{79} & 0 & 0 \\ b_1 & b_2 & b_3 & b_4 & b_5 & b_6 & b_7 & b_8 & b_9 \end{bmatrix}$$

Figure 20.2. $\tilde{A}$ for $w = 5$ and $n = 9$

Gaussian elimination without pivot search does not lead to an increase in band width. The same holds true if the search for the pivot elements is restricted to the row at hand, i.e. if there are only column exchanges. Thus the algorithm can run its course entirely within A. The number of computations is, as previously mentioned, proportional to nw^2.

The Ritz method leads to positive definite Hermitian matrices. A pivot search thus is unnecessary, as it also is with most difference equations. The number of computations without pivot search is

$$\frac{n}{2}(w^2 + 3w - 2).$$

Appendix 5 contains two FORTRAN programs, for $w = 3$ and $w \geq 3$.

In many cases the band width of a matrix A can be reduced substantially through a simultaneous exchange of rows and columns. For example, we can convert

$$\begin{bmatrix} x & & & & x \\ & x & & & \\ & & x & & \\ & & & x & \\ & & & & x \end{bmatrix}$$

into

$$\begin{bmatrix} x & x & & & \\ & x & & & \\ & & x & & \\ & & & x & \\ & & & & x \end{bmatrix}.$$

through an exchange of rows 2 and 5 and columns 2 and 5. Such a simultaneous exchange of rows and columns is a similarity transformation with a permutation matrix. In the example at hand, we have the permutation

$$(1,2,3,4,5) \rightarrow (1,5,3,4,2),$$

which we abbreviate to (1,5,3,4,2). Slightly more complicated is the example

$$\begin{bmatrix} x & x & & & & x \\ x & x & x & & & \\ & x & x & x & & \\ & & x & x & x & \\ & & & x & x & x \\ x & & & & x & x \end{bmatrix} .$$

The permutation (1,3,5,6,4,2) leads to the matrix

$$\begin{bmatrix} x & x & x & & & \\ x & x & & x & & \\ x & & x & & x & \\ & x & & x & & x \\ & & x & & x & x \\ & & & x & x & x \end{bmatrix} .$$

The band width has been reduced from 11 to 5. A further reduction of band width is not possible.

For each matrix A there exists a permutation which transforms A into a matrix with minimal band width. This permutation is not uniquely determined, as a rule. Unfortunately there is no algorithm for large n which finds a permutation of this type within reasonable bounds of computation. The algorithms used in practice produce permutations which typically have a band width close to the minimal band width. There are no theoretical predictions which indicate how far this band width deviates from the minimal one.

Among the numerous *approximating* algorithms of this type, the algorithm of Cuthill-McKee 1969 and a variation of Gibbs-Poole-Stockmeyer 1976 are used the most. Both algorithms are based on graph theoretical considerations. Sometimes the first algorithm provides the better result, and other times, the second. The difference is usually minimal. The second algorithm almost always requires less computing time. For that reason, we want to expand on it here. The corresponding FORTRAN program is to be found in Appendix 5.

The band widths of

$$A = (a_{ij}) \quad \text{and} \quad B = (|a_{ij}| + |a_{ji}|)$$

are the same. For an arbitrary permutation matrix P, the band widths of

$$P^{-1}AP \quad \text{and} \quad P^{-1}BP$$

remain the same. Thus we may assume that A is symmetric. The diagonal elements of A have no significance for band width. So, without loss of generality, we can let the diagonal entries be all ones.

The i-th and j-th row of A are called connected if $a_{ij} \neq 0$. We write $zi \sim zj$ or $zj \sim zi$. Thus, for

$$A = \begin{bmatrix} x & x & x \\ x & x & \\ x & & x \end{bmatrix}$$

we have $z1 \sim z1, z1 \sim z2, z1 \sim z3, z2 \sim z1, z2 \sim z2, z3 \sim z1, z3 \sim z3$. This leads to the picture

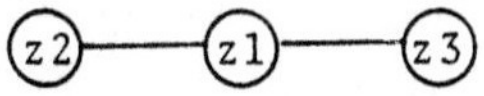

This can be regarded as an undirected graph with numbered knots. The rows of the matrix are the knots of the graph.

Conversely, the plan of the matrix can be reconstructed from a graph. Thus, the graph

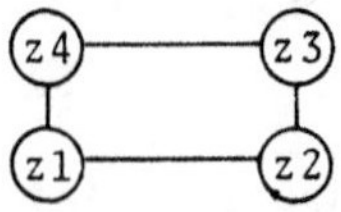

yields the matrix

$$A = \begin{bmatrix} x & x & & x \\ x & x & x & \\ & x & x & x \\ x & & x & x \end{bmatrix}.$$

Definition 20.3: $(G,\sim)$ is called an *undirected graph* if the following hold:

(1) G is a nonempty set. The elements are called *knots*.

(2) $\sim$ is a relation between certain knots g and h. Notation: "$g \sim h$" or "g and h are *connected*".

(3) For all $g \in G$, $g \sim g$. $g \sim h$ always implies $h \sim g$.

A knot g has *degree* p if g is connected with exactly $p+1$ knots in G. A graph is called *connected* if, for arbitrary knots g and h in G there always exists an $r \in \mathbb{N}$ and $k_i \in G$, $i = 0(1)r$, with the properties:

(4) $k_0 = g$, $k_r = h$.

(5) $k_{i-1} \sim k_i$ for $i = 1(1)r$. □

Let $\tilde{G}$ be an arbitrary nonempty subset of G. Then $(\tilde{G},\sim)$ is also a graph (subgraph of $(G,\sim)$). If $(\tilde{G},\sim)$ is connected, we simply say that $\tilde{G}$ is connected.

Definition 20.4: Let the knots of a finite graph $(G,\sim)$ be numbered arbitrarily, i.e. let $G = \{g_1,g_2,\ldots,g_n\}$.

$$w = 1 + 2\max\{d \mid d = |i-j| \text{ where } g_i \sim g_j\}$$

is called the *band width of the graph* $(G,\sim)$ *with respect to the given numbering*. □

The problem now becomes one of finding a numbering of the knots for which the band width w is as small as possible. As a first step, we separate the graph into levels.

Definition 20.5: $L = (L_1, L_2, \ldots, L_r)$ is called the *level structure* of the graph $(G,\sim)$ when the following hold true:

(1) The sets L_i, $i = 1(1)r$, (*levels*) are nonempty disjoint subsets of G. Their union is G.

(2) For $g \in L_i$, $h \in L_j$ and $g \sim h$, $|i-j| \leq 1$.

r is called the *depth* of the level structure. Its *width* k is the number of elements in the level with the most elements. □

Figure 20.6 shows how to separate a graph into levels.

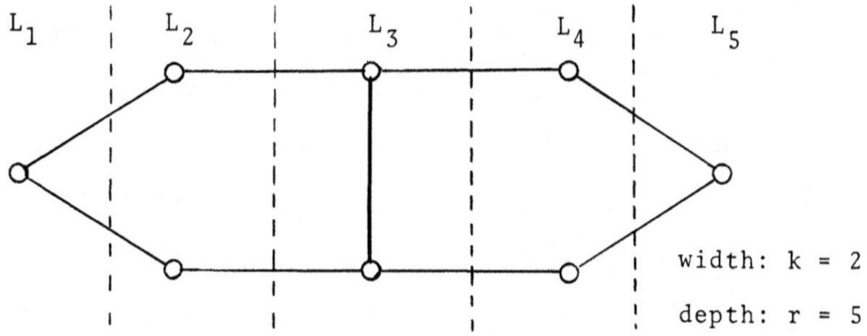

Figure 20.6. Separating a graph into levels

Theorem 20.7: Let $L = (L_1, L_2, \ldots, L_r)$ be a level structure of $(G,\sim)$ of width k. Then there exists a numbering of the knots such that the band width of the graph with respect to this numbering is less than or equal to $4k-1$.

Proof: First number the elements of L_1, then the elements of L_2, etc. For $g_\mu \sim g_\nu$, $g_\mu \in L_i$, and $g_\nu \in L_j$ it follows that $|i-j| \leq 1$, $|\mu-\nu| \leq 2k-1$, and hence, $w \leq 4k-1$. □

This theorem gives reason for constructing level structures with the smallest width possible. Usually constructing level structures of the greatest possible depth leads to the same end.

Theorem 20.8: For every knot g of a connected graph there exists exactly one level structure

$$R(g) = (L_1, L_2, \ldots, L_r)$$

satisfying:

(1) $L_1 = \{g\}$.

(2) For every $h \in L_i$ with $i > 1$, there is a $k \in L_{i-1}$ with $k \sim h$. This level structure is called the level structure with root g.

The proof is trivial. □

In the following, let the graph $(G,\sim)$ be finite and connected. In actuality, the connected components can be numbered sequentially. The algorithm for band width reduction begins with the following steps:

(A) Choose a knot $g \in G$ of minimal degree.

(B) Compute $R(g) = (L_1, L_2, \ldots, L_r)$. Let the elements of the last level L_r be k_j, $j = 1(1)p$.

(C) Set $j = 1$.

(D) Compute $R(k_j) = (M_1, M_2, \ldots, M_s)$ and set m_j = width of level structure $R(k_j)$. If $s > r$, set $g = k_j$ and return to (B).

(E) If $j < p$, increase j by 1 and repeat (D).

(F) Choose $j \in \{1, \ldots, p\}$ so that m_j is minimal. Then set $h = k_j$.

The first part of the algorithm thus determines two knots g and h and the corresponding level structures

$$R(g) = (L_1,L_2,\dots,L_r), \quad R(h) = (M_1,M_2,\dots,M_s).$$

Lemma 20.9: The following are true:

(1) $s = r$.

(2) $h \in L_r$.

(3) $g \in M_r$.

(4) $L_i \subset \bigcup_{j=1}^{i} M_{r+1-j} = A_i, \quad i = 1(1)r.$

(5) $M_i \subset \bigcup_{j=1}^{i} L_{r+1-j} = B_i, \quad i = 1(1)r.$

Proof: Conclusion (2) is trivial. For $i > s$ let M_i be the empty set. We then prove Conclusion (5) by induction. By (2), the induction begins with $M_1 = \{h\} \subset L_r$. Now the induction step. All knots in M_{i+1} are connected with knots in M_i (cf. Theorem 20.8(2)). Since the elements of L_μ are connected only to elements in $L_{\mu-1}$, L_μ, and $L_{\mu+1}$, and since M_i is a subset of B_i, we have $M_{i+1} \subset B_{i+1}$.

Conclusion (5) implies that $s \geq r$. By construction (cf. Step (D)), $s \leq r$. This proves Conclusion (1). From (5) we obtain

$$M_i \subset B_i$$

$$\bigcup_{i=1}^{r-1} M_i \subset \bigcup_{i=1}^{r-1} B_i = B_{r-1} = G-L_1$$

$$L_1 \subset M_r$$

$$g \in M_r.$$

This is conclusion (3). The proof of (4) is analogous to the proof of (5), in view of (3). □

The conclusions of the lemma need to be supplemented by experience. In most practical applications it turns out that:

(6) The depth r of the two level structures $R(g)$ and $R(h)$ is either the maximal depth which can be achieved for a level structure of $(G,\sim)$ or a good approximation thereto.

(7) The sets $L_i \cap M_{r+1-i}$, $i = 1(1)n$, in many cases contain most of the elements of $L_i \cup M_{r+1-i}$. The two level structures thus are very similar, through seldom identical.

(8) The return from step (D) of the algorithm to step (B) occurs at all in only a few examples, and then no more frequently than once per graph. This observation naturally is of great significance for computing times.

In the next part of the algorithm, the two level structures $R(g)$ and $R(h)$ are used to construct a level structure

$$S = (S_1, S_2, \ldots, S_r)$$

of the same depth and smallest possible width. In the process, every knot $k \in G$ is assigned to a level S_i by *one* of the following rules:

Rule 1: If $k \in L_i$, let $k \in S_i$.

Rule 2: If $k \in M_{r+1-i}$, let $k \in S_i$.

For the elements of $L_i \cap M_{r+1-i}$, the two rules have the same result. So in any case, $L_i \cap M_{r+1-i} \subset S_i$. The set

$$V = G - \bigcup_{i=1}^{r} (L_i \cap M_{r+1-i})$$

splits into p connected components, $V_1, V_2, \ldots, V_p$.

Unfortunately it is not possible to use one of the Rules 1 and 2 independently of each other for all the elements of V. Such an approach would generally not lead to a level structure S. But there is

Lemma 20.10: If in each connected component of V the same rule (either always 1 or always 2) is applied constantly, then S is a level structure. We leave the proof to the reader.

In the following we use $|T|$ to denote the number of elements of the set T. Let κ_1 be the width of $R(g)$ and κ_2 the width of $R(h)$. The second part of the algorithm consists of four separate steps for determining S:

(G) Compute κ_1 and κ_2, set

$$S_i = L_i \cap M_{r+1-i}, \quad i = 1(1)r$$

and determine V. If V is the empty set, this part of the algorithm is complete (and continue at (K)). Otherwise split V into connected components V_j, $j = 1(1)p$. Order the V_j so that always $|V_{j+1}| \le |V_j|$, $j = 1(1)p-1$.

(H) Set $\nu = 1$.

(I) Expand all S_i to $\tilde{S}_i$, $i = 1(1)r$, by including the knots from V_ν by rule 1. Expand S_i to $\tilde{\tilde{S}}_i$, $i = 1(1)r$, by including the knots from V_ν by rule 2. Compute

$$\kappa_3 = \max\{|\tilde{S}_i| \;|\; i = 1(1)r \text{ where } \tilde{S}_i \neq S_i\}$$

$$\kappa_4 = \max\{|\tilde{\tilde{S}}_i| \;|\; i = 1(1)r \text{ where } \tilde{\tilde{S}}_i \neq S_i\}.$$

If $\kappa_3 < \kappa_4$ or if $\kappa_3 = \kappa_4$ and $\kappa_1 \le \kappa_2$, set $S_i = \tilde{S}_i$; otherwise set $S_i = \tilde{\tilde{S}}_i$, $i = 1(1)r$.

(J) For $\nu < p$, increase ν by 1 and repeat (I).

This completes the computation of the level structure S. Figure 20.11 shows the level structures R(g), R(h), and S for a graph. The knots in G-V are denoted by x. In this case V consists of two components, each having three knots. For the left component rule 2 was used, and for the right, rule 1. In this way, one obtains the optimal width of 3. The second part of the algorithm consumes the greater part of the computing time. This is especially so when V has many components.

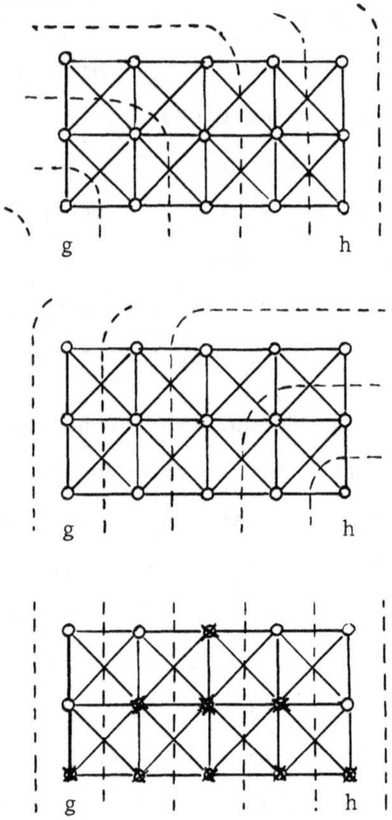

Figure 20.11. Level structures R(g), R(h), and S

Finally, in the third and last part of the algorithm, one starts with the level structure S and derives the numbering of the graph:

(K) Set $p = 1$. Go to (N).

(L) Let k run through all the elements of S_{p-1} in the order of the numbering. For fixed k, number all $\tilde{k} \in S_p$ which are connected to k and which have not yet been numbered so that the degree of $\tilde{k}$ does not decrease.

(M) Let k run through all the elements of S_p which have already been numbered, in the order of the numbering. For fixed k, number all $\tilde{k} \in S_p$ which have not been numbered and which are connected to k so that the degree of $\tilde{k}$ does not decrease.

(N) If there remain elements of S_p which are unnumbered, search for an unnumbered element of S_p of minimal degree. Assign it the next number and return to (M).

(O) Increase p by 1 and, if $p \le r$, return to (L).

This last step also completes the numbering. In some cases, the result can be improved through the use of an iterative algorithm due to Rosen 1968. Lierz 1976 contains a report on practical experience with the algorithm.

All algorithms for band width reduction are useful primarily for sparse matrices which are not too large and irregularly patterned. These arise primarily in finite element approaches where the geometry is complicated. Typical magnitudes are on the order of $n = 1000$ and $w = 50$.

21. Buneman Algorithm

Let G be a rectangle with sides parallel to the axes, and let the following problem be posed on G:

$$\begin{aligned} \Delta u(x,y) &= q(x,y), & (x,y) &\in G \\ u(x,y) &= \psi(x,y), & (x,y) &\in \partial G. \end{aligned}$$

This problem is to be solved with a five point difference method. We will assume that the distance separating neighboring lattice points from each other is always equal to h (cf. Section 13). The difference equations for the discrete function w are then uniformly the same for all internal lattice points:

$$w(x+h,y) + w(x-h,y) + w(x,y+h) + w(x,y-h) - 4w(x,y) = h^2 q(x,y).$$

If one of the points $(x+h,y)$, $(x-h,y)$, $(x,y+h)$, or $(x,y-h)$ should lie in the boundary of the rectangle, w is to be replaced by the boundary value ψ at that point. We number the points by rows or columns. In either case, we obtain a linear system of equations of the following type:

$$\begin{aligned} &Mw = z \\ &M = \begin{bmatrix} A & I & & \\ I & A & \ddots & \\ & \ddots & \ddots & I \\ & & I & A \end{bmatrix} \in \mathrm{MAT}(pn,pn,\mathbb{R}) \\ &A = \begin{bmatrix} -4 & 1 & & \\ 1 & -4 & \ddots & \\ & \ddots & \ddots & 1 \\ & & 1 & -4 \end{bmatrix} \in \mathrm{MAT}(n,n,\mathbb{R}) \end{aligned} \qquad (21.1)$$

$$w = \begin{bmatrix} w_1 \\ \vdots \\ w_p \end{bmatrix}, \quad z = \begin{bmatrix} z_1 \\ \vdots \\ z_p \end{bmatrix}; \; w_i, z_i \in \mathbb{R}^n \quad (i=1(1)p).$$

The inhomogeneity z contains $h^2 q(x,y)$ as well as the boundary value $\psi(x,y)$.

O. Buneman has developed a simple, fast algorithm for solving (21.1) for the case where

$$p = 2^{k+1} - 1, \quad k \in \mathbb{N}.$$

In the following, we always let $w_o = w_{2^{k+1}} = 0$. Three consecutive components of w will satisfy the block equations:

$$\begin{aligned} w_{j-2} + Aw_{j-1} + w_j \qquad\qquad\qquad &= z_{j-1} \\ w_{j-1} + Aw_j + w_{j+1} \qquad &= z_j \\ w_j + Aw_{j+1} + w_{j+2} &= z_{j+1}. \end{aligned}$$

Multiply the middle equation by $-A$ and then add the three equations. When restricted to $j = 2(2)2^{k+1}-2$, this results in

$$w_{j-2} + (2I-A^2)w_j + w_{j+2} = z_{j-1} + z_{j+1} - Az_j. \qquad (21.2)$$

This represents 2^k-1 block equations for the unknowns of even index. After solving this *reduced* system, we can determine the remaining unknowns of odd index by solving the n-dimensional system

$$Aw_j = z_j - w_{j-1} - w_{j+1}, \quad j = 1(2)2^{k+1}-1 \qquad (21.3)$$

The reduction process described here can now be applied again, to the 2^k-1 block equations (21.2), since these have the same structure as (21.1). Using the notation

$$A^{(1)} = 2I - A^2$$
$$z_j^{(1)} = z_{j-1} + z_{j+1} - Az_j, \qquad j = 2(2)2^{k+1}-2$$

we obtain the *twice-reduced* system of block equations

$$w_{j-4} + [2I-(A^{(1)})^2]w_j + w_{j+4} = z_{j-2}^{(1)} + z_{j+2}^{(1)} - A^{(1)} z_j^{(1)}, \quad j=4(4)2^{k+1}-4.$$

What we have here is a system with $2^{k-1}-1$ block equations. After this system is solved, the remaining unknowns of even index can be found as the solution of the n-dimensional system

$$A^{(1)} w_j = z_j^{(1)} - w_{j-2} - w_{j+2}, \qquad j = 2(4)2^{k+1}-2.$$

The remaining unknowns of odd index are then computed via (21.3).

The reduction process we have described can obviously be repeated again and again, until after k reduction steps only one system, with a single block equation, remains to be solved. After that comes the process of substituting the unknowns already computed in the remaining systems of equations.

The entire method can be described recursively as follows. Let

$$A^{(0)} = A$$
$$z_j^{(0)} = z_j, \qquad j = 1(1)2^{k+1}-1.$$

Then, for $r = 1(1)k$, define

$$\begin{aligned} A^{(r)} &= 2I - (A^{(r-1)})^2 \\ z_j^{(r)} &= z_{j-2^{r-1}}^{(r-1)} + z_{j+2^{r-1}}^{(r-1)} - A^{(r-1)} z_j^{(r-1)}, \\ &\qquad j = 2^r(2^r)2^{k+1}-2^r. \end{aligned} \tag{21.4}$$

Then, for $r = k(-1)0$, solve successively

$$A^{(r)}w_j = z_j^{(r)} - w_{j+2^r} - w_{j-2^r}, \quad j=2^r(2^{r+1})2^{k+1}-2^r. \quad (21.5)$$

This method is known as *cyclic odd/even reduction,* and frequently, simply as the *COR* algorithm. The band width of the matrices $A^{(r)}$ grows with r, so that a direct determination of the matrices requires extensive matrix multiplication. For this reason, they are explicitly represented instead as the product of simpler matrices in the following.

The recursion formula (21.4) implies that $A^{(r)}$ is a polynomial P_{2^r} of degree 2^r in A:

$$A^{(r)} = P_{2^r}(A) = \sum_{j=0}^{2^{r-1}} c_{2j}^{(r)} A^{2j}, \quad c_{2j}^{(r)} \in \mathbb{R}, \quad r = 1(1)k.$$

The highest order coefficient is

$$c_{2^r}^{(r)} = -1.$$

We want to represent the polynomial as a product of linear factors. To do that, we need to know its zeros. These are found most easily via the substitution $\xi = -2\cos\Theta$. From (21.4) we obtain the functional equation

$$P_{2^r}(\xi) = 2 - [P_{2^{r-1}}(\xi)]^2$$

which reduces inductively to

$$P_{2^r}(\xi) = -2\cos(2^r\Theta).$$

The zeros of $P_{2^r}(\xi)$ thus are

$$\xi_j = -2\cos\left(\frac{2j-1}{2^{r+1}}\pi\right), \quad j = 1(1)2^r.$$

The matrices $A^{(r)}$ therefore can be represented as a product of linear factors:

$$A^{(r)} = \begin{cases} A, & r = 0 \\ -\prod_{j=1}^{2^r}\left[A + 2\cos\left(\frac{2j-1}{2^{r+1}}\pi\right)I\right], & r=1(1)k. \end{cases} \tag{21.6}$$

This factorization can be used both in the reduction phase (21.4) for computing $z_j^{(r)}$ and in the following solution phase (21.5) for determining w_j. In this way, the matrix multiplication as well as the systems of equations are limited to tridiagonal matrices of a special kind.

This last described method is called *cyclic odd/even reduction* with *factorization*, and frequently is simply known as the *CORF* algorithm.

Various theoretical and practical investigations have demonstrated that the COR and CORF algorithms described here are numerically unstable (cf. e.g. Buzbee-Golub-Nielson 1970 and Schröder-Trottenberg-Reutersberg 1976). *Buneman* has developed a stabilization (cf. e.g. Buzbee-Golub-Nielson 1970) which is based on a mathematically equivalent reformulation for computing the vectors $z_j^{(r)}$ from (21.4).

Let

$$p_j^{(0)} = 0, \quad q_j^{(0)} = z_j, \quad j = 1(1)2^{k+1}-1.$$

Then, for $r = 1(1)k$, define

$$\begin{aligned} p_j^{(r)} &= p_j^{(r-1)} - (A^{(r-1)})^{-1}(p_{j-2^{r-1}}^{(r-1)} + p_{j+2^{r-1}}^{(r-1)} - q_j^{(r-1)}) \\ q_j^{(r)} &= q_{j-2^{r-1}}^{(r-1)} + q_{j+2^{r-1}}^{(r-1)} - 2p_j^{(r)} \\ j &= 2^r(2^r)2^{k+1} - 2^r. \end{aligned} \tag{21.7}$$

The sequences $p_j^{(r)}$, $q_j^{(r)}$, and $z_j^{(r)}$ are related as follows:

$$z_j^{(r)} = A^{(r)} p_j^{(r)} + q_j^{(r)}$$
$$r = 0(1)k, \quad j = 2^r(2^r)2^{k+1} - 2^r. \tag{21.8}$$

The proof follows by induction on r. For $r = 0$, the assertion is immediate, since $p_j^{(0)} = 0$, and $q_j^{(0)} = z_j = z_j^{(0)}$. Now suppose the conclusion (21.8) holds for r. The induction step $r \Rightarrow r+1$ is:

$$A^{(r+1)} p_j^{(r+1)} + q_j^{(r+1)} = 2p_j^{(r+1)} - (A^{(r)})^2 p_j^{(r+1)} + q_{j-2^r}^{(r)} + q_{j+2^r}^{(r)} - 2p_j^{(r+1)}$$
$$= -(A^{(r)})^2 p_j^{(r)} + A^{(r)}(p_{j-2^r}^{(r)} + p_{j+2^r}^{(r)} - q_j^{(r)}) + q_{j-2^r}^{(r)} + q_{j+2^r}^{(r)}$$
$$= A^{(r)} p_{j-2^r}^{(r)} + q_{j-2^r}^{(r)} + A^{(r)} p_{j+2^r}^{(r)} + q_{j+2^r}^{(r)} - A^{(r)}(A^{(r)} p_j^{(r)} + q_j^{(r)})$$
$$= z_{j-2^r}^{(r)} + z_{j+2^r}^{(r)} - A^{(r)} z_j^{(r)} = z_j^{(r+1)}.$$

The Buneman algorithm can be summarized as follows:

1. *Reduction phase:* Compute the sequences $p_j^{(r)}$ and $q_j^{(r)}$ from (21.7):

$$p_j^{(0)} = 0, \; q_j^{(0)} = z_j, \quad j = 1(1)2^{k+1}-1$$
$$A^{(r-1)}\left(p_j^{(r-1)} - p_j^{(r)}\right) = p_{j-2^{r-1}}^{(r-1)} + p_{j+2^{r-1}}^{(r-1)} - q_j^{(r-1)} \tag{21.9}$$
$$p_j^{(r)} = p_j^{(r-1)} - \left(p_j^{(r-1)} - p_j^{(r)}\right)$$
$$q_j^{(r)} = q_{j-2^{r-1}}^{(r-1)} + q_{j+2^{r-1}}^{(r-1)} - 2p_j^{(r)}$$
$$r = 1(1)k, \quad j = 2^r(2^r)2^{k+1} - 2^r.$$

2. *Solution phase:* Solve the system of equations (21.5), using the relation (21.8):

$$A^{(r)}(w_j - p_j^{(r)}) = q_j^{(r)} - w_{j+2^r} - w_{j-2^r}$$

$$w_j = p_j^{(r)} + (w_j - p_j^{(r)}) \qquad (21.10)$$

$$r = k(-1)0, \quad j = 2^r(2^{r+1})2^{k+1} - 2^r.$$

In both the reduction and solution phase we have systems of equations with coefficient matrix $A^{(r)}$, and these systems are solved by using the factorization (21.6). Thus, for each system it is necessary to solve 2^r systems in the special symmetric, diagonal dominant, tridiagonal matrix

$$\begin{bmatrix} c-4 & 1 & & \\ 1 & c-4 & \ddots & \\ & \ddots & \ddots & 1 \\ & & 1 & c-4 \end{bmatrix} \in \mathrm{MAT}(n,n,\mathbb{R}),$$

$$-2 < c = 2\cos\left(\frac{2j-1}{2^{r+1}}\pi\right) < 2.$$

The vectors $p_j^{(0)}$ and $q_j^{(0)}$, which are set equal to zero and z_j, respectively, at the beginning of the reduction phase, take up a segment of length $pn = (2^{k+1}-1)n$ in memory. Because of the special order of computation, the vectors $p_j^{(r)}$ and $q_j^{(r)}$ can overwrite $p_j^{(0)}$ and $q_j^{(0)}$ in their respective places, since the latter quantities are no longer needed. Similarly, in the solution phase, the successively computed solution vectors w_j can overwrite the $q_j^{(r)}$ in their corresponding places.

For $k = 2$ and $p = 7$, the sequence of computation and memory content in the reduction and solution phases is given in Table 21.11.

Table 21.11: Stages of the computation for the reduction phase (upper part) and the solution phase (lower part)

Step	r	requires	computes	memory content at end of step
1	0	initialization: $p_j^{(0)}=0,\ q_j^{(0)}=z_j$	$j=1(1)7$	
2	1	$p_1^{(0)},p_2^{(0)},p_3^{(0)},p_4^{(0)},p_5^{(0)},p_6^{(0)},p_7^{(0)}$ $q_1^{(0)},q_2^{(0)},q_3^{(0)},q_4^{(0)},q_5^{(0)},q_6^{(0)},q_7^{(0)}$	$p_2^{(1)},p_4^{(1)},p_6^{(1)}$ $q_2^{(1)},q_4^{(1)},q_6^{(1)}$	$p_1^{(0)},p_2^{(1)},p_3^{(0)},p_4^{(1)},p_5^{(0)},p_6^{(1)},p_7^{(0)}$ $q_1^{(0)},q_2^{(1)},q_3^{(0)},q_4^{(1)},q_5^{(0)},q_6^{(1)},q_7^{(0)}$
3	2	$p_2^{(1)},p_4^{(1)},p_6^{(1)}$ $q_2^{(1)},q_4^{(1)},q_6^{(1)}$	$p_4^{(2)}$ $q_4^{(2)}$	$p_1^{(0)},p_2^{(1)},p_3^{(0)},p_4^{(2)},p_5^{(0)},p_6^{(1)},p_7^{(0)}$ $q_1^{(0)},q_2^{(1)},q_3^{(0)},q_4^{(2)},q_5^{(0)},q_6^{(1)},q_7^{(0)}$
1	2	$p_4^{(2)}$ $q_4^{(2)}$	w_4	$q_1^{(0)},q_2^{(1)},q_3^{(0)},w_4,q_5^{(0)},q_6^{(1)},q_7^{(0)}$
2	1	$p_2^{(1)},p_6^{(1)}$ $q_2^{(1)},q_6^{(1)}$ w_4	w_2,w_6	$q_1^{(0)},w_2,q_3^{(0)},w_4,q_5^{(0)},w_6,q_7^{(0)}$
3	0	$p_1^{(0)},p_3^{(0)},p_5^{(0)},p_7^{(0)}$ $q_1^{(0)},q_3^{(0)},q_5^{(0)},q_7^{(0)}$ w_2,w_4,w_6	w_1,w_3,w_5,w_7	$w_1,w_2,w_3,w_4,w_5,w_6,w_7$

The Buneman algorithm described here thus requires twice as many memory locations as there are discretization points. There exists a modification, (see Buzbee-Golub-Nielson 1970), which uses the same number of locations. However, the modification requires more extensive computation.

We next determine the number of arithmetic operations in the Buneman algorithm for the case of a square G, i.e. $p = n$. Solving a single tridiagonal system requires $6n-5$ operations when a specialized Gaussian elimination is used. For the reduction phase, noting that

$$p = n = 2^{k+1} - 1, \quad k+1 = \log_2(n+1),$$

we find the number of operations to be

$$\sum_{r=1}^{k} \sum_{j\varepsilon M_r} [6n+2^{r-1}(6n-5)]$$

where

$$M_r = \{j = 2^r(2^r)2^{k+1}-2^r\}$$

and cardinality$(M_r) = 2^{k+1-r} - 1$. Therefore,

$$\sum_{r=1}^{k} (2^{k+1-r}-1)[6n+2^{r-1}(6n-5)] = 3n^2\log_2(n+1) + O[n \log_2(n+1)].$$

Similarly, for the solution phase we have

$$\sum_{r=0}^{k} \sum_{j\varepsilon\tilde{M}_r} [3n+2^r(6n-5)]$$

where

$$\tilde{M}_r = \{j = 2^r(2^{r+1})2^{k+1}-2^r\}$$

and cardinality$(\tilde{M}_r) = 2^{k-r}$. Therefore,

$$\sum_{r=0}^{k} 2^{k-r}[3n+2^r(6n-5)] = 3n^2\log_2(n+1)+3n^2+O[n \log_2(n+1)].$$

Altogether, we get that the number of operations in the Buneman algorithm is

$$6n^2 \log_2(n+1) + 3n^2 + O[n \log_2(n+1)].$$

At the beginning of the method, it is necessary to compute values of cosine, but with the use of the appropriate recursion formula, this requires only $O(n)$ operations, and thus can be ignored.

Finally, Appendix 6 contains a FORTRAN program for the Buneman algorithm.

22. The Schröder-Trottenberg reduction method

We begin by way of introduction with the ordinary differential equation

$$-u''(x) = q(x).$$

The standard discretization is

$$\frac{2u(x)-u(x+h)-u(x-h)}{h^2} = q(x)$$

for which there is the alternative notation

$$(2I-T_h-T_h^{-1})u(x) = h^2 q(x). \tag{22.1}$$

Here I denotes the identity and T_h the translation operator defined by

$$T_h u(x) = u(x+h).$$

Multiplying equation (22.1) by

$$(2I+T_h+T_h^{-1})$$

yields

$$(2I-T_h^2-T_h^{-2})u(x) = h^2(2I+T_h+T_h^{-1})q(x).$$

Set

$$q_0(x) = q(x), \quad q_1(x) = (2I+T_h+T_h^{-1})q_0(x)$$

and use the relations

$$T_h^j = T_{jh}, \quad T_h^{-j} = T_{jh}^{-1}, \quad j \in \mathbb{N}.$$

This simple notational change leads to the *simply reduced* equation

$$(2I-T_{2h}-T_{2h}^{-1})u(x) = h^2 q_1(x).$$

This process can be repeated arbitrarily often. The *m-fold reduced* equations are

$$(2I-T_{kh}-T_{kh}^{-1})u(x) = h^2 q_m(x) \tag{22.2}$$

where $k = 2^m$ and

$$q_m(x) = (2I+T_{\ell h}+T_{\ell h}^{-1})q_{m-1}(x), \quad \ell = 2^{m-1}.$$

The boundary value problem

$$-u''(x) = q(x), \quad u(0) = A, \quad u(1) = B$$

can now be solved as follows. Divide the interval $[0,1]$ into 2^n subintervals of length $h = 1/2^n$. Then compute sequentially

$q_1(x)$ at the points $2h, 4h, \ldots, 1-2h$

$q_2(x)$ at the points $4h, 8h, \ldots, 1-4h$

$\vdots$

$q_{n-1}(x)$ at the point $2^{n-1}h = 1/2$.

The $(n-1)$-fold reduced equation

$$(2I-T_{2^{n-1}h}-T^{-1}_{2^{n-1}h})u(1/2) = h^2 q_{n-1}(1/2)$$

allows $u(1/2)$ to be determined immediately, since the values of u are known at the endpoints 0 and 1. Similarly, we immediately obtain $u(1/4)$ and $u(3/4)$ from the (n-2)-fold reduced equation. Continuing with the method described leads successively to the functional values of u at the lattice points jh, $j = 1(1)2^n-1$.

In the following we generalize the one dimensional reduction method to differential or difference equations with constant coefficients.

Definition 22.3: Let $G_h = \{\nu h \mid \nu \in \mathbb{Z}\}$, $h > 0$. Then we call

$$S_h = \sum_{\nu=-r}^{r} \alpha_\nu T_h^\nu, \qquad \alpha_\nu \in \mathbb{R}$$

a *one-dimensional difference star* on the lattice G_h. S_h is called *symmetric* if $\alpha_{-\nu} = \alpha_\nu$ for $\nu = 1(1)r$. S_h is called *odd* (*even*) if $\alpha_{2\nu} = 0$ $(\alpha_{2\nu+1} = 0)$ for all ν with $-r \leq 2\nu \leq r$ $(-r \leq 2\nu+1 \leq r)$. In Schröder-Trottenberg 1973, the sum for S_h is simply abbreviated to

$$[\alpha_{-r}\alpha_{-r+1} \cdots\cdots \alpha_{r-1}\alpha_r]_h. \qquad \square$$

Each difference star obviously can be split into a sum

$$S_h = P_h + Q_h,$$

where P_h is an even difference star, and Q_h is an odd one. Since $T_h^{2\nu} = T_{2h}^\nu$, the even part can also be regarded as the difference star of the lattice $G_{2h} = \{2\nu h \mid \nu \in \mathbb{Z}\}$.

The reduction step now looks like this:

$$S_h u(x) = (P_h+Q_h)u(x) = h^2 q(x)$$

$$(P_h-Q_h)(P_h+Q_h)u(x) = (P_h^2-Q_h^2)u(x) = h^2(P_h-Q_h)q(x).$$

Since $P_h^2 - Q_h^2$ obviously is an even difference star, it can be represented as a difference star on the lattice G_{2h}. We denote it by R_{2h}. We further define

$$q_1(x) = (P_h-Q_h)q(x).$$

Thus one obtains the simply reduced equation

$$R_{2h}u(x) = h^2 q_1(x).$$

In the special case

$$S_h = \alpha_{-1}T_h^{-1} + \alpha_o T_h^o + \alpha_1 T_h^1$$

one obtains

$$P_h^2 = \alpha_o^2 T_h^o$$

$$Q_h^2 = \alpha_{-1}^2 T_h^{-2} + 2\alpha_{-1}\alpha_1 T_h^o + \alpha_1^2 T_h^2$$

$$R_{2h} = -\alpha_{-1}^2 T_{2h}^{-1} + (\alpha_o^2 - 2\alpha_{-1}\alpha_1)T_{2h}^o - \alpha_1^2 T_{2h}.$$

The reduction process can now be carried on in analogy with (22.2). In general, the difference star will change from step to step. The number of summands, however, remains constant.

The reduction method described also surpasses the Gauss algorithm in one dimension in numerical stability. Our main concern however is with two dimensional problems. We begin once again with a simple example.

The standard discretization of the differential equation

$$-\Delta u(x) = q(x), \qquad x \in \mathbb{R}^2$$

is

$$4u(x)-u(x+he_1)-u(x-he_1)-u(x+he_2)-u(x-he_2) = h^2 q(x).$$

The translation operators

$$T_{\nu,h}u(x) = u(x+he_\nu)$$

lead to the notation

$$(4I-T_{1,h}-T_{1,h}^{-1}-T_{2,h}-T_{2,h}^{-1})u(x) = h^2 q(x).$$

Multiplying the equation by

$$(4I+T_{1,h}+T_{1,h}^{-1}+T_{2,h}+T_{2,h}^{-1})$$

yields

$$\{16I-2(T_{1,h}T_{2,h}+T_{1,h}T_{2,h}^{-1}+T_{1,h}^{-1}T_{2,h}+T_{1,h}^{-1}T_{2,h}^{-1}) - T_{1,h}^{2} - T_{1,h}^{-2}-T_{2,h}^{2}-T_{2,h}^{-2}-4I\}u(x)$$

$$= h^2(4I+T_{1,h}+T_{1,h}^{-1}+T_{2,h}+T_{2,h}^{-1})q(x).$$

We set

$$q_1(x) = (4I+T_{1,h}+T_{1,h}^{-1}+T_{2,h}+T_{2,h}^{-1})q(x)$$

and obtain the simply reduced equations

$$\{12I-2(T_{1,h}T_{2,h}+T_{1,h}T_{2,h}^{-1}+T_{1,h}^{-1}T_{2,h}+T_{1,h}^{-1}T_{2,h}^{-1}) - (T_{1,h}^{2}+T_{1,h}^{-2}+T_{2,h}^{2}+T_{2,h}^{-2})\}u(x) = h^2 q_1(x).$$

In contrast to the reduction process for a one dimensional difference equation, here the number of summands has increased from 5 to 9. The related lattice points have spread farther apart. The new difference star is an even polynomial in the four translation operators $T_{1,h}$, $T_{1,h}^{-1}$,

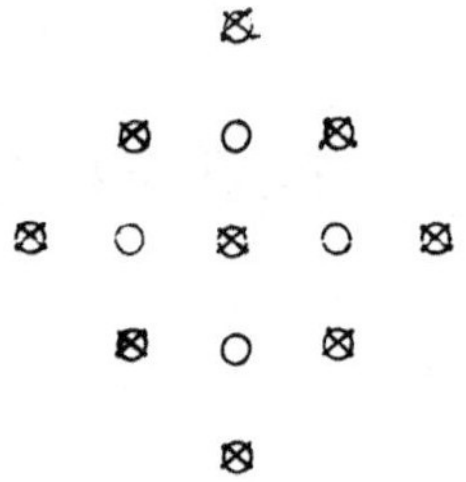

Figure 22.4: Related lattice points after one reduction step.

$T_{2,h}$, and $T_{2,h}^{-1}$. The related lattice points are shown in Figure 22.4 with an "x". Such an even polynomial can be rewritten as a polynomial in two 'larger' translation operators. Let

$$\begin{aligned} e_3 &= (e_1 + e_2)/\sqrt{2} \\ e_4 &= (e_2 - e_1)/\sqrt{2} \end{aligned} \tag{22.5}$$

be the unit vectors rotated by $\pi/4$ counterclockwise, and let

$$\begin{aligned} k &= h\sqrt{2} \\ T_{3,k}u(x) &= u(x+ke_3) = u(x+he_1+he_2) \\ T_{4,k}u(x) &= u(x+ke_4) = u(x+he_2-he_1). \end{aligned}$$

Then we have

$$\begin{aligned} T_{3,k} &= T_{1,h}\cdot T_{2,h} \\ T_{4,k} &= T_{1,h}^{-1}\cdot T_{2,h} \\ T_{3,k}^{-1} &= T_{1,h}^{-1}\cdot T_{2,h}^{-1} \\ T_{4,k}^{-1} &= T_{1,h}\cdot T_{2,h}^{-1}. \end{aligned} \tag{22.6}$$

The simply reduced equation therefore can be written in the following form:

$$\{12I-2(T_{3,k}+T_{4,k}+T_{3,k}^{-1}+T_{4,k}^{-1})-(T_{3,k}T_{4,k}+T_{3,k}T_{4,k}^{-1}+T_{3,k}^{-1}T_{4,k}$$
$$+ T_{3,k}^{-1}T_{4,k}^{-1}\}u(x) = h^2q_1(x).$$

The difference star on the left side once again can be split into an even part,

$$12I - (T_{3,k}T_{4,k}+T_{3,k}T_{4,k}^{-1}+T_{3,k}^{-1}T_{4,k}+T_{3,k}^{-1}T_{4,k}^{-1})$$

and an odd part,

$$-2(T_{3,k}+T_{4,k}+T_{3,k}^{-1}+T_{4,k}^{-1})$$

and reduced anew. One then obtains a polynomial in the translation operators

$$T_{1,2h},\ T_{1,2h}^{-1},\ T_{2,2h},\ T_{2,2h}^{-1}.$$

Thus, beginning with a polynomial in

$$T_{1,h},\ T_{1,h}^{-1},\ T_{2,h},\ T_{2,h}^{-1}$$

we have obtained, after one reduction step, a polynomial in

$$T_{3,h\sqrt{2}},\ T_{3,h\sqrt{2}}^{-1},\ T_{4,h\sqrt{2}},\ T_{4,h\sqrt{2}}^{-1}$$

and after two reduction steps, a polynomial in

$$T_{1,2h},\ T_{1,2h}^{-1},\ T_{2,2h},\ T_{2,2h}^{-1}$$

In particular, this means that the $\pi/4$ rotation $e_1 \to e_3$ and $e_2 \to e_4$ has been undone after the second reduction. This process as described can be repeated arbitrarily often. The amount of computation required, however, grows substantially with the repetitions. For this reason we have not carried the second reduction step through explicitly.

We now discuss the general case of two-dimensional reduction methods for differential and difference equations with constant coefficients. We preface this with some general results on polynomials and difference stars.

Definition 22.7: Let $P(x_1,\ldots,x_p)$ be a real polynomial in p variables $x_1,\ldots,x_p$. P is called *even* if

$$P(-x_1,\ldots,-x_p) = P(x_1,\ldots,x_p),$$

and P is called *odd* if

$$P(-x_1,\ldots,-x_p) = -P(x_1,\ldots,x_p). \qquad \square$$

It is obvious that the product of two even or of two odd polynomials is even, and that the product of an even polynomial with an odd one is odd. Further, every polynomial can be written as the sum of one even and one odd polynomial.

Lemma 22.8: Let $P(x_1,x_2,x_3,x_4)$ be an even polynomial. Then there exist polynomials $\tilde{P}$ and $\tilde{\tilde{P}}$ with the properties:

$$P(x,x^{-1},y,y^{-1}) = \tilde{P}(xy,(xy)^{-1},yx^{-1},xy^{-1})$$

$$P(xy,(xy)^{-1},yx^{-1},xy^{-1}) = \tilde{\tilde{P}}(x^2,x^{-2},y^2,y^{-2}).$$

Proof: We have

$$P(x,x^{-1},y,y^{-1}) = \sum_{i,j\in\mathbb{Z}} \alpha_{ij}x^iy^j, \qquad \alpha_{ij} \in \mathbb{R}.$$

Since P is even, we need only sum over those $i,j \in \mathbb{Z}$ with $i+j$ even. For such i,j we have

$$x^iy^j = (xy)^r(yx^{-1})^s, \quad r = (i+j)/2, \quad s = (j-i)/2.$$

We obtain

$$P(x,x^{-1},y,y^{-1}) = \sum_{r,s\in\mathbb{Z}} \alpha_{r-s,r+s}(xy)^r(yx^{-1})^s$$

and therefore, the first part of the conclusion. Similarly,

we obtain

$$P(xy,(xy)^{-1},yx^{-1},xy^{-1}) = \sum_{i,j\in\mathbb{Z}} \alpha_{ij}(xy)^i(yx^{-1})^j$$

$$(xy)^i(yx^{-1})^j = x^{2r}y^{2s}, \quad r = (i-j)/2, \quad s = (i+j)/2$$

$$P(xy,(xy)^{-1},yx^{-1},xy^{-1}) = \sum_{r,s\in\mathbb{Z}} \alpha_{r+s,s-r}x^{2r}y^{2s}. \qquad \square$$

Definition 22.9: For fixed $h > 0$, define the lattice G_ν by

$$G_\nu = \begin{cases} \{x\in\mathbb{R}^2 \mid x=2^{\nu/2}h(ie_1+je_2);\ i,j\in\mathbb{Z}\} & \text{when } \nu \text{ is even} \\ \{x\in\mathbb{R}^2 \mid x=2^{\nu/2}h(ie_3+je_4);\ i,j\in\mathbb{Z}\} & \text{when } \nu \text{ is odd.} \end{cases}$$

A *difference star* S_ν on the lattice G_ν is a polynomial in

$$T_{1,k},\ T_{1,k}^{-1},\ T_{2,k},\ T_{2,k}^{-1} \quad \text{for even } \nu$$

or in

$$T_{3,k},\ T_{3,k}^{-1},\ T_{4,k},\ T_{4,k}^{-1} \quad \text{for odd } \nu.$$

Here $k = 2^{\nu/2}h$. S_ν is called *even* (*odd*) if the corresponding polynomial is even (odd). $\square$ It follows that $G_0 \supset G_1 \supset G_2 \supset \ldots$. Figure 22.10 shows G_0, G_1, G_2, and G_3.

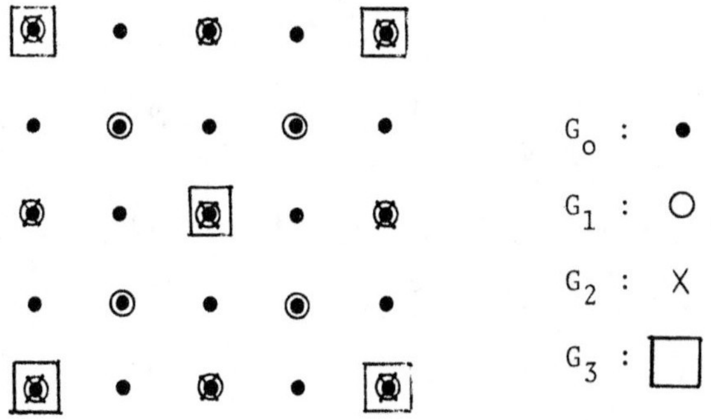

Figure 22.10: A representation of lattices G_0, G_1, G_2 and G_3.

<u>Theorem 22.11</u>: Let S_ν be a difference star on the lattice G_ν. Let S_ν be partitioned into a sum

$$S_\nu = P_\nu + Q_\nu$$

where P_ν is an even difference star and Q_ν is an odd difference star. Then

$$S_{\nu+1} = (P_\nu - Q_\nu)S_\nu$$

may be regarded as a difference star on the coarser lattice $G_{\nu+1} \subset G_\nu$.

Proof: The difference star

$$S_{\nu+1} = P_\nu^2 - Q_\nu^2$$

is even since P_ν^2 and Q_ν^2 are even. Now let ν be even. By Lemma 22.8, there exists a polynomial $\tilde{S}_{\nu+1}$ with

$$S_{\nu+1}(T_{1,k},T_{1,k}^{-1},T_{2,k},T_{2,k}^{-1})$$
$$= \tilde{S}_{\nu+1}(T_{1,k}T_{2,k},(T_{1,k}T_{2,k})^{-1},T_{2,k}T_{1,k}^{-1},T_{1,k}T_{2,k}^{-1}).$$

Since

$$T_{1,k}\cdot T_{2,k} = T_{3,k\sqrt{2}}$$
$$T_{2,k}\cdot T_{1,k}^{-1} = T_{4,k\sqrt{2}}$$

one obtains, where $m = k\sqrt{2} = 2^{(\nu+1)/2}h$,

$$S_{\nu+1}(T_{1,k},T_{1,k}^{-1},T_{2,k},T_{2,k}^{-1}) = \tilde{S}_{\nu+1}(T_{3,m},T_{3,m}^{-1},T_{4,m},T_{4,m}^{-1}).$$

The right side of this equation is a difference star on the lattice $G_{\nu+1}$. For odd ν, one obtains, correspondingly and taking note of (22.6):

$$S_{\nu+1}(T_{3,k},T_{3,k}^{-1},T_{4,k},T_{4,k}^{-1})$$
$$= S_{\nu+1}(T_{1,\ell}T_{2,\ell},(T_{1,\ell}T_{2,\ell})^{-1},T_{1,\ell}^{-1}T_{2,\ell},T_{1,\ell}T_{2,\ell}^{-1})$$
$$= \tilde{\tilde{S}}_{\nu+1}(T_{1,\ell}^{2},T_{1,\ell}^{-2},T_{2,\ell}^{2},T_{2,\ell}^{-2}) = \tilde{\tilde{S}}_{\nu+1}(T_{1,m},T_{1,m}^{-1},T_{2,m},T_{2,m}^{-1})$$
$$\ell = k/\sqrt{2}, \quad m = 2\ell = k\sqrt{2} = 2^{(\nu+1)/2}h. \quad \square$$

We may summarize these results as follows:

Reduction step for equations with constant coefficients:

initial equation:	$S_\nu u(x) = h^2 q_\nu(x)$,	$x \in G_\nu$
partition:	$S_\nu = P_\nu + Q_\nu$, P_ν even, Q_ν odd	
compute:	$q_{\nu+1}(x) = (P_\nu - Q_\nu)q_\nu(x)$,	$x \in G_{\nu+1}$
reduced equation:	$S_{\nu+1}u(x) = h^2 q_{\nu+1}(x)$,	$x \in G_{\nu+1}$.

The reduction process can be repeated arbitrarily often. The difference stars then depend only on the differential equation and not on its right side. Thus, for a particular type of differential equation, the stars can be computed once for all time and then stored.

Our representation of the difference stars becomes rather involved for large ν. Instead of

$$S_\nu = \begin{cases} \sum_{i,j\in\mathbb{Z}} a_{ij}T_{1,k}^{i}\cdot T_{2,k}^{j} & \text{for } \nu \text{ even} \\ \sum_{i,j\in\mathbb{Z}} a_{ij}T_{3,k}^{i}\cdot T_{4,k}^{j} & \text{for } \nu \text{ odd} \end{cases}$$
$$k = 2^{\nu/2}h$$

we can use the Schröder-Trottenberg abbreviation and write

$$S_\nu = \begin{bmatrix} & \vdots & \vdots & \vdots & \\ \cdots\cdots & a_{-1,1} & a_{0,1} & a_{1,1} & \cdots\cdots \\ \cdots\cdots & a_{-1,0} & a_{0,0} & a_{1,0} & \cdots\cdots \\ \cdots\cdots & a_{-1,-1} & a_{0,-1} & a_{1,-1} & \cdots\cdots \\ & \vdots & \vdots & \vdots & \end{bmatrix}_\nu$$

With the differential equation

$$-\Delta u(x) = q(x),$$

the standard discretization leads to the star

$$S_0 = \begin{bmatrix} 0 & -1 & 0 \\ -1 & 4 & -1 \\ 0 & -1 & 0 \end{bmatrix}_0 .$$

The first three reductions are:

$$S_1 = \begin{bmatrix} -1 & -2 & -1 \\ -2 & 12 & -2 \\ -1 & -2 & -1 \end{bmatrix}_1$$

$$S_2 = \begin{bmatrix} 0 & 0 & 1 & 0 & 0 \\ 0 & -2 & -32 & -2 & 0 \\ 1 & -32 & 132 & -32 & 1 \\ 0 & -2 & -32 & -2 & 0 \\ 0 & 0 & 1 & 0 & 0 \end{bmatrix}_2$$

$$S_3 = \begin{bmatrix} 1 & -4 & 6 & -4 & 1 \\ -4 & -752 & -2584 & -752 & -4 \\ 6 & -2584 & 13348 & -2584 & 6 \\ -4 & -752 & -2584 & -752 & -4 \\ 1 & -4 & 6 & -4 & 1 \end{bmatrix}_3 .$$

The even components P_ν of the stars S_ν are:

$$P_0 = \begin{bmatrix} 0 & 0 & 0 \\ 0 & 4 & 0 \\ 0 & 0 & 0 \end{bmatrix}_0 , \quad P_1 = \begin{bmatrix} -1 & 0 & -1 \\ 0 & 12 & 0 \\ -1 & 0 & -1 \end{bmatrix}_1$$

$$P_2 = \begin{bmatrix} 0 & 0 & 1 & 0 & 0 \\ 0 & -2 & 0 & -2 & 0 \\ 1 & 0 & 132 & 0 & 1 \\ 0 & -2 & 0 & -2 & 0 \\ 0 & 0 & 1 & 0 & 0 \end{bmatrix}_2$$

$$P_3 = \begin{bmatrix} 1 & 0 & 6 & 0 & 1 \\ 0 & -752 & 0 & -752 & 0 \\ 6 & 0 & 13348 & 0 & 6 \\ 0 & -752 & 0 & -752 & 0 \\ 1 & 0 & 6 & 0 & 1 \end{bmatrix}_3 .$$

When the reduction method is applied to boundary value problems for elliptic differential equations, a difficulty arises in that the formulas are valid only on a restricted region whenever the lattice point under consideration is sufficiently far from the boundary. In certain special cases, however, one is able to evade this constraint.

Let the unit square

$$G = (0,1) \times (0,1)$$

be given. Let the boundary conditions either be periodic, i.e.

$$\left.\begin{aligned} u(0,x_2) &= u(1,x_2) \\ u(x_1,0) &= u(x_1,1) \\ \partial_1 u(0,x_2) &= \partial_1 u(1,x_2) \\ \partial_2 u(x_1,0) &= \partial_2 u(x_1,1), \end{aligned}\right. \qquad x_1, x_2 \in [0,1]$$

or homogeneous, i.e. $u(x_1,x_2) = 0$ for $(x_1,x_2) \in \partial G$. Let the difference star approximating the differential equation be a symmetric nine point formula:

$$S_o = \begin{bmatrix} \gamma & \beta & \gamma \\ \beta & \alpha & \beta \\ \gamma & \beta & \gamma \end{bmatrix}_o .$$

The extension of the domain of definition of the difference equation

$$S_o u(x) = q_o(x)$$

to $x \in G_o$ is accomplished in these special cases by a continuation of the lattice function $u(x)$ to all of G_o. In the case of periodic boundary conditions, $u(x)$ and $q_o(x)$ are continued periodically on G_o. In the case of homogeneous boundary conditions, $u(x)$ and $q_o(x)$ are continued antisymmetrically with respect to the boundaries of the unit square:

$$u(-x_1,x_2) = -u(x_1,x_2), \quad q_o(-x_1,x_2) = -q_o(x_1,x_2)$$
$$u(x_1,-x_2) = -u(x_1,x_2), \quad q_o(x_1,-x_2) = -q_o(x_1,x_2)$$
$$u(1+x_1,x_2) = -u(1-x_1,x_2), \quad q_o(1+x_1,x_2) = -q_o(1-x_1,x_2)$$
$$u(x_1,1+x_2) = -u(x_1,1-x_2), \quad q_o(x_1,1+x_2) = -q_o(x_1,1-x_2)$$
$$q_o(x_1,x_2) = 0 \quad \text{for} \quad (x_1,x_2) \in \partial G.$$

This leads to doubly periodic functions. Figure 22.12 shows their relationship.

-	+	-	+	-
+	-	+	-	+
-	+	-	+	-

Figure 22.12: Antisymmetric continuation.

The homogeneous boundary conditions and the symmetry of the difference star assure the validity of the extended difference equations at the boundary points of G, and therefore, on all of G_0. An analogous extension of the exact solution of the differential equation, however, is not normally possible, since the resulting function will not be twice differentiable.

We present an example in which we carry out the reduction method in the case of a homogeneous boundary condition and for $h = 1/n$ and $n = 4$. After three reduction steps, we obtain a star, S_3, which is a polynomial in the translation operators

$$T_{3,k}, T_{3,k}^{-1}, T_{4,k}, T_{4,k}^{-1}, \quad k = 2^{3/2}h = 2h\sqrt{2}.$$

The corresponding difference equation holds on the lattice G_3 (cf. Figure 22.10). Because of the special nature of the extension of u, it only remains to satisfy the equation

$$S_3 u(1/2,1/2) = q_3(1/2,1/2).$$

By appealing to periodicity and symmetry, we can determine all summands from $u(1/2,1/2)$, $u(0,0)$, $u(1,0)$, $u(0,1)$, and $u(1,1)$. But the values of u on the boundary are zero. Thus $u(1/2,1/2)$ can be computed immediately. The equations resulting from the difference star S_2 on the lattice G_2 no longer contain new unknowns (cf. Figure 22.10). The equations

$$S_1 u(1/4,1/4) = q_1(1/4,1/4), \quad S_1 u(3/4,1/4) = q_1(3/4,1/4),$$
$$S_1 u(1/4,3/4) = q_1(1/4,3/4), \quad S_1 u(3/4,3/4) = q_1(3/4,3/4),$$

for the values of u at the lattice points $(1/4,1/4)$, $(3/4,1/4)$, $(1/4,3/4)$, and $(3/4,3/4)$ are still coupled; otherwise only the boundary values of u and the by now determined value $u(1/2,1/2)$ are involved. Thus the 4×4

system can be solved. As it is strictly diagonally dominant, it can be solved, for example, in a few steps (about 3 to 5) with SOR. All remaining unknowns are then determined from

$$S_0 u(1/2,1/4) = q_0(1/2,1/4),\ S_0 u(1/4,1/2) = q_0(1/4,1/2),$$
$$S_0 u(3/4,1/2) = q_0(3/4,1/2),\ S_0 u(1/2,3/4) = q_0(1/2,3/4).$$

In all cases arising in practice, this system too is strictly diagonally dominant. The method of solution described can generally always be carried out when n is a power of 2, say $n = 2^m$, $m \in \mathbb{N}$, $h = 1/n$. The $(2m-1)$-times reduction equation

$$S_{2m-1}u(1/2,1/2) = q_{2m-1}(1/2,1/2)$$

is then simply an equation for $u(1/2,1/2)$. The values of u at the remaining lattice points then follow one after another from strictly diagonally dominant systems of equations.

By looking at the difference stars S_1, S_2, S_3 formed from

$$S_0 = \begin{bmatrix} 0 & -1 & 0 \\ -1 & 4 & -1 \\ 0 & -1 & 0 \end{bmatrix}$$

one can see that the number of coefficients differing from zero increases rapidly. In addition, the coefficients differ greatly in order of magnitude. This phenomenon generally can be observed with all following S_ν, and it is independent of the initial star in all practical cases. As a result, one typically does not work with a complete difference star S_ν, but rather with an appropriately *truncated* star. Thus, after a truncation parameter σ has been specified, all coefficients a_{ij} of S_ν with $|a_{ij}| < \sigma|a_{oo}|$ are replaced by zeros. For sufficiently small σ this has no great influence on the accuracy of the computed approximation values of u. As one example, the choice of $\sigma = 10^{-8}$ for the case of the initial star

$$S_o = \begin{bmatrix} 0 & -1 & 0 \\ -1 & 4 & -1 \\ 0 & -1 & 0 \end{bmatrix}$$

leads to the discarding of all coefficients a_{ij} with

$$|i| + |j| > 4, \quad |i| > 3, \quad |j| > 3.$$

In conclusion, we would like to compare the number of operations required for the Gauss-Seidel method (single-step method), the SOR interation, the Buneman algorithm, and the reduction method. We arrived at Table 22.13 by restricting ourselves to the Poisson equation on the unit square (model problem) together with the classic five point difference formula. All lower order terms are discarded. N^2 denotes the number of interior lattice points, respectively the number of unknowns in the system. The computational effort for the iterative method depends additionally on the factor ε by which the initial error is to be reduced. For the reduction method, we assume that all stars are truncated by $\sigma = 10^{-8}$. More exact comparisons are undertaken in Dorr 1970 and Schröder-Trottenberg 1976. Appendices 4 and 6 contain remarks on computing times. In looking at this comparison, note that iterative methods are also applicable to more general problems.

Gauss-Seidel	$\frac{4}{\pi^2} N^4 \lvert\log \varepsilon\rvert$
SOR	$\frac{7}{2\pi} N^3 \lvert\log \varepsilon\rvert$
Buneman	$[6\log_2(N+1)+3]N^2$
Reduction	$36\ N^2$

Table 22.13: The number of operations for the model problem

APPENDICES: FORTRAN PROGRAMS

Appendix 0: Introduction.

The path from the mathematical formulation of an algorithm to its realization as an effective program is often difficult. We want to illustrate this propostion with six typical examples from our field. The selections are intended to indicate the multiplicity of methods and to provide the reader with some insight into the technical details involved. Each appendix emphasizes a different perspective: computation of characteristics (Appendix 1), problems in nonlinear implicit difference methods (Appendix 2), storage for more than two independent variables (Appendix 3), description of arbitrary regions (Appendix 4), graph theoretical aids (Appendix 5), and a comprehensive program for a fast direct method (Appendix 6). Some especially difficult questions, such as step size control, cannot be discussed here.

As an aid to readability we have divided each problem into a greater number of subroutines than is usual. This is an approach we generally recommend, since it greatly simplifies the development and debugging of programs. Those who are intent on saving computing time can always create a less structured formulation afterwards, since it will only be necessary to integrate the smaller subroutines. However, with each modification, one should start anew with the highly structured original program. An alternative approach to reduced computing time is to rewrite frequently called subroutines in assembly language. This will not affect the readability of the total program so long as programs equivalent in content are available in a higher language.

The choice of FORTRAN as the programming language was a hard one for us, since we prefer to use PL/1 or PASCAL. However, FORTRAN is still the most widespread language in the technical and scientific domain. The appropriate compiler is resident in practically all installations. Further, FORTRAN programs generally run much faster than programs in the more readable languages, which is a fact of considerable significance in the solution of partial differential equations.

The programs presented here were debugged on the CDC-CYBER 76 installation at the University of Köln and in part on the IBM 370/168 installed at the nuclear research station Jülich GmbH. They should run on other installations without any great changes.

We have been careful with all nested loops, to avoid any unnecessary interchange of variables in machines with virtual memory or buffered core memory. In such installations, for example, the loop

```
      DO 10 I = 1,100
      DO 10 J = 1,100
   10 A(J,I) = 0
```

is substantially faster than

```
    DO 10 I = 1,100
    DO 10 J = 1,100
 10 A(I,J) = 0.
```

This is so because when FORTRAN is used, the elements of A appear in memory in the following order:

A(1,1), A(2,1), ..., A(100,1), A(1,2), A(2,2), ...

For most other programming languages, the order is the contrary one:

A(1,1), A(1,2), ..., A(1,100), A(2,1), A(2,2), ...

For this reason, a translation of our programs into ALGOL, PASCAL, or PL/1 requires an interchange of all indices.

There is no measure of computing time which is independent of the machine or the compiler. If one measures the time for very similar programs running on different installations, one finds quite substantial differences. We have observed differences with ratios as great as 1:3, without any plausible explanations to account for this. It is often a pure coincidence when the given compiler produces the optimal translation for the most important loops. Therefore, we use the number of equally weighted floating point operations as a measure of computing time in these appendices. This count is more or less on target only with the large installations. On the smaller machines, multiplication and division consume substantially more time than addition and subtraction.

Appendix 1: Method of Massau

This method is described in detail in Section 3. It deals with an initial value problem for a quasilinear hyperbolic system of two equations on a region G in $\mathbb{R}^2$:

$$u_y = A(x,y,u)u_x + g(x,y,u), \qquad (x,y) \in G$$
$$u(x,0) = \psi(x,0), \qquad (x,0) \in G$$

where $A \in C^1(G \times \tilde{G}, \mathrm{MAT}(2,2,\mathbb{R}))$, and $g \in C^1(G \times \tilde{G}, \mathbb{R}^2)$. $\tilde{G}$ is an arbitrary subset of $\mathbb{R}^2$.

The initial values $\psi(x) = (\psi_1(x),\psi_2(x))$ are given at the equidistant points $(x_i,0)$, insofar as these points belong to G:

$$x_i = x_{n_1} + 2h(i-n_1), \quad i = n_1(1)n_2.$$

The lattice points in this interval which do not belong to G must be marked by B(I) = .FALSE..

Throughout the complete computation, U(*,I) contains the following four components:

$$u_1(x_i,y_i), \quad u_2(x_i,y_i), \quad x_i, \quad y_i.$$

The corresponding characteristic coordinates (cf. Figure 3.10) are:

SIGMA = SIGMAO + 2hi, TAU.

At the start, the COMMON block MASS has to be filled:

N1 = n_1

N2 = n_2

H2 = 2h

SIGMAO = $x_{n_1} - 2hn_1$

TAU = 0.

For those i for which $(x_i,0)$ belongs to G, we set

B(I) = .TRUE.

U(1,I) = $\psi_1(x_i)$

U(2,I) = $\psi_2(x_i)$

U(3,I) = x_i

U(4,I) = 0,

and otherwise, simply set

B(I) = .FALSE.

Each time MASSAU is called, a new level of the characteristic lattice (TAU = h, TAU = 2h, ...) is computed. The program also alters N1, N2, and SIGMAO. The number of lattice points in each level reduces each time by at least one. At the end, N2 $\leq$ N1. The results for a level can be printed out between two calls of MASSAU.

To describe A and g, it is necessary in each concrete case to write two subroutines, MATRIX and QUELL. The initial parameter is a vector U of length 4, with components

$$u_1(x,y), \quad u_2(x,y), \quad x, \; y.$$

Program MATRIX sets one more logical variable, L, as follows: L = .FALSE. if U lies outside the common domain of definition $G \times \tilde{G}$ of A and g; L = .TRUE. otherwise.

To determine the eigenvalues and eigenvectors of A, MASSAU calls the subroutine EIGEN. Both programs contain a number of checks:

(1) Are A and g defined?

(2) Are the eigenvalues of A real and distinct?

(3) Is the orientation of the (x,y)-coordinate system and of the (σ,τ)-system the same?

The lattice point under consideration is deleted as necessary (by setting B(I) = .FALSE.). Consider the example

$$A = -\begin{pmatrix} u_1 & 2u_2 \\ u_2/2 & u_1 \end{pmatrix}, \quad g = 0, \quad \psi = \begin{pmatrix} \frac{1}{10}e^x\sin(2\pi x) \\ 1 \end{pmatrix}, \quad G = (0,2) \times \mathbb{R}.$$

Figures 1 and 2 show the characteristic net in the (σ,τ)-system and the (x,y)-system. For $2h = 1/32$, we have the sector

$$0.656250 \leq \sigma \leq 1.406250$$

$$0.312500 \leq \tau \leq 0.703125 \qquad \text{20th to 45th level.}$$

In the (x,y)-coordinate system, different characteristics of the type $\sigma + \tau = \mathit{constant}$ will intersect each other, so in this region the computation cannot be carried out. If one drops the above checks, one obtains results for the complete "determinancy triangle". In the region where the solution exists, there cannot be any intersection of characteristics in the same family.

The method of Massau is particularly well suited to global extrapolation, as shown with the following example:

$$A = \begin{pmatrix} 0 & 1 \\ -\dfrac{1-u_1^2}{1-u_2^2} & \dfrac{2u_1u_2}{1-u_2^2} \end{pmatrix}, \quad g = \begin{pmatrix} 0 \\ -\dfrac{4u_1e^{2x}}{1-u_2^2} \end{pmatrix}, \quad \psi = \begin{pmatrix} 0 \\ 2e^x \end{pmatrix},$$

$$G = (0,1) \times \mathbb{R}.$$

The exact solution is

$$u(x,y) = 2e^x \begin{pmatrix} \sin y \\ \cos y \end{pmatrix}$$

The corresponding program and the subroutine MATRIX and QUELL are listed below.

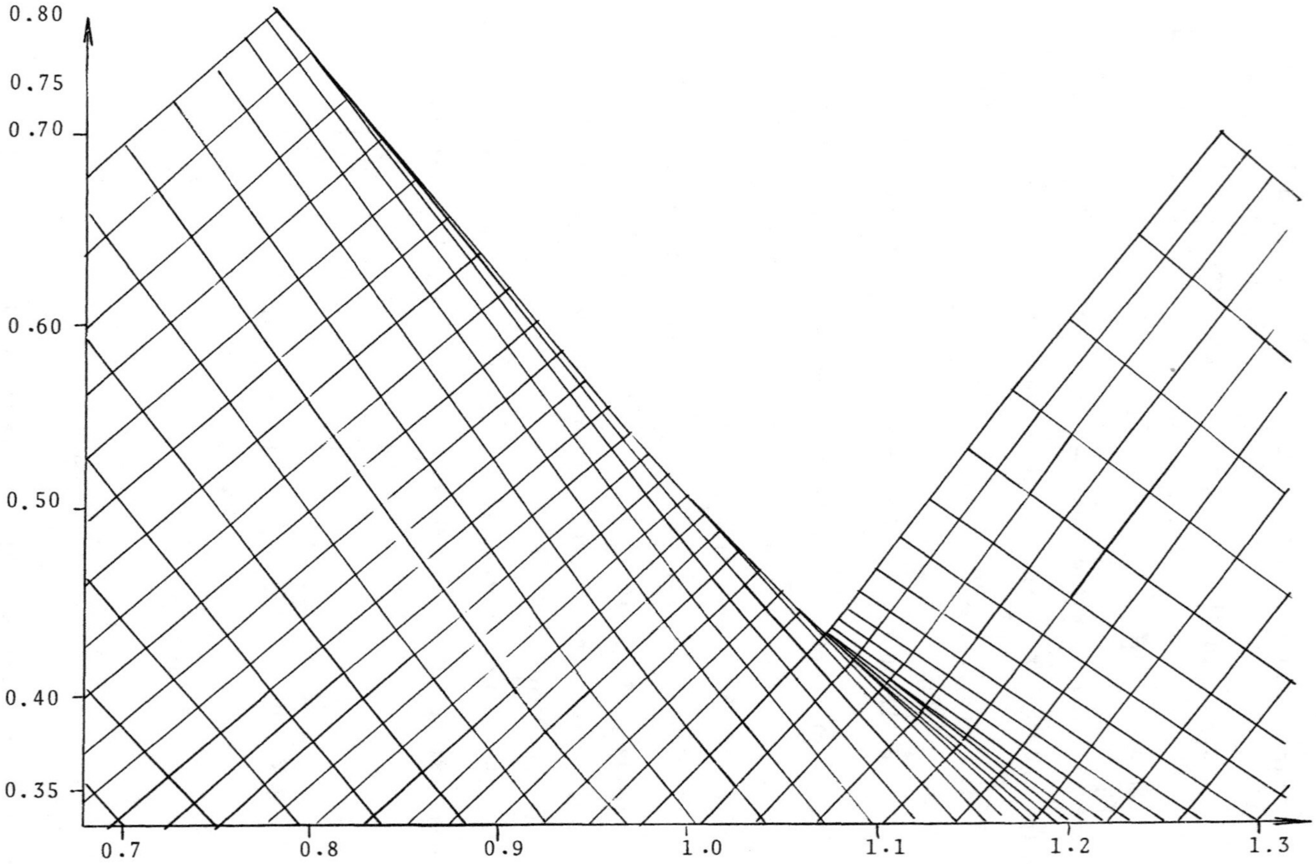

Figure 1: Characteristic net in the (x,y)-coordinate system

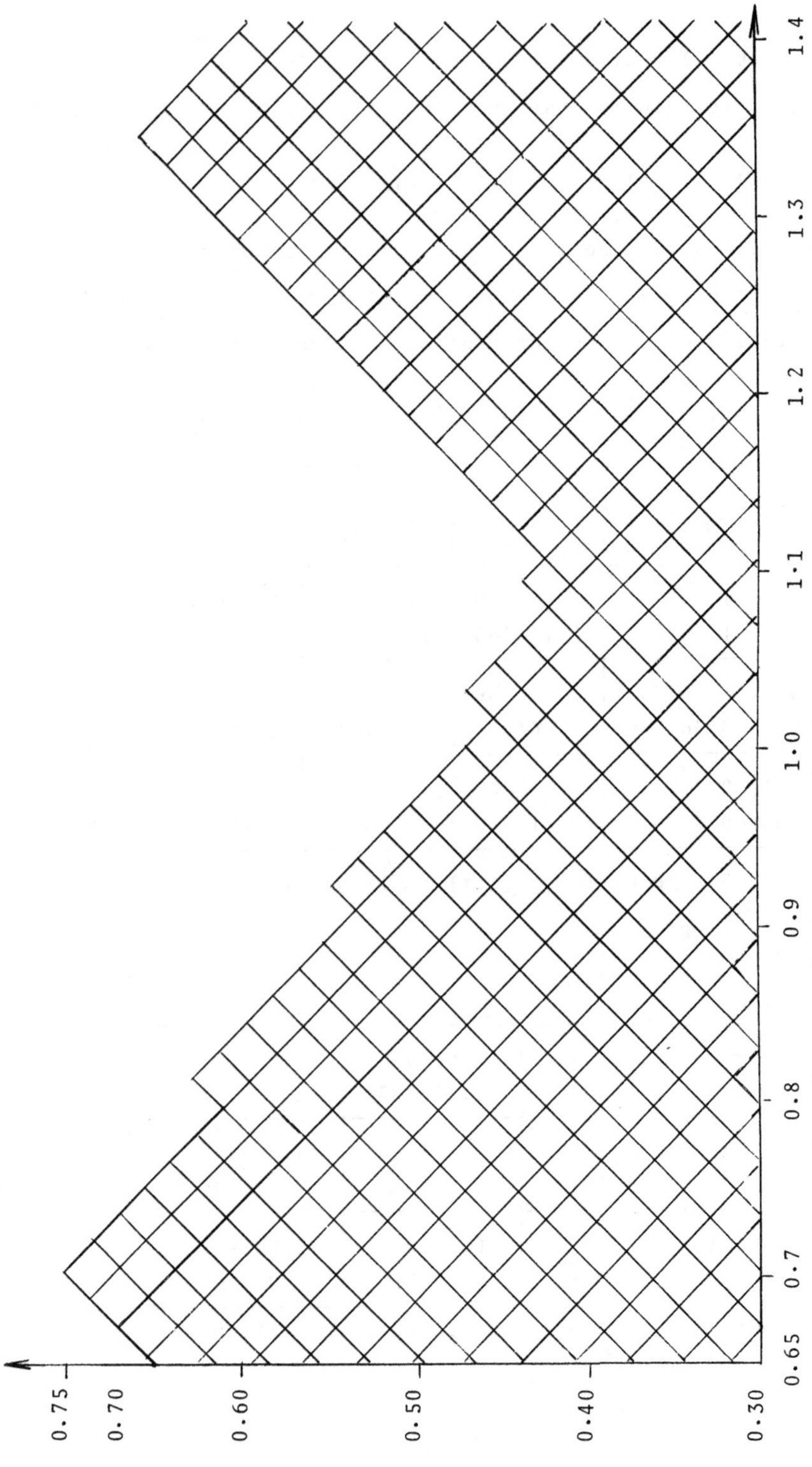

Figure 2: Characteristic net in the (σ,τ)-coordinate system

Table 3 contains the results for $(\sigma,\tau) = (1/2,1/2)$. The discrete solution has an asymptotic series of the form

$$\tau_0(x,y) + \tau_1(x,y)h + \tau_2(x,y)h^2 + \dots .$$

The results after the first and second extrapolation are found in Tables 4 and 5. The errors Δu_1 and Δu_2 in all three tables were computed from the values in a column:

$$\Delta u_1 = 2e^x \sin y - u_1, \quad \Delta u_2 = 2e^x \cos y - u_2.$$

H2	1/32	1/64	1/128	1/256
x y	1.551750 1.122827	1.582640 1.138263	1.598355 1.147540	1.606317 1.152624
U1 U2	8.804323 3.361230	9.004039 3.664041	9.104452 3.838782	9.154850 3.932681
ΔU1 ΔU2	-0.296282 0.727335	-0.165040 0.416843	-0.087380 0.223264	-0.044999 0.115574

Table 3: Results for $(\sigma,\tau) = (1/2,1/2)$.

x y	1.613530 1.153699	1.614070 1.156817	1.614279 1.157708
U1 U2	9.203755 3.966852	9.204865 4.013523	9.205248 4.026580
ΔU1 ΔU2	-0.023579 0.100843	-0.007085 0.027710	-0.001948 0.007299

Table 4: Results after the first extrapolation

x y	1.614250 1.157856	1.614349 1.158005
U1 U2	9.205235 4.029080	9.205376 4.030932
ΔU1 ΔU2	-0.001605 0.003320	-0.000234 0.000496

Table 5: Results after the second extrapolation

```
      SUBROUTINE MASSAU
C
C     VARIABLES OF THE COMMON BLOCK
C     U(1,I) AND U(2,I) COMPONENTS OF U,
C     U(3,I)=X, U(4,I)=Y,
C     WHERE N1.LE.I.LE.N2.
C     B(I)=.FALSE. MEANS THAT THE POINT DOES NOT BELONG
C     TO THE GRID.
C     THE CHARACTERISTIC COORDINATES OF THE
C     POINT (U(3,I),U(4,I)) ARE (SIGMAO+I*H2,TAU).
C     THE BLOCK MUST BE INITIALIZED BY THE MAIN PROGRAMME.
C
      REAL U(4,500),SIGMAO,TAU,H2
      INTEGER N1,N2
      LOGICAL B(500)
      COMMON /MASS/ U,SIGMAO,TAU,H2,N1,N2,B
C
      REAL E1(2,2),E2(2,2),LAMB1(2),LAMB2(2),G1(2),G2(2)
      REAL C1,C2,D,XO,X1,X2,YO,Y1,Y2,RD
      REAL DX01,DX21,DY01,DY02,DY21,U11,U12
      INTEGER N3,N4,I
      LOGICAL L,LL
C
      N3=N2-N1
   10 IF(N3.LE.0) RETURN
      IF(B(N1)) GOTO 20
   11 N1=N1+1
      N3=N3-1
      GOTO 10
   20 LL=.FALSE.
      N4=N2-1
C
C     BEGINNING OF THE MAIN LOOP
C
      DO 100 I=N1,N4
      IF(.NOT.B(I+1)) GOTO 90
      IF(LL) GOTO 30
      IF(.NOT.B(I)) GOTO 90
      CALL EIGEN(U(1,I),E1,LAMB1,L)
      IF(.NOT.L) GOTO 90
      CALL QUELL(U(1,I),G1)
```

```
   30 CALL EIGEN(U(1,I+1),E2,LAMB2,L)
      IF(.NOT.L) GOTO 90
      CALL QUELL(U(1,I+1),G2)
C
C     SOLUTION OF THE FOLLOWING EQUATIONS
C     (X0-X1)+LAMB1(1)*(Y0-Y1)=0
C     (X0-X1)+LAMB2(2)*(Y0-Y1)=(X2-X1)+LAMB2(2)*(Y2-Y1)
C
      C1=LAMB1(1)
      C2=LAMB2(2)
      D=C2-C1
      IF(D.LT.1.E-6*AMAX1(ABS(C1),ABS(C2))) GOTO 80
      X1=U(3,I)
      X2=U(3,I+1)
      Y1=U(4,I)
      Y2=U(4,I+1)
      DX21=X2-X1
      DY21=Y2-Y1
      RD=(DX21+C2*DY21)/D
      DX01=-C1*RD
      DY01=RD
      X0=X1+DX01
      Y0=Y1+DY01
      DY02=Y0-Y2
C
C     CHECK WHETHER THE TRANSFORMATION FROM (SIGMA,TAU)
C     TO (X,Y) IS POSSIBLE
C
      IF((DX21*DY01-DX01*DY21).LE.0.) GOTO 80
C
C     SOLUTION OF THE FOLLOWING EQUATIONS
C
C     E1(1,1)*(U(1,I)-U11)+E1(1,2)*(U(2,I)-U12)=
C     DY01*(E1(1,1)*G1(1)+E1(1,2)*G1(2))
C     E2(2,1)*(U(1,I)-U11)+E2(2,2)*(U(2,I)-U12)=
C     E2(2,1)*(DY02*G2(1)+U21-U11)+E2(2,2)*(DY02*G2(2)+U22-U12)
C
C     U11=OLD VALUE OF U(1,I)
C     U12=OLD VALUE OF U(2,I)
C     U21=OLD VALUE OF U(1,I+1)
C     U22=OLD VALUE OF U(2,I+1)
C
      D=E1(1,1)*E2(2,2)-E2(2,1)*E1(1,2)
      IF(ABS(D).LT.1.E-6) GOTO 80
      U11=U(1,I)
      U12=U(2,I)
      C1=DY01*(E1(1,1)*G1(1)+E1(1,2)*G1(2))
      C2=E2(2,1)*(DY02*G2(1)+U(1,I+1)-U11) +
     F   E2(2,2)*(DY02*G2(2)+U(2,I+1)-U12)
      U(1,I)=U11+(C1*E2(2,2)-C2*E1(1,2))/D
      U(2,I)=U12+(E1(1,1)*C2-E2(2,1)*C1)/D
C
C
      U(3,I)=X0
      U(4,I)=Y0
```

```
   70 LAMB1(1)=LAMB2(1)
      E1(1,1)=E2(1,1)
      E1(1,2)=E2(1,2)
      G1(1)=G2(1)
      G1(2)=G2(2)
      LL=.TRUE.
      GOTO 100
   80 B(I)=.FALSE.
      GOTO 70
   90 B(I)=.FALSE.
      LL=.FALSE.
  100 CONTINUE
C
C     END OF THE MAIN LOOP
C
      B(N2)=.FALSE.
  110 N2=N2-1
      IF(.NOT.B(N2).AND.N2.GT.N1) GOTO 110
      SIGMAO=SIGMAO+H2*0.5
      TAU=TAU+H2*0.5
      RETURN
      END
```

```
      SUBROUTINE EIGEN(U,E,LAMBDA,L)
C
      REAL U(4),E(2,2),LAMBDA(2)
      LOGICAL L
C
C     INPUT PARAMETERS
C     U CONTAINS U(1),U(2),X,Y
C
C     OUTPUT PARAMETERS
C     EIGENVALUES LAMBDA(1).LT.LAMBDA(2)
C     MATRIX E (IN THE TEXT DENOTED BY E**-1)
C     L=.FALSE. INDICATES THAT THE COMPUTATION IS NOT POSSIBLE
C
      REAL A(2,2),C,D,C1,C2,C3,C4
      LOGICAL SW
      L=.TRUE.
      CALL MATRIX(U,A,L)
      IF(.NOT.L) RETURN
C
C     COMPUTATION OF THE EIGENVALUES OF A
```

```
C
      C=A(1,1)+A(2,2)
      D=A(1,1)-A(2,2)
      D=D*D+4.*A(1,2)*A(2,1)
      IF(D.LE.0) GO TO 101
      D=SQRT(D)
      IF(D.LT.1.E-6*ABS(C)) GO TO 101
      LAMBDA(1)=0.5*(C-D)
      LAMBDA(2)=0.5*(C+D)
C
C     SOLUTION OF THE FOLLOWING HOMOGENEOUS EQUATIONS
C     E(1,1)*(A(1,1)-LAMBDA(1))+E(1,2)*A(2,1)=0
C     E(1,1)*A(1,2)+E(1,2)*(A(2,2)-LAMBDA(1))=0
C     E(2,1)*(A(1,1)-LAMBDA(2))+E(2,2)*A(2,1)=0
C     E(2,1)*A(1,2)+E(2,2)*(A(2,2)-LAMBDA(2))=0
C
      C=LAMBDA(1)
      SW=.FALSE.
   10 C1=ABS(A(1,1)-C)
      C2=ABS(A(2,1))
      C3=ABS(A(1,2))
      C4=ABS(A(2,2)-C)
      IF(AMAX1(C1,C2).LT.AMAX1(C3,C4)) GO TO 30
      IF(C2.LT.C1) GO TO 20
      C1=1.
      C2=(C-A(1,1))/A(2,1)
      GO TO 50
   20 C2=1.
      C1=A(2,1)/(C-A(1,1))
      GO TO 50
   30 IF(C3.LT.C4) GO TO 40
      C2=1.
      C1=(C-A(2,2))/A(1,2)
      GO TO 50
   40 C1=1.
      C2=A(1,2)/(C-A(2,2))
   50 IF(SW) GO TO 60
      E(1,1)=C1
      E(1,2)=C2
      C=LAMBDA(2)
      SW=.TRUE.
      GO TO 10
   60 E(2,1)=C1
      E(2,2)=C2
      RETURN
  101 L=.FALSE.
      RETURN
      END
```

EXAMPLE (MENTIONED IN THE TEXT)

```
      MAIN PROGRAMME:

C     DESCRIPTION OF THE COMMON BLOCK
C     IN THE SUBROUTINE MASSAU
      REAL U(4,500),SIGMAO,TAU,H2
      INTEGER N1,N2
      LOGICAL B(500)
      COMMON /MASS/ U,SIGMAO,TAU,H2,N1,N2,B
C
      REAL X,DU1,DU2,SIGMA
      INTEGER I,J
C
C     INITIALIZATION OF THE COMMON BLOCK
C
      TAU=0.
      N1=1
      N2=65
      H2=1./32.
      PI = 4.*ATAN(1.)
      SIGMAO=-H2
      X=0.
      DO 10 I=1,N2
      U(1,I)=0.1*SIN(2.*PI*X)*EXP(X)
      U(2,I)=1.
      U(3,I)=X
      U(4,I)=0.
      B(I)=.TRUE.
   10 X=X+H2
C
C     LOOP FOR PRINTING AND EXECUTING THE SUBROUTINE
C
      DO 40 I=1,65
      DO 39 J=N1,N2
      IF(.NOT.B(J)) GOTO 39
      SIGMA=SIGMAO+J*H2
      WRITE(6,49) SIGMA,TAU,U(3,J),U(4,J),U(1,J),U(2,J)
   39 CONTINUE
      WRITE(6,50)
C
      IF(N2.LE.N1) STOP
      CALL MASSAU
   40 CONTINUE
      STOP
C
   49 FORMAT(1X,2F8.5,1X,6F13.9)
   50 FORMAT(1H1)
      END
```

SUBROUTINES:

```
      SUBROUTINE QUELL(U,G)
C
C     INPUT PARAMETER U CONTAINS U(1),U(2),X,Y
C     OUTPUT PARAMETERS ARE G(1),G(2)
C
      REAL U(4),G(2)
      G(1)=0.
      G(2)=0.
      RETURN
      END
```

```
      SUBROUTINE MATRIX(U,A,L)
C
C     INPUT PARAMETER U CONTAINS U(1),U(2),X,Y
C     OUTPUT PARAMETERS ARE THE MATRIX A AND L
C     L=.TRUE. IF U BELONGS TO THE DOMAIN OF THE COEFFICIENT
C     MATRIX A AND OF THE TERM G. OTHERWISE, L=.FALSE.
C
      REAL U(4),A(2,2)
      LOGICAL L
C
      REAL U1,U2
C
      U1=U(1)
      U2=U(2)
      L=.TRUE.
      A(1,1)=-U1
      A(1,2)=-2.*U2
      A(2,1)=-0.5*U2
      A(2,2)=-U1
      RETURN
      END
```

Appendix 2: Total implicit difference method for solving a nonlinear parabolic differential equation.

The total implicit difference method has proven itself useful for strongly nonlinear parabolic equations. With it one avoids all the stability problems which so severely complicate the use of other methods. In the case of one (space) variable, the amount of effort required to solve the system of equations is often overestimated.

The following programs solve the problem

$$\begin{aligned} &u_t = \alpha(u)u_{xx} - q(u), && x \in (r,s),\ t > 0\\ &u(x,0) = \phi(x), && x \in [r,s]\\ &u(r,t) = \phi_r(t),\ u(s,t) = \phi_s(t), && t > 0. \end{aligned}$$

Associated with this is the difference method

$$\begin{aligned} u(x,t+h)-u(x,t) = \lambda\alpha(u(x,t+h))[&u(x+\Delta x,t+h)+u(x-\Delta x,t+h)\\ &- 2u(x,t+h)] - hq(u(x,t+h)) \end{aligned}$$

where $\Delta x > 0$, $h > 0$, and $\lambda = h/(\Delta x)^2$. When specialized to

$$\begin{aligned} &\Delta x = (s-r)/(n+1), && n \in \mathbb{N} \text{ fixed}\\ &x = r + j\Delta x, && j = 1(1)n \end{aligned}$$

this becomes a nonlinear system in n unknowns. It is solved with Newton's method. For each iterative step, we have to solve a linear system with a tridiagonal matrix. The linear equations are

$$a_{1j}\tilde{\tilde{u}}_{j-1} + a_{2j}\tilde{\tilde{u}}_j + a_{3j}\tilde{\tilde{u}}_{j+1} = a_{4j}, \quad j = 1(1)n$$

where

$$a_{1j} = -\lambda\alpha(\tilde{u}_j)$$

$$a_{2j} = 1+2\lambda\alpha(\tilde{u}_j)-\lambda\alpha'(\tilde{u}_j)[\tilde{u}_{j+1}+\tilde{u}_{j-1}-2\tilde{u}_j]+hq'(\tilde{u}_j)$$

$$a_{3j} = -\lambda\alpha(\tilde{u}_j)$$

$$a_{4j} = u_j-[1+2\lambda\alpha(\tilde{u}_j)]\tilde{u}_j+\lambda\alpha(\tilde{u}_j)[\tilde{u}_{j+1}+\tilde{u}_{j-1}]-hq(\tilde{u}_j)$$

u_j = solution of the difference equation at the point $(r+j\Delta x,t+h)$.

$\tilde{u}_j$ = corresponding Newton approximation for $u(r+j\Delta x,t+h)$.

When this sytem has been solved, the $\tilde{u}_j$ are replaced by $\tilde{u}_j + \tilde{\tilde{u}}_j$; the a_{ij} are recomputed; etc., until there is no noticeable improvement in the $\tilde{u}_j$. Usually two to four Newton steps suffice.

Since the subscript 0 is invalid in FORTRAN, the quantities $u(x+j\Delta x,t)$ are denoted in the programs by U(J+1). For the same reason, the Newton approximation $\tilde{u}_j$ is called U1(J+1).

The method consists of eight subroutines:

HEATTR, AIN, RIN, GAUBD3, ALPHA, DALPHA, QUELL, DQUELL.

HEATTR is called once by the main program for each time increment. Its name is an abbreviation for heat transfer. The other subroutines are used indirectly only. The last four subroutines must be rewritten for each concrete case. They are REAL FUNCTIONs with one scalar argument of REAL type, which describe the functions $\alpha(u)$, $\alpha'(u)$, $q(u)$, and $q'(u)$.

The other subroutines do not depend on the particulars of the problem. AIN computes the coefficients a_{ij} of the linear system of equations. GAUBD3 solves the equations. This program is described in detail along with the programs dealing with band matrices in Appendix 5. The coefficients a_{1j}, a_{2j}, and a_{3j} are recomputed only at every third Newton's step. In the intervening two steps, the old values are reused, and the subroutine RIN is called instead of AIN. RIN only computes a_{4j}. Afterwards, GAUBD3 runs in a simplified form. For this reason, the third variable is .TRUE.. We call these iterative steps abbreviated Newton's steps.

Before HEATTR can be called the first time, it is necessary to fill the COMMON block /HEAT/:

$$N = n$$

$$DX = \Delta x = (s-r)/(n+1)$$

$$U(J+1) = \phi(r+j\Delta x) \qquad j = 0(1)n+1$$

$$H = h.$$

H can be changed from one time step to another. If $u(r,t)$ and $u(s,t)$ depend on t, it is necessary to set the new boundary values $U(1) = \phi_r(t+h)$ and $U(N+2) = \phi_s(t+h)$ before each call of HEATTR by the main program.

An abbreviated Newton's step uses approximately 60% of the floating point operations of a regular Newton's step:

(1) Regular Newton's step:

n	calls of ALPHA, DALPHA, QUELL, DQUELL
21n+4	operations in AIN
8n-7	operations in GAUBD3
4n	operations in HEATTR.

(2) Abbreviated Newton's step:

n	calls of ALPHA, QUELL
10n+3	operations in RIN
5n-4	operations in GAUBD3
4n	operations in HEATTR.

This sequence of different steps--a regular step followed by two abbreviated steps--naturally is not optimal in every single case. Our error test for a relative accuracy of 10^{-5} is also arbitrary. If so desired, it suffices to make the necessary changes in HEATTR, namely at

```
IF(AMAX.LE.0.00001*UMAX) GO TO 70
```

and

```
IF(ITER1.LT.3) GO TO 21.
```

As previously noted, two to four Newton's iterations usually suffice. This corresponds to four to eight times this effort with a naive explicit method. If u and $\alpha(u)$ change substantially, the explicit method allows only extremely small incrementations h. This can reach such extremes that the method is useless from a practical standpoint. However, if $q'(u) < 0$, then even for the total implicit method one should have $hq'(u) > -1$, i.e. $h < 1/|q'(u)|$.

For very large n, to reduce the rounding error in AIN and RIN we recommend the use of double precision when executing the instruction

```
      A(4,J)=U(J+1)-(1.+LAMBD2*AJ)*UJ+LAMBDA*AJ*
     *         (U1(J+2)+U1(J))-H*QJ.
```

This is done by declaring

```
      DOUBLE PRECISION LAMBDA, LAMBD2, AJ, UJ, U12, U10
```

and replacing the instructions above by the following three instructions:

```
      U12 = U1(J+2)
      U10 = U1(J)
      A(4,J) = U(J+1)-(1.+LAMBD2*AJ)*UJ
     +           +LAMBDA*AJ*(U12+U10).
```

All remaining floating point variables remain REAL. Programs other than AIN and RIN do not have to be changed.

```
      SUBROUTINE HEATTR(ITER)
C
C     U(I) VALUES OF U AT X=X0+(I-1)*DX, I=1(1)N+2
C     U(1), U(N+2) BOUNDARY VALUES
C     H STEP SIZE WITH RESPECT THE TIME COORDINATE
C
      REAL U(513),H,DX
      INTEGER N
      COMMON/HEAT/U,H,DX,N
C
      REAL U1(513),AJ,UJ,AMAX,UMAX,A(4,511)
      INTEGER ITER,I,ITER1,N1,N2,J
      N1=N+1
      N2=N+2
C
C     FIRST STEP OF THE NEWTON ITERATION
C
      CALL AIN(A,U)
      CALL GAUBD3(A,N,.FALSE.)
      DO 20 J=2,N1
   20 U1(J)=U(J)+A(4,J-1)
      U1(1)=U(1)
      U1(N2)=U(N2)
      ITER=1
      ITER1=1
C
C     STEP OF THE MODIFIED NEWTON ITERATION
C
   21 CALL RIN(A,U1)
      CALL GAUBD3(A,N,.TRUE.)
      GO TO 30
C
C     STEP OF THE USUAL NEWTON ITERATION
C
   25 CALL AIN(A,U1)
      CALL GAUBD3(A,N,.FALSE.)
      ITER1=0.
C
C     RESTORING AND CHECKING
C
   30 AMAX=0.
      UMAX=0.
      DO 40 J=2,N1
      AJ=A(4,J-1)
      UJ=U1(J)+AJ
      AJ=ABS(AJ)
      IF(AJ.GT.AMAX) AMAX=AJ
      U1(J)=UJ
      UJ=ABS(UJ)
      IF(UJ.GT.UMAX) UMAX=UJ
   40 CONTINUE
C
```

```
C
      ITER=ITER+1
      ITER1=ITER1+1
      IF(AMAX.LE.0.00001*UMAX) GO TO 70
      IF(ITER.GT.20) GO TO 110
      IF(ITER1.LT.3) GO TO 21
      GO TO 25
C
C     U=U1
C
   70 DO 80 J=2,N1
   80 U(J)=U1(J)
      RETURN
C
C
  110 WRITE(6,111)
  111 FORMAT(15H NO CONVERGENCE)
      STOP
      END
```

```
      SUBROUTINE AIN(A,U1)
C
C     EVALUATION OF THE COEFFICIENTS AND OF THE RIGHT-HAND
C     SIDE OF THE SYSTEM OF LINEAR EQUATIONS
C
      REAL A(4,511),U1(513)
C
C     COMMON BLOCK COMPARE HEATTR
C
      REAL U(513),H,DX
      INTEGER N
      COMMON/HEAT/U,H,DX,N
C
      REAL LAMBDA,LAMBD2,LAMBDM,UJ,AJ,DAJ,QJ,DQJ
      INTEGER J
      REAL Z
      LAMBDA=H/(DX*DX)
      LAMBD2=2.*LAMBDA
      LAMBDM=-LAMBDA
      DO 10 J=1,N
      UJ=U1(J+1)
      AJ=ALPHA(UJ)
      DAJ=DALPHA(UJ)
      QJ=QUELL(UJ)
      DQJ=DQUELL(UJ)
      Z=LAMBDM*AJ
```

```
      A(1,J)=Z
      A(2,J)=1.+LAMBD2*(AJ+DAJ*UJ)-LAMBDA*DAJ*
     *       (U1(J+2)+U1(J))+H*DQJ
      A(3,J)=Z
      A(4,J)=U(J+1)-(1.+LAMBD2*AJ)*UJ+LAMBDA*AJ*
     *       (U1(J+2)+U1(J))-H*QJ
   10 CONTINUE
      RETURN
      END
```

```
      SUBROUTINE RIN(A,U1)
C
C     EVALUATION OF THE RIGHT-HAND SIDE OF THE
C     LINEAR EQUATIONS
C
      REAL A(4,511),U1(513)
C
C     COMMON BLOCK COMPARE HEATTR
C
      REAL U(513),H,DX
      INTEGER N
      COMMON/HEAT/U,H,DX,N
C
      REAL LAMBDA,LAMBD2,UJ,AJ,QJ
      INTEGER J
      LAMBDA=H/(DX*DX)
      LAMBD2=2.*LAMBDA
      DO 10 J=1,N
      UJ=U1(J+1)
      AJ=ALPHA(UJ)
      QJ=QUELL(UJ)
      A(4,J)=U(J+1)-(1.+LAMBD2*AJ)*UJ+LAMBDA*AJ*
     *       (U1(J+2)+U1(J))-H*QJ
   10 CONTINUE
      RETURN
      END
```

EXAMPLE

MAIN PROGRAMME:

```
      REAL U(513),H,DX
      INTEGER N
      COMMON/HEAT/U,H,DX,N
      REAL PI,T
      INTEGER I,J,ITER
      PI=4.*ATAN(1.)
C
      N=7
      DX=1./8.
      H=1./64.
      DO 10 I=1,9
   10 U(I)=SIN(PI*DX*FLOAT(I-1))
C
      T=0.
      DO 20 I=1,10
      CALL HEATTR(ITER)
      WRITE(6,22) ITER
      T=T+H
      WRITE(6,21) T
      WRITE(6,21)(U(J),J=1,9)
   20 CONTINUE
      STOP
C
   21 FORMAT(1X,9F12.9/1X,9F12.9)
   22 FORMAT(6H ITER=,I2)
      END
```

SUBROUTINES:

```
REAL FUNCTION ALPHA(U)
REAL U
ALPHA=(1.-0.5*U)/(4.*ATAN(1.))**2
RETURN
END
```

```
REAL FUNCTION DALPHA(U)
REAL U
DALPHA=-.5/(4.*ATAN(1.))**2
RETURN
END
```

```
REAL FUNCTION QUELL(U)
REAL U
QUELL=U*U*0.5
RETURN
END
```

```
REAL FUNCTION DQUELL(U)
REAL U
DQUELL=   U
RETURN
END
```

Appendix 3: Lax-Wendroff-Richtmyer method for the case of two space variables.

The subroutines presented here deal with the initial value problem

$$u_t = A_1 u_x + A_2 u_y + Du + q, \qquad x,y \in \mathbb{R},\ t > 0$$
$$u(x,y,0) = \phi(x,y), \qquad x,y \in \mathbb{R}.$$

Here $A_1, A_2, D \in C^1(\mathbb{R}^2, \mathrm{MAT}(n,n,\mathbb{R}))$, $D(x,y) = \mathrm{diag}(d_{jj}(x,y))$, $q \in C^1(\mathbb{R}^2 \times [0,\infty), \mathbb{R}^n)$. We require that the problem be properly posed in $L^2(\mathbb{R}^2, \mathbb{C}^n)$. For this it suffices, e.g., that (1) $A_1(x,y)$, $A_2(x,y)$ are always symmetric, and (2) A_1, A_2, D, q have compact support.

Because of the terms $Du + q$, the differential equation in the problem considered here is a small generalization of Example 10.9. There is one small difficulty in extending the difference scheme to this slightly generalized differential equation. One can take care of the terms $Du + q$ in the differential equation with the additional summand

$$h[D(x,y)u(x,y,t) + q(x,y,t)]$$

or better yet, with

$$h\{\tfrac{1}{2}D(x,y)[u(x,y,t) + u(x,y,t+h)] + q(x,y,t+\tfrac{1}{2}h)\}$$

This creates no new stability problems (cf. Theorem 5.13). Consistency is also preserved. However, the order of consistency is reduced from 2 to 1. The original consistency proof considered only solutions of the differential equation ("consistency in the class of solutions") and not arbitrary sufficiently smooth functions; here we have a different differential equation.

We use the difference method

$$u(x,y,t+h) = [I-\tfrac{h}{2}D(x,y)]^{-1}[I+S(h)\circ K(h)+\tfrac{h}{2}D(x,y)]u(x,y,t)$$

$$+ h[I-\tfrac{h}{2}D(x,y)]^{-1}q(x,y,t+\tfrac{h}{2})+\tfrac{h}{2}(I-\tfrac{h}{2}D(x,y))^{-1}S(h)q(x,y,t)$$

where

$$K(h) = \tfrac{1}{2}S(h) + (\tfrac{1}{4}I+\tfrac{h}{8}D(x,y))\ (T_{\Delta,1}+T_{\Delta,2}+T_{\Delta,1}^{-1}+T_{\Delta,2}^{-1})$$

$$S(h) = \tfrac{1}{2}\lambda[A_1(x,y)(T_{\Delta,1}-T_{\Delta,1}^{-1}) + A_2(x,y)(T_{\Delta,2}-T_{\Delta,2}^{-1})].$$

For $D(x,y) = 0$ and $q(x,y,t) = 0$, this is the original method $(r = 1)$. But then the order of convergence is 2 in every case. Naturally, this presupposes that the coefficients, inhomogeneities, and solution are all sufficiently often differentiable.

The computation procedes in two steps.

1st Step (SUBROUTINE STEP1):

$$v(x,y) = K(h)u(x,y,t) + \tfrac{1}{2}hq(x,y,t).$$

2nd Step (SUBROUTINE STEP2):

$$u(x,y,t+h) = [I - \tfrac{h}{2}D(x,y)]^{-1}\circ$$

$$\{[I+\tfrac{h}{2}D(x,y)]u(x,y,t)+S(h)v(x,y)+hq(x,y,t+\tfrac{h}{2})\}.$$

The last instruction is somewhat less susceptible to rounding error in the following formulation:

$$u(x,y,t+h) = u(x,y,t) + [I-\tfrac{h}{2}D(x,y)]^{-1}\circ$$

$$\{S(h)v(x,y)+h[D(x,y)u(x,y,t)+q(x,y,t+\tfrac{h}{2})]\}.$$

If $u(x,y,t)$ is given at the lattice points

$$(x,y) = (\mu\Delta,\nu\Delta) \qquad (\mu,\nu \in \mathbb{Z},\ \mu+\nu \ \text{even})$$

then $v(x,y)$ can be computed at the following points:

$$(x,y) = (\mu\Delta,\nu\Delta), \quad \mu,\nu \in \mathbb{Z}, \ \mu+\nu \ \text{odd}.$$

From these values and the old values $u(x,y)$, one obtains $u(x,y,t+h)$ at the points

$$(x,y) = (\mu\Delta,\nu\Delta), \quad \mu,\nu \in \mathbb{Z}, \ \mu+\nu \ \text{even}.$$

This completes the computation for one time increment.

If steps 1 and 2 follow each other directly, then u and v have to be stored. Therefore, we divide each time step into substeps, in which the u-values are computed only for the lattice points on a line

$$x + y = 2\alpha\Delta = \textit{constant}.$$

For this one only needs the v-values for

$$x + y = (2\alpha-1)\Delta \quad \text{and} \quad x + y = (2\alpha+1)\Delta$$

(as shown in Figure 1). At first, only these v-values are stored, and in the next substep, half of these are overwritten. Thus we alternately compute v on a line and u on a line. SUBROUTINE STEP1 computes only the v values on a line. STEP2 does compute all of u, but in passing from one line to the next, STEP2 calls STEP1 to compute v.

The program starts with the lattice points in the square

$$\{(x,y) \mid -1 \le x+y \le 1 \ \text{and} \ -1 \le x-y \le 1\}.$$

Because of the difference star of the Lax-Wendroff method, we lose fewer lattice points at the boundary per time step than with an equally large square with sides parallel to

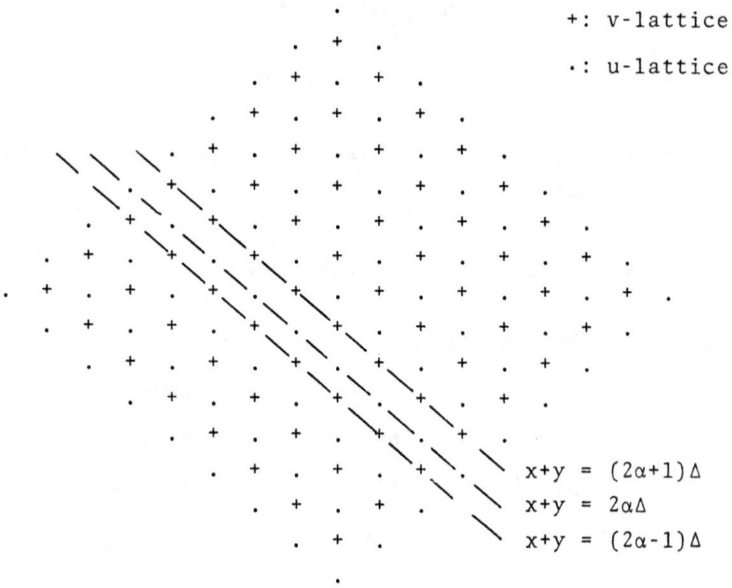

Figure 1: Lattice points for $m_0 = 3$, $m = 8$, $= 1/8$.

the axes. We set $m = 2^{m_0}$ and $\Delta = 1/m$.

Altogether, the solution of the initial value problem requires six subroutines:

INITIO, STEP1, STEP2, PRINT, COEFF, FUNC.

The last two subroutines have to be rewritten for each application. COEFF computes the matrices A_1, A_2, and D and the inhomogeneity q. FUNC provides the initial values.

The first program called by the user is INITIO. It defines the lattice, the number of components of u, computes the initial values with the help of FUNC, and enters all this in COMMON. Additionally, INITIO computes

$$\text{DMAX} = \max(0, d_{jj}(x,y)).$$

When calling STEP2, one should choose h no larger than

1.0/DMAX. In each time step, STEP2 is called first to carry out the computation, and then PRINT is called to print out the results.

After s_2 time steps, only $(m+1-2s_2)^2$ lattice points remain. Thus, at most $m/2$ steps can be carried out. After an incorrect call of INITIO or of STEP2, IERR > 0 is set as follows:

in INITIO:

IERR = 1: m_0 outside the boundaries.

IERR = 2: n_0 outside the boundaries.

in STEP2:

IERR = 1: s_0 too large.

STEP1 and STEP2 each contain only one computation intensive loop:

STEP1	STEP2
`DO 100 K=1,MS`	`DO 100 J=J1,J2`
⋮	⋮
`100 Y=Y-DELTA`	`100 Y1=Y1+DELTA`

In the following accounting of the number of floating point calls we ignore all operations outside these loops, with the exception of calls of STEP1 in STEP2.

STEP1:

$(m-2s_2)$	calls of COEFF
$(m-2s_2)(4n^2+12n+2)$	operations.

STEP2:

$(m-2s_2)$	calls of STEP1
$(m-2s_2-1)^2$	calls of COEFF
$(m-2s_2-1)^2(4n^2+11n+2)$	operations

Each time step therefore consumes approximately

$2(m-2s_2)^2$ calls of COEFF

$2(m-2s_2)^2(4n^2+11n)$ operations.

The total effort required for all $m/2$ time steps thus is

σ calls of COEFF

$\sigma(4n^2+11n)$ operations

where

$$\sigma = 8\sum_{\mu=1}^{m/2} \mu^2 = \tfrac{1}{3}m(m+1)(m+2).$$

If the matrices A_1 and A_2 contain many zeros (as in the wave equation for example) then the term $4n^2$ can be reduced substantially. To accomplish this it is enough, in STEP1 and STEP2, to reprogram only the loops beginning with

```
DO 20 LL = 1,N.
```

If enough memory is available, A_1, A_2, and D can be computed in advance, and CALL COEFF can be replaced by the appropriate reference. If q is t-dependent, however, it will have to be computed for each time step. In this way, the computing time can be reduced to a tolerable level in many concrete cases.

For the case of the wave equation

$$A_1(x,y) = \begin{pmatrix} 0 & 1 & 0 \\ 1 & 0 & 0 \\ 0 & 0 & 0 \end{pmatrix}, \quad A_2(x,y) = \begin{pmatrix} 0 & 0 & 1 \\ 0 & 0 & 0 \\ 1 & 0 & 0 \end{pmatrix}$$

$$D(x,y) = 0, \qquad q(x,y,t) = 0$$

we have tried to verify experimentally the theoretical stability bound $\lambda \le \sqrt{2}$. The initial values

$$\phi(x,y) = \begin{pmatrix} 0 \\ \cos x \\ \cos y \end{pmatrix}$$

have the exact solution

$$u(x,y,t) = \begin{pmatrix} -\sin t(\sin x + \sin y) \\ \cos t \ \cos x \\ \cos t \ \cos y \end{pmatrix}.$$

We chose $m_0 = 7$, $m = 128$, $\Delta = 1/128$, $h = \lambda\Delta$ where $\lambda = 1.3(0.1)1.7$. After 63 steps, we compared the numerical results with the exact solutions at the remaining 9 lattice points. The absolute error for the first component of u is generally smaller than the absolute error for the other components (cf. Table 2). The instability is not really noticeable until $\lambda = 1.7$, where it is most likely due to the still small number of steps.

λ	max. absolute error	
	1st comp.	2nd & 3rd comp.
≤ 1.5	$1.0 \cdot 10^{-7}$	$4.0 \cdot 10^{-6}$
1.6	$4.4 \cdot 10^{-5}$	$5.3 \cdot 10^{-5}$
1.7	$2.0 \cdot 10^{0}$	$1.2 \cdot 10^{0}$

Table 2

Nevertheless, the computations already become problematical with $\lambda \geq 1.5$. A multiplication of h by $1 + 10^{-12}$ creates a perturbation, for $\lambda = 1.3$ and $\lambda = 1.4$, of the same order of magnitude as the perturbation of h. For $\lambda = 1.5$, however, the relative changes in the results are greater up to a factor of 1000, and for $\lambda = 1.6$, this amplification can reach 10^9 for some points.

We have tested consistency among other places in an

example wherein A_1 and A_2 are full, and all elements of A_1 and A_2, as well as the diagonal elements of D, are essentially space-dependent. In this case, q depends on x, y, and t. The corresponding subroutines COEFF and FUNC are listed below. The initial value problem has the exact solution

$$u(x,y,t) = e^{-t}\begin{pmatrix} \cos x + \cos y \\ \cos x + \sin y \end{pmatrix}.$$

The computation was carried out for the four examples:

(1) $\Delta = 1/8$, $h = 1/32$, $\lambda = 1/4$, $s_2 = 1$
(2) $\Delta = 1/16$, $h = 1/64$, $\lambda = 1/4$, $s_2 = 2$
(3) $\Delta = 1/32$, $h = 1/128$, $\lambda = 1/4$, $s_2 = 4$
(4) $\Delta = 1/64$, $h = 1/256$, $\lambda = 1/4$, $s_2 = 8$.

The end results thus all belong to the same time $T = s_2h = 1/32$. Therefore, at lattice points with the same space coordinates, better approximations can be computed with the aid of a global extrapolation. Our approach assumes an asymptotic expansion of the type

$$\tau_0(x,y) + h^2\tau_2(x,y) + h^3\tau_3(x,y) + h^4\tau_4(x,y) + \dots .$$

The first and third extrapolation do in fact improve the results substantially. The summand $h^3\tau_3(x,y)$ should be very small in our example relative to the terms $h^2\tau_2(x,y)$ and $h^4\tau_4(x,y)$. The absolute error of the unextrapolated values decreases with h from about 10^{-3} to 10^{-5}. After the third extrapolation, the errors at all 49 points (and for both components of u) are less than 10^{-9}.

We do not intend to recommend the Lax-Wendroff-Richtmyer method as a basis for an extrapolation method as a result of these numerical results. For that it is too complicated and too computation intensive. However, global extrapolation is a far-reaching method for testing a program for hidden programming errors and for susceptibility to rounding error.

```
      SUBROUTINE INITIO (COEFF,FUNC,TO,MO,NO,DMAX,IERR)
C
C     FOR THE DESCRIPTION OF COEFF COMPARE STEP2.
C     THE SUBROUTINE FUNC YIELDS THE INITIAL VALUES F(N) AT THE
C     POINTS X,Y. THE USER HAS TO DECLARE THIS SUBROUTINE AS
C     EXTERNAL.
C     T=TO, N=NO, M=2**MO, FOR DMAX COMPARE TEXT.
C
      INTEGER I,IERR,I1,I2,J,MMAX1,MO,NN,NO
      REAL DMAX,MINUS,TO,XO,X1,YO,Y1
      REAL A1(4,4),A2(4,4),D(4),Q(4),F(4)
C
C     MEANING OF THE VARIABLES OF THE COMMON BLOCK
C
C     M        NUMBER OF THE PARTS OF THE INTERVAL (O,1),
C     DELTA    =1./M,
C     MMAX     UPPER BOUND FOR M,
C     N        NUMBER OF THE COMPONENTS OF THE SOLUTION
C              (1.LE.N.LE.4),
C     S2       NUMBER OF CALLS OF STEP2 DURING THE EXECUTION
C              OF INITIO,
C     T        TIME AFTER S2 STEPS,
C     H        STEP SIZE WITH RESPECT TO THE TIME,
C     LAMBDA   =H/DELTA (LAMBDA.GT.O),
C     U        SOLUTION. U(*,I,J) BELONGS TO THE POINT
C                                            X=DELTA*(J+I-MMAX-2)
C                                            Y=DELTA*(J-I),
C     V        INTERMEDIATE VALUES (COMPARE TEXT)
C              V(*,2,I) BELONGS TO THE POINT X=DELTA*(J+I-MMAX-1)
C                                            Y=DELTA*(J-I)
C              J IS THE RESPECTIVE PARAMETER OF STEP1
C              V(*,1,I) BELONGS TO THE POINT X1=X-DELTA
C                                            Y1=Y-DELTA
C     MMAX AND THE BOUNDS OF THE ARRAYS U(4,DIM2,DIM2) AND
C     V(4,2,DIM1) ARE RELATED AS FOLLOWS
C     MMAX     DIM1     DIM2
C      32       32       33
C      64       64       65
C     128      128      129
C     ...      ...      ...
C
      INTEGER MMAX,M,N,S2
      REAL U(4,65,65),V(4,2,64),H,DELTA,LAMBDA,T
      COMMON  U,V,H,DELTA,LAMBDA,T,MMAX,M,N,S2
      DATA MINUS /-1.E50/
      MMAX=64
C
      MMAX1=MMAX+1
      M=2**MO
      IF( MO.LT.1 .OR. M .GT.MMAX ) GOTO 998
      IF( NO.LT.1 .OR. NO.GT.4    ) GOTO 997
```

```
C
C     SET V(*,2,*)=0 AND ASSIGN MINUS INFINITY (HERE -1E50)
C     TO U(*,*,*).
C
      DO 10 J=1,MMAX
       DO 10 NN=1,N
   10   V(NN,2,J)=0.
      DO 20 I=1,MMAX1
       DO 20 J=1,MMAX1
   20   U(1,I,J)=MINUS
C
      T=TO
      N=NO
      S2=0
      DMAX=0.
      IERR=0
      DELTA=1./FLOAT(M)
      I1=(MMAX-M)/2+1
      I2=I1+M
      XO=-1.
      YO=0.
      DO 40 J=I1,I2
       X1=XO
       Y1=YO
       DO 30 I=I1,I2
        CALL FUNC  (X1,Y1,F)
        CALL COEFF (X1,Y1,TO,A1,A2,D,Q)
        X1=X1+DELTA
        Y1=Y1-DELTA
        DO 30 NN=1,N
         U(NN,I,J)=F(NN)
         IF( D(NN).GT.DMAX ) DMAX=D(NN)
   30    CONTINUE
       XO=XO+DELTA
   40  YO=YO+DELTA
      RETURN
  997 IERR=1
      RETURN
  998 IERR=2
      RETURN
      END
```

```
      SUBROUTINE  STEP1 (COEFF,J)
      INTEGER I1,I2,J,J1,J2,K,L,LL,MS
      REAL H2,H8,LAM4,SUM,X,Y
      REAL A1(4,4),A2(4,4),D(4),Q(4),UX(4),UY(4)
C
C     FOR VARIABLES IN COMMON COMPARE INITIO
C
      INTEGER MMAX,M,N,S2
      REAL U(4,65,65), V(4,2,64),H,DELTA,LAMBDA,T
      COMMON  U,V,H,DELTA,LAMBDA,T,MMAX,M,N,S2
C
      H2=H*.5
      H8=H*.125
      LAM4=LAMBDA*.25
      MS=M-2*S2
      I1=(MMAX-MS)/2+1
      J1=J
      I2=I1+1
      J2=J1+1
      DO 10 K=1,MS
       DO 10 L=1,N
   10   V(L,1,K)=V(L,2,K)
      X=DELTA*FLOAT(J1+I1-MMAX-1)
      Y=DELTA*FLOAT(J1-I1)
      DO 100 K=1,MS
       DO 15 LL=1,N
        UX(LL)=U(LL,I2,J2)-U(LL,I1,J1)
   15   UY(LL)=U(LL,I1,J2)-U(LL,I2,J1)
       CALL  COEFF (X,Y,T,A1,A2,D,Q)
       DO 30 L=1,N
        SUM=0.
        DO 20 LL=1,N
   20    SUM=SUM+A1(L,LL)*UX(LL)+A2(L,LL)*UY(LL)
        V(L,2,K)=LAM4*SUM+H2*Q(L)+
     +            (0.25+H8*D(L))*(U(L,I2,J2)+U(L,I1,J1)+
     +                           U(L,I1,J2)+U(L,I2,J1))
   30   CONTINUE
       I1=I1+1
       I2=I2+1
       X=X+DELTA
  100  Y=Y-DELTA
      RETURN
      END
```

```
      SUBROUTINE  STEP2 (COEFF,HO,IERR)
C
C     THE SUBROUTINE COEFF EVALUATES THE COEFFICIENTS A1, A2, D
C     AND THE SOURCE TERM Q OF THE DIFFERENTIAL EQUATIONS. COEFF
C     IS TO BE DECLARED AS EXTERNAL.
C     A1(N,N), A2(N,N), D(N) MAY DEPEND ON X AND Y, Q(N) MAY
C     DEPEND ON X,Y, AND T.
C     HO IS THE SIZE WITH RESPECT TO TIME. HO MAY CHANGE FROM
C     ONE CALL STEP2 TO THE NEXT CALL ACCORDING TO THE
C     STABILITY CONDITION.
C
      EXTERNAL COEFF
      INTEGER I,IERR,I1,I2,J,J1,J2,K,KK,L,LL,MS
      REAL HO,H2,LAM2,MINUS,SUM,T2,X,X1,Y,Y1
      REAL A1(4,4),A2(4,4),D(4),Q(4),VX(4),VY(4)
C
C     FOR VARIABLES IN COMMON COMPARE INITIO
C
      INTEGER MMAX,M,N,S2
      REAL U(4,65,65),V(4,2,64),H,DELTA,LAMBDA,T
      COMMON  U,V,H,DELTA,LAMBDA,T,MMAX,M,N,S2
      DATA MINUS /-1.E50/
C
      MS=M-2*S2
      IF( MS.LT.1 ) GOTO 99
      IERR=0
      H=HO
      LAMBDA=H/DELTA
      LAM2=LAMBDA*.5
      H2=H*.5
      T2=T+H2
      I1=(MMAX-MS)/2+1
      I2=I1+MS
      J1=I1+1
      J2=I2-1
      CALL  STEP1 (COEFF,I1)
      X1=DELTA*FLOAT(I1+I1-MMAX)
      Y1=0.
C
      DO 100 J=J1,J2
       X=X1
       Y=Y1
       K=1
       KK=2
       CALL  STEP1 (COEFF,J)
       DO 50 I=J1,J2
        DO 15 LL=1,N
         VX(LL)=V(LL,2,KK)-V(LL,1,K )
   15    VY(LL)=V(LL,2,K )-V(LL,1,KK)
        CALL  COEFF (X,Y,T2,A1,A2,D,Q)
        DO 30  L=1,N
         SUM=0.
```

```
      DO 20 LL=1,N
 20    SUM=SUM+A1(L,LL)*VX(LL)+A2(L,LL)*VY(LL)
      U(L,I,J)=U(L,I,J)+(LAM2*SUM+H*(D(L)*U(L,I,J)+Q(L)))/
    /                         (1.-H2*D(L))
 30   CONTINUE
     X=X+DELTA
     Y=Y-DELTA
     K=K+1
 50  KK=KK+1
    X1=X1+DELTA
100 Y1=Y1+DELTA
    DO 110 J=I1,I2
     U(1,I1,J)=MINUS
     U(1,I2,J)=MINUS
     U(1,J,I1)=MINUS
110  U(1,J,I2)=MINUS
    T=T+H
    S2=S2+1
    RETURN
 99 IERR=1
    RETURN
    END
```

```
      SUBROUTINE PRINT
      INTEGER I,J,L,MMAX1
      REAL MINUS,X,Y
C
C     FOR VARIABLES IN COMMON COMPARE INITIO
C
      INTEGER MMAX,M,N,S2
      REAL U(4,65,65),V(4,2,64),H,DELTA,LAMBDA,T
      COMMON U,V,H,DELTA,LAMBDA,T,MMAX,M,N,S2
      DATA MINUS /-1.E50/
C
      MMAX1=MMAX+1
      DO 30 J=1,MMAX1
       DO 20 I=1,MMAX1
        IF( U(1,I,J).LE.MINUS ) GOTO 20
        X=DELTA*FLOAT(J+I-MMAX-2)
        Y=DELTA*FLOAT(J-I)
        DO 10 L=1,N
 10      WRITE(6,800) L,I,J,U(L,I,J),X,Y
 20     CONTINUE
 30    CONTINUE
      RETURN
800   FORMAT(1H ,10X,2HU(,I2,1H,,I2,1H,,I2,1H),5X,E20.14,
     F              5X,2HX=,F10.6,2X,2HY=,F10.6)
      END
```

EXAMPLE (MENTIONED IN THE TEXT)

```
      SUBROUTINE  COEFF (X,Y,T,A1,A2,D,Q)
      REAL A1(4,4),A2(4,4),D(4),Q(4)
      SINX=SIN(X)
      COSY=COS(Y)
      CX1=COS(X)+1.
      CX2=CX1+1.
      SY1=SIN(Y)-1.
      SS1=SY1*SY1-1.
C
C     I SIN(X)   , COS(X)+1 I     I COS(Y)   , SIN(Y)-1             I
C  A1=I                     I  A2=I                                 I
C     I COS(X)+1, COS(X)+2 I     I SIN(Y)-1, SIN(Y)*(SIN(Y)-2) I
C
      A1(1,1)=SINX
      A1(1,2)=CX1
      A1(2,1)=CX1
      A1(2,2)=CX2
      A2(1,1)=COSY
      A2(1,2)=SY1
      A2(2,1)=SY1
      A2(2,2)=SS1
      D(1)=0.
      D(2)=SY1-CX1
      Q(1)=0.
      Q(2)=-EXP(-T)*(COSY*SS1-SINX*CX2)
      RETURN
      END




      SUBROUTINE FUNC (X,Y,F)
      REAL F(4)
      F(1)=SIN(X)+COS(Y)
      F(2)=COS(X)+SIN(Y)
      RETURN
      END
```

Appendix 4: Difference methods with SOR for solving the Poisson equation on nonrectangular regions.

Let $G \subset \mathbb{R}^2$ be a bounded region and let

$$\begin{aligned} \Delta u(x,y) &= q(x,y) \qquad & (x,y) &\in G \\ u(x,y) &= \psi(x,y) \qquad & (x,y) &\in \partial G. \end{aligned}$$

Furthermore, let one of the following four conditions be satisfied:

(1) $G \subset (-1,1) \times (-1,+1) = Q_1$

(2) $G \subset (-1,3) \times (-1,+1) = Q_2$

G,q,ψ are symmetric with respect to the line $x = 1$.

(3) $G \subset (-1,+1) \times (-1,3) = Q_3$ (2)

G,q,ψ are symmetric with respect to the line $y = 1$.

(4) $G \subset (-1,3) \times (-1,3) = Q_4$

G,q,ψ are symmetric with respect to the lines $x = 1$ and $y = 1$.

The symmetry conditions imply that the normal derivative of u vanishes on the lines of symmetry. This additional boundary condition results in a modified boundary value problem for u on the region $(-1,1) \times (-1,1) \cap G$.

The program uses the five point difference formula of Section 13. The linear system of equations is solved by SOR (cf. Section 18). Because of the symmetries, computation is restricted to the lattice points in the square $[-1,1] \times [-1,1]$. This leads to a substantial reduction in computing time for each iteration. The optimal overrelaxation parameter ω_b and the number of required iterations m, however, remain as large as with a computation over the entire region.

Altogether, nine subroutines are used:

POIS, SOR, SAVE, QNORM, NEIGHB, CHARDL,
CHAR, QUELL, RAND.

The last three named programs depend on the concrete problem and describe G, q, and ψ. Formally, we have REAL FUNCTIONs of two arguments of type REAL. CHAR is a characteristic function of the region G:

$$\mathrm{CHAR(X,Y)} \begin{cases} > 0 & \text{if } (X,Y) \in G \\ = 0 & \text{if } (X,Y) \in \partial G \\ < 0 & \text{otherwise.} \end{cases}$$

This function should be continuous, but need not be differentiable. If

ABS(CHAR(X,Y)) .LT. 1.E-4

it is assumed that the distance from the point to the boundary G is at most 10^{-3}. Each region is truncated by the program so as to lie in the appropriate rectangle Q_i, $i \in \{1,2,3,4\}$. For $G = (-1,1) \times (-1,1)$, therefore, CHAR(X,Y) = 1 suffices. If a region G is the intersection (union) of two regions G_1 and G_2, then the minimum (maximum) of the characteristic functions of G_1 and G_2 is a characteristic function for G.

POIS is called by the main program. The first two parameters are the names of the function programs RAND and QUELL. The name of CHAR is fixed. The remaining parameters of POIS are

BR, BO
M, EPSP, OMEGAP
BR = .TRUE. $\iff$ x = 1 is a line of symmetry
BO = .TRUE. $\iff$ y = 1 is a line of symmetry.

The mesh of the lattice is H = 1./2**M. EPSP is the absolute error up to which the iteration is continued. When EPSP = 0., the computation defaults to 10^{-3}. OMEGAP is the overrelaxation parameter. When OMEGAP = 0., ω_b is determined numerically by the program. Note that ω_b does depend on G, but does not depend on q or ψ. The numerical approximation also depends on EPSP. It improves as EPSP gets smaller. In case OMEGAP = 0, POIS should be called sequentially with M = 2, 3, 4, ... The program then uses as its given initial value for determining ω_b the approximate value from the preceding coarser lattice. OMEGAP remains zero.

In each iteration, SOR uses 7e+11f floating point operations, where

e : number of boundary distant lattice points

f : number of boundary close lattice points.

In the composition of the system of equations, the following main terms should be distinguished:

calls of QUELL : proportional to 1/H**2

calls of RAND : proportional to 1/H

calls of CHAR : proportional to |log EPSP|/H.

The program is actually designed for regions which are not rectangular. For rectangles, the Buneman algorithm (cf. Appendix 6) is substantially faster. It is nevertheless enticing to compare the theoretical results for the model problem (Example 18.15) with the numerical results. Let

$$\Delta u(x,y) = -20 \quad \text{in} \quad G = (0,1) \times (0,1)$$
$$u(x,y) = 0 \quad \text{on} \quad \partial G.$$

Since the iteration begins with $u(x,y) \equiv 0$, the norm of the initial error $\|\varepsilon^{(0)}\|_2$ is coarsely approximated by 1. The following results were obtained first of all with EPSP = 1./1000. The error reduction of the iterative method is thus about 1/1000.

Table 1 contains the theoretical values of ω_b and the numerical approximations $\tilde{\omega}_b$. The approximations, as suspected, are always too large.

h	ω_b	$\tilde{\omega}_b$
1/8	1.447	1.527
1/16	1.674	1.721
1/32	1.822	1.847
1/64	1.907	1.920
1/128	1.952	1.959

Table 1. ω_b and its numerical approximations

Table 2 contains the number of iterations m_1, m_2, and m_3, and the computing times t_1, t_2, and t_3, for OMEGAP = $\omega = 0$, $\omega = \tilde{\omega}_b$, and $\omega = \omega_b$. Column 2 contains the theoretically required number of steps m_{ω_b} from Example 18.15. m_1 and t_1 describe the computational effort involved in determining $\tilde{\omega}_b$. The times were measured on a CDC CYBER 76 system (in units of 1 second).

h	m_{ω_b}	m_1	m_2	m_3	t_1	t_2	t_3
1/8	8	22	14	14	0.021	0.016	0.016
1/16	17	36	28	32	0.098	0.077	0.085
1/32	35	52	56	64	0.487	0.506	0.751
1/64	70	132	112	128	4.484	3.785	4.298

Table 2. Iterative steps and computing time for EPSP = 10^{-3}

Surprisingly, the number of iterations is smaller for $\omega = \tilde{\omega}_b$ than for $\omega = \omega_b$. This does not contradict the theory, since the spectral radius $\rho(\mathscr{L}_\omega)$ only describes asymptotic convergence behavior. In fact, the relationship reverses for EPSP $= 10^{-9}$. Table 3 contains the number, Δm_{ω_b}, Δm_2, Δm_3, of *additional* iterations required to achieve this degree of accuracy.

Δm_{ω_b} : theoretical number, as in Example 18.5
Δm_2 : computation with $\omega = \tilde{\omega}_b$
Δm_3 : computation with $\omega = \omega_b$.

In both cases, i.e. for EPSP $= 10^{-3}$ and EPSP $= 10^{-9}$, it is our experience that

$$\omega = \omega_b + (2-\omega_b)/50$$

is better than ω_b.

h	Δm_{ω_b}	Δm_2	Δm_3
1/8	17	20	20
1/16	35	44	36
1/32	70	88	72
1/64	141	176	144

Table 3. Additional iterations for EPSP $= 10^{-9}$

```
      SUBROUTINE POIS(RAND,QUELL,BR,BO,M,EPSP,OMEGAP)
C
C     PARAMETERS OF THE SUBROUTINE
C
      REAL EPSP,OMEGAP
      INTEGER M
      LOGICAL BR,BO
C     RAND AND QUELL ARE FUNCTIONS FOR THE BOUNDARY VALUES
C     AND THE SOURCE TERM, RESPECTIVELY.
C
C     VARIABLES OF THE COMMON BLOCK
C
      REAL W(66,66),W1(65,65),WPS(65),Q(65,65),COEFF(1600)
      INTEGER NR(66,66),MITTE,N1,N2,ITER,NSYM,L1(65),L2(65),NO
      EQUIVALENCE (L1(1),W1(1,1)),(L2(1),Q(1,1))
      EQUIVALENCE (NO,W1(1,1)),(MITTE,Q(1,1)),(WPS(1),W1(1,2))
      COMMON W,W1,Q,COEFF,NR
      COMMON N1,N2,ITER,NSYM
C
C     LOCAL VARIABLES
C
      REAL D(4),PUNKT,STERN,BET2,EPS,EPS1,EPS2,H,H2,
     ,     OALT,OMEGAB,OMEGA,X,XN,Y,Z1,Z2,Z3
      INTEGER I,J,K,K1,K2,LCOEFF,N,NN,N3,N4,
     ,        MALT,MITTEO,MITTE1,MMAX,LCMAX
      LOGICAL BRN,BON,HBIT
      DATA PUNKT/1H./
      DATA STERN/1H*/
      DATA MALT /0/
C
C
C     MEANING OF THE VARIABLES
C
C     W(I,J)      VALUE OF THE UNKNOWN FUNCTION AT (X,Y),
C                 WHERE X=(I-MITTE)*H, Y=(J-MITTE)*H
C     W1(I,J),    AUXILIARY STORAGE
C     WPS(I)
C     COEFF(L)    HERE THE COEFFICIENTS OF THE DIFFERENCE EQUATION
C                 BELONGING TO A POINT NEAR THE BOUNDARY ARE
C                 STORED. COEFF(L), COEFF(L+1), COEFF(L+2), AND
C                 COEFF(L+3) ARE RELATED TO ONE POINT.
C     Q(I,J)      RIGHT-HAND SIDE OF THE DIFFERENCE EQUATION
C
C     N1,N2,      THE INTERIOR POINTS OF THE REGION SATISFY THE
C     L1(J),      INEQUALITIES
C     L2(J),               N1.LE.J.LE.N2          N1.GT.1
C                       L1(J).LE.I.LE.L2(J)
C     THIS SET OF INDICES MAY ALSO CONTAIN OUTER POINTS
C     INDICATED BY NR(I,J)=-1. L1(J) IS EVEN IN ORDER TO
C     SIMPLIFY THE RED-BLACK ORDERING OF THE SOR ITERATION.
```

```
C
C     THE ARRAY BOUNDS OF (1) W,NR, OF (2) W1,WPS,Q,L1,L2, AND
C     OF (3) COEFF CAN BE CHANGED SIMULTANEOUSLY WITH MMAX:
C        MMAX=4    BOUNDS    (1)  34    (2)  33    (3)  800
C        MMAX=5    BOUNDS    (1)  66    (2)  65    (3) 1600
C        MMAX=6    BOUNDS    (1) 130    (2) 129    (3) 3200
C     THE REAL VARIABLES MAY BE REPLACED BY DOUBLE PRECISION
C     VARIABLES.
C
C     D(K)    DISTANCES OF A BOUNDARY CLOSE POINT FROM THE
C             NEIGHBOURING POINTS.
C     H       DISTANCES OF THE BOUNDARY DISTANT POINTS (GRID SIZE)
C     M       H=1./2**M
C     N       =2**M
C     H2      =H*H
C     NN      =N/2
C     MALT    =0 IN THE CASE OF THE FIRST RUN OF THIS
C             SUBROUTINE; OTHERWISE, MALT COINCIDES WITH M
C             FROM THE FOREGOING RUN.
C     OALT    OMEGAB OF THE LAST RUN; OTHERWISE UNDEFINED
C     BR      =.TRUE. IN THE CASE OF SYMMETRIE WITH RESPECT
C             TO THE LINE X=1
C     BO      =.TRUE. IN THE CASE OF SYMMETRIE WITH RESPECT
C             TO THE LINE Y=1
C     BRN     =.NOT.BR
C     BON     =.NOT.BO
C     EPS     RELATIVE ACCURACY. THE SOR ITERATION IS CONTINUED
C             UNTIL EPS IS REACHED.
C     EPS1    RELATIVE ACCURACY FOR DETERMINING THE DIFFERENCE
C             2-OMEGAB
C     OMEGA   PRELIMINARY SOR PARAMETER THAT IS USED FOR
C             COMPUTING OMEGAB (OMEGA.LT.OMEGAB)
C     ITER    NUMBER OF STEPS OF THE SOR ITERATION
C     NSYM    LENGTH OF THE LINES OF SYMMETRIE IN W(I,J)
C
C
C     IF THE PARAMETERS "EPSP" AND "OMEGAP" EQUAL ZERO, THE
C     PROGRAMME DEFINES "EPS=0.001" AND COMPUTES THE OPTIMAL
C     "OMEGAB". IN THE CASE OF "OMEGAP.GT.0.", THE PARAMETER
C     OMEGAB=OMEGAP IS USED DURING THE WHOLE ITERATION.
C
C     COMPONENTS OF REAL ARRAYS AND INTEGER VARIABLES EQUATED
C     BY AN EQUIVALENCE STATEMENT ARE USED ONLY AS INTEGERS.
C
      OMEGAB=OMEGAP
      EPS=EPSP
      MMAX=5
      LCMAX=50*(2**MMAX)
      MITTE0=2**MMAX
      MITTE =MITTE0+1
      MITTE1=MITTE +1
```

```
C     M MUST SATISFY 2.LE.M.LE.MMAX
      IF(2.LE.M .AND. M.LE.MMAX) GO TO 1
      PRINT 97, M,MMAX
   97 FORMAT(4H1 M=,I1,11H NOT IN (2,,I1,1H))
      STOP
    1 N=2**M
C     PRELIMINARY "OMEGA" IS 1.. ONLY IN THE CASE OF "M=MALT+1"
C     THE VALUE OF "OMEGAB" OF THE FOREGOING RUN IS ASSIGNED
C     TO "OMEGA".
      OMEGA=1.
      IF(M.EQ.MALT+1) OMEGA=OALT
      MALT=M
      IF(EPS.LE.0.) EPS=0.001
      EPS1=-1./ALOG(EPS)
      EPS2=0.1*EPS
C     THE NUMBER NO OF BISECTION STEPS IN NACHB
C     IS ABOUT -LOG2(EPS).
      NO=-1.5*ALOG(EPS)+0.5
      ITER=0
      NN=N/2
      IF(BR.OR.BO) NN=N
      XN = N
      H = 1./XN
      H2 = H*H
      N1 = MITTE1-N
      N2 = MITTE +N
      N3=N1-1
      N4=N2+1
      NSYM=N2
      BRN=.NOT.BR
      BON=.NOT.BO
      N2R=N2
      N2O=N2
      IF(BRN) N2R=N2R-1
      IF(BON) N2O=N2O-1
C     THE VALUES 0. AND -1 ARE ASSIGNED TO "W" AND "NR", RESP.,
C     AT ALL POINTS OF THE SQUARE  -1..LE.Y.LE.+1.
C                                  -1..LE.X.LE.+1.
      DO 3 J=N3,N4
       DO 3 I=N3,N4
        W(I,J)=0.
    3   NR(I,J)=-1
C
C     THE FOLLOWING NESTED LOOP DETERMINES ALL INTERIOR POINTS.
C     AT THE SAME TIME "K1", "K2", "L1", "L2" ARE DEFINED
C     SUCH THAT
C             K1.LE.J.LE.K2
C         L1(J).LE.I.LE.L2(J)
C     HOLDS FOR ALL INTERIOR POINTS.
      K1=N2
      K2=N1
```

```
      DO 5 J=N1,N20
       Y = J-MITTE
       Y = Y*H
       L1(J)=N2
       L2(J)=N1
       DO 4 I=N1,N2R
        X = I-MITTE
        X = X*H
C       (X,Y) IS AN INTERIOR POINT IF "CHAR(X,Y).GT.0.". IN ORDER
C       TO AVOID TOO SMALL DISTANCES FROM THE BOUNDARY,
C       "CHAR(X,Y).GT.EPS2" IS REQUIRED FOR INTERIOR POINTS.
C       THEREFORE IT IS PERMITTED THAT SOME POINTS HAVE A
C       DISTANCE GREATER THAN "H" FROM THE BOUNDARY. WE ALLOW A
C       DISTANCE UP TO 1.1*H. THIS HARDLY INFLUENCES THE ACCURACY.
        IF (CHAR(X,Y) .LE. EPS2) GOTO 4
        K1=MIN0(K1,J)
        K2=MAX0(K2,J)
        L1(J)=MIN0(L1(J),I)
        L2(J)=MAX0(L2(J),I)
C       "NR=0" AND "Q=QUELL(X,Y)*H*H" IS USED AT INTERIOR POINTS.
C       THESE VALUES WILL BE CHANGED ONLY AT POINTS NEAR THE
C       BOUNDARY.
        NR(I,J) = 0
        Q(I,J) = QUELL(X,Y)*H2
    4   CONTINUE
       NR(NSYM+1,J)=NR(NSYM-1,J)
C      ROUNDING OFF L1(J) TO THE PRECEDING EVEN INTEGER.
       L1(J)=L1(J)-MOD(L1(J),2)
    5  CONTINUE
      DO 6 I=N1,N2R
       NR(I,NSYM+1)=NR(I,NSYM-1)
    6  CONTINUE
C
C     NEW PAGE. AFTERWARDS THE GRID IS PRINTED OUT PROVIDED
C     THAT M.LE.5.
C               "STAR"=INTERIOR POINT
C             "PERIOD"=EXTERIOR POINT
C     "X" IS ORIENTATED FROM THE LEFT TO THE RIGHT,
C     "Y" FROM BOTTOM TO TOP.
      IF (M .GT. 5) GO TO 10
      PRINT99
   99 FORMAT (1H1)
      J=N20
      DO 9 K = N1,N20
       DO 7 I = N1,N2R
        WPS(I) = PUNKT
        IF(NR(I,J).GE.0) WPS(I) = STERN
    7   CONTINUE
       PRINT 8, (WPS(I), I=N1,N2R)
    8  FORMAT(3X,64A2)
       J = J-1
    9  CONTINUE
```

```
C     IN THE CASE OF "K1.GT.K2" THERE ARE NO INTERIOR POINTS.
      IF(K1.LE.K2) GOTO 10
      PRINT 98
   98 FORMAT(30H1 THERE ARE NO INTERIOR POINTS)
      STOP
C     HENCEFORTH THE INTERIOR POINTS SATISFY
C                   N1.LE.J.LE.N2
C                L1(J).LE.I.LE.L2(J)
   10 N1=K1
      N2=K2
C
C
C     IT FOLLOWS THE DETERMINATION OF THE BOUNDARY CLOSE
C     POINTS AND THE COMPUTATION OF THE COEFFICIENTS OF THE
C     DIFFERENCE EQUATION IN THESE POINTS. THE COEFFICIENTS ARE
C     ASSIGNED TO
C               COEFF(LCOEFF), COEFF(LCOEFF+1),
C               COEFF(LCOEFF+2), COEFF(LCOEFF+3)
C
      LCOEFF = 1
      DO 30 J=N1,N2
       Y = J-MITTE
       Y = Y*H
       K1=L1(J)
       K2=L2(J)
       DO 29 I=K1,K2
C       NO CHECKS FOR EXTERIOR POINTS.
        IF (NR(I,J) .LT. 0) GOTO 29
C       THE SUBROUTINE "NEIGHB" DEFINES THE ARRAY "D".
C          "D(1),D(2),D(3),D(4) = DISTANCES FROM THE NEIGHBOURS"
C          FOR BOUNDARY CLOSE POINTS;
C          "D(1)=-H", OTHERWISE.
        CALL NEIGHB(D, I, J, H)
C       ONLY BOUNDARY CLOSE POINTS ARE TREATED IN THE FOLLOWING
        IF (D(1).LT.0.) GO TO 29
C       IF "LCOEFF.GT.LCMAX" THE ARRAY "COEFF" IS FILLED. THE
C       PROGRAMME MUST BE TERMINATED. LCMAX (=LENGTH OF LCOEFF)
C       CAN BE ENLARGED. IN THIS CASE THE COMMON BLOCKS ARE
C       TO BE CHANGED.
        IF(LCOEFF.GT.LCMAX) GOTO 100
        X = I-MITTE
        X = X*H
        Z1=D(1)+D(3)
        Z2=D(2)+D(4)
        Z3 = 1./(Z1*D(1))+1./(Z2*D(2))+1./(Z1*D(3))+1./(Z2*D(4))
        Q(I,J) = Q(I,J)*2./(Z3*H2)
        Z1=Z1*Z3
        Z2=Z2*Z3
        HBIT=.TRUE.
```

```
       IF (NR(I+1,J)) 11, 12, 12
 11    Q(I,J) = Q(I,J) - 4./(D(1)*Z1)*RAND(X+D(1),Y)
       COEFF(LCOEFF) = 0.
       HBIT=HBIT.AND.D(1).EQ.H
       GOTO 13
 12    COEFF(LCOEFF) = 4./(D(1)*Z1)
 13    IF (NR(I,J+1)) 14, 15, 15
 14    Q(I,J) = Q(I,J) - 4./(D(2)*Z2)*RAND(X,Y+D(2))
       COEFF(LCOEFF+1) = 0.
       HBIT=HBIT.AND.D(2).EQ.H
       GOTO 16
 15    COEFF(LCOEFF+1) = 4./(D(2)*Z2)
 16    IF (NR(I-1,J)) 17, 18, 18
 17    Q(I,J) = Q(I,J) - 4./(D(3)*Z1)*RAND(X-D(3),Y)
       COEFF(LCOEFF+2) = 0.
       HBIT=HBIT.AND.D(3).EQ.H
       GOTO 19
 18    COEFF(LCOEFF+2) = 4./(D(3)*Z1)
 19    IF (NR(I,J-1)) 20, 21, 21
 20    Q(I,J) = Q(I,J) - 4./(D(4)*Z2)*RAND(X,Y-D(4))
       COEFF(LCOEFF+3) = 0.
       HBIT=HBIT.AND.D(4).EQ.H
       GOTO 22
 21    COEFF(LCOEFF+3) = 4./(D(4)*Z2)
 22    NR(I,J) = 0
       IF(HBIT) GOTO 29
       NR(I,J) = LCOEFF
       LCOEFF = LCOEFF + 4
 29    CONTINUE
 30  CONTINUE
C
C
      LCOEFF = LCOEFF/4
      PRINT 40,LCOEFF
 40   FORMAT(1X/ 34H   NUMBER OF BOUNDARY CLOSE POINTS,
                 14H (D(L).NE.H) =, I4)
C
C
C     THE NEXT LOOP ENDING WITH STATEMENT NUMBER "59" COMPUTES
C     THE OPTIMAL "OMEGAB". THE COMPUTATION IS OMITTED IF
C     "OMEGAB.GT.0.".
      IF(OMEGAB.GT.0.) GOTO 60
C     "OMEGAB" WILL BE IMPROVED ITERATIVELY, STARTING WITH
C     A UNSUITABLE VALUE.
      OMEGAB=2.
C     AT FIRST "NN" STEPS OF THE SOR ITERATION ARE EXECUTED.
C     AFTER NEARLY NN STEPS THE INFLUENCE OF THE BOUNDARY
C     VALUES BEARS UPON THE MIDDLE OF THE REGION.
 31   DO 32 I = 1,NN
 32    CALL SOR(OMEGA)
C     "W1=W"
      CALL SAVE
```

```
C     A FURTHER STEP OF THE SOR ITERATION.
      CALL SOR(OMEGA)
C     "Z1"=SUMME"(W1(I,J)-W(I,J))**2"
      CALL QNORM(Z1)
      CALL SAVE
      CALL SOR(OMEGA)
      CALL QNORM(Z2)
C     IF THE COMPONENT OF THE INITIAL ERROR BELONGING TO
C     THE LARGEST EIGENVALUE OF THE MATRIX IS TOO SMALL,
C     THE STARTING VALUES MUST BE CHANGED.
      IF(Z2.GE.Z1) GOTO 110
C     "Z3"=APPROXIMATION TO THE SPECTRAL RADIUS OF THE ITERATION
C     MATRIX OF THE SOR ITERATION WITH PARAMETER "OMEGA".
      Z3 = SQRT(Z2/Z1)
C     "BET2"=APPROXIMATION TO THE SQUARED SPECTRAL RADIUS OF
C     THE ITERATION MATRIX OF THE JACOBI ITERATION.
      BET2 = (Z3+OMEGA-1.)**2/(Z3*OMEGA**2)
C     "Z3"=NEW APPROXIMATION OF "OMEGAB"
      Z3 = 2./(1.+SQRT(1.-BET2))
C     THE DIFFERENCE "2.-OMEGAB" IS TO BE DETERMINED UP TO
C     THE RELATIVE ACCURACY "EPS1". IN CASE OF WORSE ACCURACY
C     THE WHOLE PROCESS IS TO BE REPEATED WITH "OMEGAB=Z3".
      IF (ABS(Z3-OMEGAB) .LT. (2.-Z3)*EPS1) GO TO 59
      OMEGAB=Z3
      GOTO 31
C     SINCE IT IS MORE ADVANTAGEOUS TO USE A LARGER THAN
C     A SMALLER VALUE, THE APPROXIMATION OF "OMEGAB" IS
C     SOMEWHAT ENLARGED. THEN "OMEGAB" IS ROUNDED UP.
   59 Z3=Z3+EPS1*(2.-Z3)+16.
      OMEGAB=Z3-16.
C     "OMEGAB" IS ASSIGNED TO "OALT". THIS VALUE IS KEPT SINCE
C     IT CAN BE USED IN A FOLLOWING RUN WITH "M=M+1".
      OALT=OMEGAB
C     PRINTING OF "OMEGAB" AND OF THE NUMBER OF ITERATION
C     STEPS NEEDED UP TO NOW.
      PRINT 61, OMEGAB, ITER
   61 FORMAT (1X/ 11H   OMEGAB =,F6.3,7H ITER =,I4)
   62 FORMAT(29H   TOTAL NUMBER OF ITERATIONS,I6/1H )
C
C
C     "W1=W"
   60 CALL SAVE
C     NN STEPS OF THE SOR ITERATION
      DO 80 I = 1,NN
   80  CALL SOR(OMEGAB)
C     CHECK OF ACCURACY. THE ACCURACY IS NEARLY INDEPENDENT OF
C     "M" AND "H", SINCE "NN" ITERATIONS ARE USED.
      DO 90 J = N1,N2
       K1=L1(J)
       K2=L2(J)
       DO 90 I = K1,K2
        IF (ABS(W1(I,J)-W(I,J)) .GT. EPS) GOTO 60
   90   CONTINUE
```

```
C
C     PRINT-OUT OF "W" AND OF THE TOTAL NUMBER OF SOR ITERATIONS.
C
      IF (M.GT.3) PRINT 99
      PRINT 62, ITER
      N3 = N3 + 1
      J = N2
      Y=(N2-MITTE)*H
      DO 93 K=N1,N2
       PRINT 92, Y,(W(I,J),I=N3,N2R)
   92  FORMAT(1X,F9.7,4X,8F7.4,4X,8F7.4/7(14X,8F7.4,4X,8F7.4/))
       J = J - 1
       Y=Y-H
   93  CONTINUE
      RETURN
C
  100 PRINT 101
  101 FORMAT(33H   TOO MANY BOUNDARY CLOSE POINTS)
      STOP
C
C     CHANGE OF STARTING VALUES AT THE INTERIOR POINTS.
C
  110 DO 120 J=N1,N2
       K1= L1(J)
       K2=L2(J)
       DO 119 I=K1,K2
        IF(NR(I,J).LT.0) GO TO 119
        W(I,J)=W(I,J)-1.
  119   CONTINUE
  120  CONTINUE
      GOTO 31
      END
```

```
      SUBROUTINE SOR(OMEGA)
C
C     THE SUBROUTINE "SOR" PERFORMS ONE ITERATION STEP.
C     THE NUMBER OF ITERATIONS IS COUNTED BY "ITER".
C     THE RED-BLACK ORDERING IS USED FOR THE INTERIOR POINTS.
C     SINCE "L1(J)" IS EVEN,
C          M=1  , MOD(I+J,2)=MOD(N1,2)
C     HOLDS FOR THE FIRST RUN, WHILE
C          M=2  , MOD(I+J,2)=MOD(N1+1,2)
C     HOLDS FOR THE SECOND RUN.
C
C     PARAMETER
C
      REAL OMEGA
C
C     VARIABLES OF THE COMMON BLOCK
C
      REAL W(66,66),W1(65,65),WPS(65),Q(65,65),COEFF(1600)
      INTEGER NR(66,66),MITTE,N1,N2,ITER,NSYM,L1(65),L2(65),NO
      EQUIVALENCE (L1(1),W1(1,1)),(L2(1),Q(1,1))
      EQUIVALENCE (NO,W1(1,1)),(MITTE,Q(1,1))
      COMMON W,W1,Q,COEFF,NR
      COMMON N1,N2,ITER,NSYM
C
C     LOCAL VARIABLES
C
      REAL OM,OM1
      INTEGER I,J,K,K2,L,M,N
C
      ITER=ITER+1
      OM = OMEGA*0.25
      OM1=1.-OMEGA
      M = 1
      N = 0
    5 DO 50 J=N1,N2
       K=L1(J)+N
       K2=L2(J)
       DO 40 I = K,K2,2
        IF (NR(I,J)) 40, 10, 20
   10   W(I,J) = W(I,J)*OM1 + OM*(W(I+1,J)+W(I,J+1)+W(I-1,J)
     +                           +W(I,J-1)-Q(I,J))
        GOTO 40
   20   L = NR(I,J)
        W(I,J) = W(I,J)*OM1 + OM*(COEFF(L)*W(I+1,J)+COEFF(L+1)*
     *      W(I,J+1)+COEFF(L+2)*W(I-1,J)+COEFF(L+3)*W(I,J-1)-Q(I,J))
   40   CONTINUE
       W(NSYM+1,J)=W(NSYM-1,J)
       N = 1-N
   50  CONTINUE
      IF(N2.LT.NSYM) GOTO 54
      K=L1(NSYM-1)
      K2=L2(NSYM-1)
```

```
      DO 53 I=K,K2
       W(I,NSYM+1)=W(I,NSYM-1)
   53  CONTINUE
   54 IF (M-2) 55,56,56
   55 N = 1
      M = 2
      GOTO 5
   56 RETURN
      END
```

```
      SUBROUTINE NEIGHB(D, I, J, H)
C
C     "NEIGHB" COMPUTES THE DISTANCE OF THE INTERIOR POINT
C     (X,Y) WITH
C          X=(I-MITTE)*H
C          Y=(J-MITTE)*H
C     FROM THE NEIGHBOURS. IN THE CASE OF A BOUNDARY DISTANT
C     POINT, THE RESULT IS
C          D(1)=-H , D(2)=D(3)=D(4)=H.
C     FOR BOUNDARY CLOSE POINTS FOUR POSITIVE NUMBERS ARE
C     COMPUTED. THE DISTANCE FROM THE BOUNDARY IS DETERMINED
C     BY A BISECTION METHOD, SINCE THE CHARACTERISTIC FUNCTION
C     OF THE REGION IS NOT REQUIRED TO BE DIFFERENTIABLE.
C
C     PARAMETERS
C
      REAL D(4),H
      INTEGER I,J
C
C     VARIABLES OF THE COMMON BLOCKS
C
      REAL W(66,66),W1(65,65),WPS(65),Q(65,65),COEFF(1600)
      INTEGER NR(66,66),MITTE,N1,N2,ITER,NSYM,L1(65),L2(65),NO
      EQUIVALENCE (L1(1),W1(1,1)),(L2(1),Q(1,1))
      EQUIVALENCE (NO,W1(1,1)),(MITTE,Q(1,1))
      COMMON W,W1,Q,COEFF,NR
      COMMON N1,N2,ITER,NSYM
C
C     LOCAL VARIABLES
C
      REAL DELTA,H1,H11,X,Y
      INTEGER I1(4),J1(4),I2,J2,K,L
      LOGICAL B
      DATA I1/1,0,-1,0/
      DATA J1/0,1,0,-1/
```

```
C
      B = .TRUE.
      H1=H/8.
      X = (I-MITTE)*H
      Y = (J-MITTE)*H
      DO 100 K = 1,4
       I2 = I + I1(K)
       J2 = J + J1(K)
       IF (NR(I2,J2).GE.0.) GO TO 90
       B = .FALSE.
       H11 = H1
       DELTA = H1
C      IN THIS LOOP THE SIGN OF "CHARDL" IS CHECKED IN STEPS OF
C      "H/8" IN ORDER TO DETECT THE FIRST CHANGE OF SIGN
       DO 10 L=1,9
        IF (CHARDL(K, DELTA, X, Y).LT.0.) GO TO 11
   10   DELTA = DELTA + H11
       DELTA=H
       GO TO 80
C      HERE THE BISECTION METHOD STARTS. "NO" IS THE REQUIRED
C      NUMBER OF STEPS. IT IS DEFINED IN POIS.
   11  H11 = H11*0.5
       DELTA = DELTA - H11
       DO 20 L = 1,NO
        H11 = H11*0.5
        IF (CHARDL(K, DELTA, X, Y).LE.0.) GO TO 15
        DELTA = DELTA + H11
        GO TO 20
   15   DELTA = DELTA - H11
   20   CONTINUE
       IF(DELTA.GT.1.1*H) DELTA=H
C      THE RESULTING DISTANCE MAY BE SOMEWHAT LARGER THAN "H".
C      BUT NOTE THE BOUNDARY POINT IS LOCATED IN
C           ABS(X).LE.1.
C           ABS(Y).LE.1.
C      SINCE EVERY REGION IS CUT IN THIS WAY.
       IF (DELTA.LE.H) GO TO 80
       IF (X + I1(K)*DELTA.GT.1. .OR.
     .     X + I1(K)*DELTA.LT.-1.) DELTA = H
       IF (Y + J1(K)*DELTA.GT.1. .OR.
     .     Y + J1(K)*DELTA.LT.-1.) DELTA = H
   80  D(K) = DELTA
       GO TO 100
   90  D(K) = H
  100  CONTINUE
      IF (B)  D(1) = -H
      RETURN
      END
```

```
      REAL FUNCTION CHARDL (K, DELTA, X, Y)
C
C     "CHARDL" TRANSFORMS THE ARGUMENTS OF "CHAR" AS SUITED TO
C     "NACHB". "CHAR" SHOULD BE PROGRAMMED AS SIMPLE AS POSSIBLE
C     TO ENABLE EASY CHANGES.
C
C     PARAMETERS
C
      REAL DELTA,X,Y
      INTEGER K
C
C     LOCAL VARIABLES
C
      REAL X1(4),Y1(4),X2,Y2
      DATA X1/1.,0.,-1.,0./
      DATA Y1/0.,1.,0.,-1./
C
      X2 = X + DELTA*X1(K)
      Y2 = Y + DELTA*Y1(K)
      CHARDL = CHAR(X2,Y2)
      RETURN
      END
```

```
      SUBROUTINE SAVE
C
C     "SAVE" STORES "W" ON "W1"
C
C     VARIABLES OF THE COMMON BLOCKS
C
      REAL W(66,66),W1(65,65),WPS(65),Q(65,65),COEFF(1600)
      INTEGER NR(66,66),MITTE,N1,N2,ITER,NSYM,L1(65),L2(65),NO
      EQUIVALENCE (L1(1),W1(1,1)),(L2(1),Q(1,1))
      EQUIVALENCE (NO,W1(1,1)),(MITTE,Q(1,1))
      COMMON W,W1,Q,COEFF,NR
      COMMON N1,N2,ITER,NSYM
C
C     LOCAL VARIABLES
C
      INTEGER I,J,K1,K2
C
      DO 10 J=N1,N2
       K1=L1(J)
       K2=L2(J)
       DO 10 I=K1,K2
   10   W1(I,J)=W(I,J)
      RETURN
      END
```

```
      SUBROUTINE QNORM(Z)
C
C     "QNORM" COMPUTES THE SQUARED EUCLIDEAN NORM OF THE
C     DIFFERENCE
C          Z=SUMME(W1(I,J)-W(I,J))**2
C
C     PARAMETER
C
      REAL Z
C
C     VARIABLES OF THE COMMON BLOCKS
C
      REAL W(66,66),W1(65,65),WPS(65),Q(65,65),COEFF(1600)
      INTEGER NR(66,66),MITTE,N1,N2,ITER,NSYM,L1(65),L2(65),NO
      EQUIVALENCE (L1(1),W1(1,1)),(L2(1),Q(1,1))
      EQUIVALENCE (NO,W1(1,1)),(MITTE,Q(1,1))
      COMMON W,W1,Q,COEFF,NR
      COMMON N1,N2,ITER,NSYM
C
C     LOCAL VARIABLES
C
      REAL SUM
      INTEGER I,J,K1,K2
C
      SUM = 0.
      DO 10 J = N1,N2
       K1=L1(J)
       K2=L2(J)
       DO 9 I = K1,K2
        IF (NR(I,J).LT. 0) GOTO 9
        SUM = SUM + (W1(I,J)-W(I,J))**2
    9   CONTINUE
   10  CONTINUE
      Z = SUM
      RETURN
      END
```

EXAMPLE:

```
      REAL FUNCTION CHAR(X,Y)
C
C     ELLIPSE
C
      Z1=X/0.9
      Z2=Y/0.6
      CHAR=1.-Z1*Z1-Z2*Z2
      RETURN
      END
```

```
      REAL FUNCTION QUELL (X,Y)
C
      QUELL=4.
      RETURN
      END
```

```
      REAL  FUNCTION RAND (X,Y)
C
      RAND=1.
      RETURN
      END
```

Appendix 5: Programs for band matrices.

We present programs for the following methods:

GAUBD3: Gaussian elimination without pivot search for tridiagonal matrices.

GAUBD: Gaussian elimination without pivot search for band matrices of arbitrary band width $w \geq 3$.

REDUCE: Band width reduction by the Gibbs-Poole-Stockmeyer method. (This includes the subroutines LEVEL, KOMPON, SSORT1, SSORT2).

In the program we use $K = (w-1)/2$ as a measure of band width (cf. Definition 20.1 and 20.4).

We first consider calls of GAUBD3 and GAUBD. A is the matrix $\tilde{A}$ from Section 20 (cf. also Figure 20.2), $N \geq 2$ is the number of equations, $K \leq N$ is the measure of bandwidth just mentioned. If the only change since the last call of the program is on the right side of the system of equations A(4,*) [or A(2*K+2,*)], set B = .TRUE., otherwise, B = .FALSE.. After the call of the program, the solution vector is in A(4,*) [or A(2*K+2,*)]. For $K > 10$, the statement for A in GAUBD has to be replaced by

```
REAL A(m,N).
```

Here m is some number greater than or equal to 2*K+2.

The number of floating point operations in one call of GAUBD3 or GAUBD is:

GAUBD3	B = .FALSE.:	$8N-7$
	B = .TRUE.:	$5N-4$
GAUBD	B = .FALSE.:	$(2K^2+5K+1)(N-1)+1$
	B = .TRUE.:	$(4K+1)(N-1)+1$.

The program REDUCE contains four explicit parameters:

N: number of rows in the matrix

M: the number of matrix elements different from zero above the main diagonal

KOLD: K before band width reduction

KNEW: K after band width reduction.

N and M are input parameters, and KOLD and KNEW are output parameters. The pattern of the matrix is described by the vector A in the COMMON block. Before a call, one enters here the row and column indices of the various matrix elements different from zero above the main diagonal. The entry order is: row index, corresponding column index, row index, corresponding column index, etc.; altogether there are M pairs of this sort. REDUCE writes into this vector that permutation which leads to band width reduction. Either KOLD > KNEW or KOLD = KNEW. In the second case, the permutation is the identity, since no band width reduction can be accomplished with this program.

The next example should make the use of REDUCE more explicit. The pattern of the matrix is given as

$$\begin{bmatrix}
x & x & & & & x & x & & & & & & & & \\
x & x & x & & & x & x & x & & & & & & & \\
& x & x & x & & & x & x & x & & & & & & \\
& & x & x & x & & & x & x & x & & & & & \\
& & & x & x & & & & x & x & & & & & \\
x & x & & & & x & x & & & & x & x & & & \\
x & x & x & & & x & x & x & & & x & x & x & & \\
& x & x & x & & & x & x & x & & & x & x & x & \\
& & x & x & x & & & x & x & x & & & x & x & x \\
& & & x & x & & & & x & x & & & & x & x \\
& & & & & x & x & & & & x & x & & & \\
& & & & & x & x & x & & & x & x & x & & \\
& & & & & & x & x & x & & & x & x & x & \\
& & & & & & & x & x & x & & & x & x & x \\
& & & & & & & & x & x & & & & x & x
\end{bmatrix}$$

This corresponds to the graph in Figure 20.11.

Input:

N = 15, M = 38,

A = 1,2,1,6,1,7,2,3,2,6,2,7,2,8,3,4,3,7,3,8,3,9,
4,5,4,8,4,9,4,10,5,9,5,10,6,7,6,11,6,12,7,8,7,11,
7,12,7,13,8,9,8,12,8,13,8,14,9,10,9,13,9,14,9,15,
10,14,10,15,11,12,12,13,13,14,14,15.

Output:

KOLD = 6, KNEW = 4,

A = 1,4,7,10,13,2,5,8,11,14,3,6,9,12,15.

The program declarations are sufficient for $N \le NMAX = 650$ and $M \le MMAX = 2048$. For large N or M, only the bounds of the COMMON variables A, VEC, IND, LIST, GRAD, and NR have to be changed. However, N must be less than 10,000 in any case. On IBM or Siemens installations with data types INTEGER*2 and INTEGER*4, two bytes suffice for IND, LIST, GRAD, and NR. All other variables should be INTEGER*4.

For a logical run of the program, it is immaterial whether the graph is connected or not. However, if the graph decomposes into very many connected components (not counting knots of degree zero), the computing times become extremely long. We have attached no special significance to this fact, since the graph in most practical cases has only one or two connected components. REDUCE is described in nine sections.

<u>Section 1</u>: Computation of KOLD and numbering of the degree zero knots. NUM contains the last number given.

<u>Section 2</u>: Building a *data base* for the following sections. During this transformation of the input values, matrix

elements entered in duplicate are eliminated. The output is the knots

A(J), J = LIST(I) to LIST(I+1)-1

which are connected to the knot I. They are ordered, in order of increasing degree. GRAD(I) gives the degree of knot I. NR(I) is the new number of the knot, or is zero if the knot does not yet have a new number. In our example, after Section 2 we obtain:

```
A = 2,6,7,                     1,3,6,7,8,
    2,4,7,8,9,                 5,3,10,8,9,
    4,10,9,                    1,11,2,12,7,
    1,11,2,6,13,3,12,          2,3,4,13,14,12,7,9,
    15,5,3,14,4,13,10,8,       5,15,4,14,9,
    6,12,7,                    11,6,13,7,8,
    12,14,7,8,9,               15,10,13,8,9,
    10,14,9.

LIST = 1,4,9,14,19,22,27,35,43,51,56,59,64,69,74,77.

GRAD = 3,5,5,5,3,5,8,8,8,5,3,5,5,5,3.

NR   = 0,0,0,0,0,0,0,0,0,0,0,0,0,0,0.
```

If the graph (disregarding knots of degree zero) consists of several connected components, Sections 3 through 8 will run the corresponding number of times.

Section 3: Step (A) and (B) of the algorithm, computation of κ_1.

Section 4: Steps (C) through (F), computation of κ_2. The return from (D) to (B) contains the instruction

```
IF(DEPTHB.GT.DEPTHF) GOTO 160
```

Section 5: Preliminary enumeration of the elements in the levels S_i (Step (G)).

Section 6: Determination and sorting of the components V_i (Step (G)).

Section 7: Steps (H) through (J). The loop on ν begins with

```
DO 410 NUE = 1, K2
```

Steps (L) and (M) are combined in the program.

Section 8: Steps (K) through (O). The loop on ℓ ends with

```
IF(L.LE.DEPTHF) GOTO 450
```

Section 9: Computation of KNEW and transfer of the new enumeration from NR to A.

LEVEL computes one level with the root START (cf. Theorem 20.8), KOMPON computes the components of V (cf. Lemma 20.10) beginning with an arbitrary starting element. SSORT1 and SSORT2 are sorting programs. To save time, we use a method of Shell 1959 (cf. also Knuth 1973), but this can be replaced just as easily with *Quicksort*.

Section 7 determines the amount of working memory required. If the graph is connected and the return from Step (D) to Step (B) occurs at most once, then the computing time is

$$O(c_1 n) + O((c_1 n)^{5/4}) + O(c_1\ c_2 n)$$

where

n = number of knots in the graph

c_1 = maximum degree of the knots

c_2 = maximum number of knots in the last level of $R(g)$.

The second summand contains the computing time for the sorts. If Quicksort is used, this term becomes $O(c_1 n \log(c_1 n))$ in the mean (statistically). The third summand corresponds to Section 4 of the program.

Suppose a boundary value problem in $\mathbb{R}^2$ is to be solved with a difference method or a finite element method. We consider the various systems of equations which result from a decreasing mesh h of the lattice. Then it is usually true that

$$n = O(1/h^2), \quad c_1 = O(1), \quad c_2 = O(1/h),$$
$$k = O(1/h) \quad \text{(band width measure)}.$$

The computing time for REDUCE thus grows at most in proportion to $1/h^3$, and for GAUBD, to $1/h^4$.

The program was tested with 166 examples. Of these, 28 are more or less comparable, in that they each had a connected graph and the number of knots was between 900 and 1000 and M was between 1497 and 2992. For this group, the computing time on a CDC-CYBER 76 varied between 0.16 and 0.37 seconds.

```
      SUBROUTINE  GAUBD3(A,N,B)
      REAL A(4,N)
      INTEGER N
      LOGICAL B
C
C     SOLUTION OF A SYSTEM OF LINEAR EQUATIONS WITH TRIDIAGONAL
C     MATRIX. THE I-TH EQUATION IS
C     A(1,I)*X(I-1)+A(2,I)*X(I)+A(3,I)*X(I+1)=A(4,I)
C     ONE TERM IS MISSING FOR THE FIRST AND LAST EQUATION.
C     THE SOLUTION X(I) WILL BE ASSIGNED TO A(4,I).
C
      REAL Q
      INTEGER I,I1
C
      IF(N.LE.1)STOP
      IF(B) GOTO 20
      DO 10 I=2,N
       Q=A(1,I)/A(2,I-1)
       A(2,I)=A(2,I)-A(3,I-1)*Q
       A(4,I)=A(4,I)-A(4,I-1)*Q
   10  A(1,I)=Q
      GOTO 40
C
   20 Q=A(4,1)
      DO 30 I=2,N
       Q=A(4,I)-A(1,I)*Q
   30  A(4,I)=Q
C
   40 Q=A(4,N)/A(2,N)
      A(4,N)=Q
      I1=N-1
      DO 50 I=2,N
       Q=(A(4,I1)-A(3,I1)*Q)/A(2,I1)
       A(4,I1)=Q
   50  I1=I1-1
      RETURN
      END
```

```
      SUBROUTINE GAUBD(A,N,K,B)
      REAL A(22,N)
      INTEGER N,K
      LOGICAL B
C
C     SOLUTION OF A SYSTEM OF LINEAR EQUATIONS WITH BAND MATRIX.
C     THE I-TH EQUATION IS
C     A(1,I)*X(I-K)+A(2,I)*X(I-K+1)+...+A(K+1,I)*X(I)+...+
C     A(2*K,I)*X(I+K-1)+A(2*K+1,I)*X(I+K) = A(2*K+2,I)
C     FOR I=1(1)K AND I=N-K+1(1)N SOME TERMS ARE MISSING.
C     THE SOLUTION X(I) WILL BE ASSIGNED TO A(2*K+2,I).
C
      REAL Q
      INTEGER K1,K2,K21,K22,I,II,II1,J,JJ,L,LL,LLL
C
      IF((K.LE.0).OR.(K.GE.N))STOP
      K1=K+1
      K2=K+2
      K21=2*K+1
      K22=K21+1
      IF(B) GO TO 100
C
      JJ=K21
      II=N-K+1
      DO 20 I=II,N
       DO 10 J=JJ,K21
   10   A(J,I)=0.
   20  JJ=JJ-1
      DO 50 I=2,N
       II=I-K
       DO 40 J=1,K
        IF(II.LE.0) GO TO 40
        Q=A(J,I)/A(K1,II)
        J1=J+1
        JK=J+K
        LLL=K2
        DO 30 L=J1,JK
         A(L,I)=A(L,I)-A(LLL,II)*Q
   30    LLL=LLL+1
        A(K22,I)=A(K22,I)-A(K22,II)*Q
        A(J,I)=Q
   40   II=II+1
   50  CONTINUE
      GO TO 200
C
  100 DO 150 I=2,N
       II=I-K
       DO 140 J=1,K
        IF(II.LE.0) GO TO 140
        A(K22,I)=A(K22,I)-A(K22,II)*A(J,I)
  140   II=II+1
  150  CONTINUE
```

```
C
 200  A(K22,N)=A(K22,N)/A(K1,N)
      II=N-1
      DO 250 I=2,N
       Q=A(K22,II)
       JJ=II+K
       IF(JJ.GT.N) JJ=N
       II1=II+1
       LL=K2
       DO 240 J=II1,JJ
        Q=Q-A(LL,II)*A(K22,J)
 240    LL=LL+1
       A(K22,II)=Q/A(K1,II)
 250   II=II-1
      RETURN
      END

      SUBROUTINE REDUCE(N,M,KOLD,KNEW)
C
C     PROGRAMME FOR REDUCING THE BANDWIDTH OF A SPARSE SYMMETRIC
C     MATRIX BY THE METHOD OF GIBBS, POOLE, AND STOCKMEYER.
C
C     INPUT
C     N                  NUMBER OF ROWS
C     M                  NUMBER OF NONVANISHING ENTRIES ABOVE
C                        THE DIAGONAL
C     A(I), I=1(1)2*M    INPUT VECTOR
C                        CONTAINING THE INDICES OF THE
C                        NONVANISHING ENTRIES ABOVE THE
C                        DIAGONAL. THE INDICES ARE ARRANGED
C                        IN THE SEQUENCE
C                        I1, J1, I2, J2, I3, J3, ...
C     OUTPUT
C     A(I), I=1(1)N      NEW NUMBERS OF I-TH ROW AND COLUMN.
C     KOLD               BANDWIDTH OF THE INPUT MATRIX
C     KNEW               BANDWIDTH AFTER PERMUTATION OF THE
C                        INPUT MATRIX ACCORDING TO A(I), I=1(1)N
C     THE ARRAY BOUNDS MAY BE CHANGED, PROVIDED THAT
C     NMAX.LT.10000
C     A(2*MMAX), VEC(NMAX),
C     IND(NMAX+1,8), LIST(NMAX+1), GRAD(NMAX), NR(NMAX)
C
      INTEGER N,M,KOLD,KNEW
C
      INTEGER A(4096),VEC(650)
      INTEGER IND(651,8),LIST(651),GRAD(650),NR(650)
      COMMON A,VEC,IND
      EQUIVALENCE (LIST(1),IND(1,1)),(GRAD(1),IND(1,2)),
                  (NR(1),IND(1,3))
C
      INTEGER NMAX,MMAX,NN,M2,N1,NUE,N1M,NUM,IS,OLD,NEW
      INTEGER F,L,L1,L2,L10,I,J,III,K,KNP1,K1,K1N,K2,K2P1
      INTEGER G,H,START,DEPTHF,DEPTHB,LEVWTH,GRDMIN,C,C2
      INTEGER KAPPA,KAPPA1,KAPPA2,KAPPA3,KAPPA4
      INTEGER IND1,IND2,INDJ2,INDJ5,INDJ6,INDI7,INDI8,VECJ
      DATA C/10000/
      C2=2*C
      NMAX=650
```

```
      MMAX=2048
      IF(N.LT.2.OR.N.GT.NMAX.OR.M.GT.MMAX) STOP
C
C     SECTION 1
C
      M2=M+M
      KOLD=0
      KNEW=N
      DO 10 I=1,8
       DO 10 J=1,N
   10   IND(J,I)=0
      IF(M.EQ.0) GOTO 680
      DO 15 I=1,M2,2
       J=IABS(A(I)-A(I+1))
       IF(J.GT.KOLD) KOLD=J
   15  CONTINUE
C
      DO 20 I=1,M2
       K1=A(I)
   20  IND(K1,7)=1
      NUM=1
      DO 30 I=1,N
       IF(IND(I,7).GT.0) GOTO 30
       NR(I)=NUM
       NUM=NUM+1
   30  CONTINUE
C
C     SECTION 2 (NEW DATA STRUCTURE)
C
      DO 40 I=1,M2,2
       K1=A(I)
       K2=A(I+1)
       A(I)=K1+C*K2
   40  A(I+1)=K2+C*K1
      CALL SSORT1(1,M2)
      J=1
      OLD=A(1)
      DO 70 I=2,M2
       NEW=A(I)
       IF(NEW.GT.OLD) J=J+1
       A(J)=NEW
   70  OLD=NEW
      M2=J
      IND(1,2)=1
      J=1
      L10=A(1)/C
      DO 90 I=1,M2
       K=A(I)
       L1=K/C
       L2=K-L1*C
       A(I)=L2
       IF(L1.EQ.L10) GOTO 90
       L10=L1
       J=J+1
       IND(J,2)=I
   90  CONTINUE
      IND(J+1,2)=M2+1
      LIST(1)=1
      J=1
```

```
      DO 110 I=1,N
       IF(IND(I,7).GT.0) J=J+1
  110  LIST(I+1)=IND(J,2)
      DO 120 I=1,N
  120  GRAD(I)=LIST(I+1)-LIST(I)
      DO 130 I=1,N
       F=LIST(I)
       L=LIST(I+1)-1
  130  CALL SSORT2(A,2,F,L)
C
C     SECTION 3 (COMPUTATION OF R(G))
C     STEPS (A) AND (B), COMPUTATION OF KAPPA 1
C     IND(I,7)     LEVEL NUMBER OF R(G)
C     VEC(I)       ELEMENTS OF THE LAST LEVEL
C
  140 GRDMIN=N
      DO 150 I=1,N
       IF(NR(I).GT.0) GOTO 150
       IF(GRDMIN.LE.GRAD(I)) GOTO 150
       START=I
       GRDMIN=GRAD(I)
  150  CONTINUE
C
  160 G=START
      NN=N
      CALL LEVEL(G,NN,DEPTHF,K1,KAPPA1)
      J=NN-K1
      DO 180 I=1,K1
       III=I+J
  180  VEC(I)=IND(III,6)
      DO 190 I=1,N
  190  IND(I,7)=IND(I,8)
C
C     SECTION 4 (COMPUTATION OF R(H))
C     STEPS (C) TO (F), COMPUTATION OF KAPPA 2
C     IND(I,8)     LEVEL NUMBERS OF R(H)
C
      LEVWTH=N
      DO 210 I=1,K1
       START=VEC(I)
       N1=N
       CALL LEVEL(START,N1,DEPTHB,K1N,KAPPA2)
       IF(DEPTHB.GT.DEPTHF) GOTO 160
       IF(KAPPA2.GE.LEVWTH) GOTO 210
       LEVWTH=KAPPA2
       VECJ=I
  210  CONTINUE
      H=VEC(VECJ)
      N1=N
      CALL LEVEL(H,N1,DEPTHB,K1N,KAPPA2)
```

```
C
C     SECTION 5 (PRELIMINARY NUMBERING OF THE ELEMENTS OF S(I))
C     STEP (G)
C     IND(I,4)     PRELIMINARY NUMBER OF ELEMENTS OF S(I)
C     IND(I,5)     LEVEL NUMBERS FOR NODES WITH SAME
C                  NUMBERING; ZERO OTHERWISE
C
      DO 230 I=1,N
       IND(I,4)=0
  230  IND(I,5)=0
      J=0
      KNP1=DEPTHF+1
      DO 260 I=1,N
       INDI8=IND(I,8)
       IF(INDI8.EQ.0) GOTO 260
       K2=KNP1-INDI8
       IF(IND(I,7).NE.K2) GOTO 250
       IND(K2,4)=IND(K2,4)+1
       K2=-K2
       J=J+1
  250  CONTINUE
       IND(I,5)=K2
  260  CONTINUE
C
C     SECTION 6 (DETERMINATION AND SORTING OF V(I))
C     STEP (G)
C     VEC(I)       STARTING VALUES OF V(I) SORTED WITH RESPECT
C                  TO IABS(V(I))
C
      K2=0
      IF(J.EQ.NN) GOTO 412
      DO 290 I=1,N
       IF(IND(I,5).LE.0) GOTO 290
       START=I
       CALL KOMPON(START,N1)
       K2=K2+1
       VEC(K2)=START
       IND(START,8)=N1
  290  CONTINUE
      DO 310 I=1,N
       IF(IND(I,5).LT.-C) IND(I,5)=IND(I,5)+C2
  310  CONTINUE
      CALL SSORT2(VEC,8,1,K2)
      N1M=VEC(K2)
      N1M=IND(N1M,8)
      DO 315 I=1,K2
  315  IND(I,8)=VEC(I)
```

```
C
C     SECTION 7 (COMPUTATION OF S)
C     STEPS (H) TO (J)
C     IND(I,4)     NUMBER OF ELEMENTS OF S(I)
C     IND(I,5)     LEVEL NUMBER OF S
C     IND(I,6)     ALL NODES V(I)
C
      DO 319 J=1,DEPTHF
  319  VEC(J)=0
      K2P1=K2+1
      DO 410 NUE=1,K2
       III=K2P1-NUE
       START=IND(III,8)
       CALL KOMPON(START,N1)
       IND1=7
       IS=0
  325  DO 330 J=1,N1
        INDJ6=IND(J,6)
        III=IND(INDJ6,IND1)+IS
        VEC(III)=VEC(III)+1
  330   IND(J,2)=III
       KAPPA=0
       DO 340 J=1,N1
        INDJ2=IND(J,2)
        III=VEC(INDJ2)
        IF(III.LE.0) GOTO 340
        III=III+IND(INDJ2,4)
        VEC(INDJ2)=0
        IF(III.GT.KAPPA) KAPPA=III
  340   CONTINUE
       IF(IS.GT.0) GOTO 346
       KAPPA3=KAPPA
       IND1=5
       IS=C2
       GOTO 325
C
  346  KAPPA4=KAPPA
       IF(KAPPA3-KAPPA4) 350,347,380
  347  IF(KAPPA1.GT.KAPPA2) GOTO 380
  350  DO 370 J=1,N1
        INDJ6=IND(J,6)
        III=IND(INDJ6,7)
        IND(INDJ6,5)=III-C2
  370   IND(III,4)=IND(III,4)+1
       GOTO 410
  380  DO 400 J=1,N1
        INDJ2=IND(J,2)
  400   IND(INDJ2,4)=IND(INDJ2,4)+1
  410  CONTINUE
      DO 411 J=1,N1M
  411  GRAD(J)=LIST(J+1)-LIST(J)
```

```
  412 DO 415 I=1,N
       IF(IND(I,5).LT.-C) IND(I,5)=IND(I,5)+C2
  415  IND(I,5)=IABS(IND(I,5))
C
C     SECTION 8 (NUMBERING)
C     STEPS (K) TO (O)
C
      DO 420 I=1,N
  420  VEC(I)=I
      CALL SSORT2(VEC,5,1,N)
      IND1=1
      L=1
      OLD=0
      NEW=1
      IND(1,7)=G
      NR(G)=NUM
      NUM=NUM+1
C
  450 I=0
C
  460 I=I+1
      IF(I.GT.NEW) GOTO 490
  470 INDI7=IND(I,7)
      L1=LIST(INDI7)
      L2=LIST(INDI7+1)-1
      DO 480 J=L1,L2
       START=A(J)
       IF(NR(START).GT.0) GOTO 480
       IF(IND(START,5).NE.L) GOTO 480
       NR(START)=NUM
       NUM=NUM+1
       NEW=NEW+1
       IND(NEW,7)=START
  480  CONTINUE
      GOTO 460
C
  490 IF(NEW-OLD.GE.IND(L,4)) GOTO 510
      GRDMIN=N
      IND2=IND1
      DO 500 J=IND1,N
       VECJ=VEC(J)
       INDJ5=IND(VECJ,5)
       IF(INDJ5-L) 499,491,501
  491  IF(NR(VECJ).GT.0) GOTO 500
       IF(GRAD(VECJ).GE.GRDMIN) GOTO 500
       GRDMIN=GRAD(VECJ)
       START=VECJ
       GOTO 500
  499  IND2=J+1
  500  CONTINUE
```

```
  501 IND1=IND2
      NR(START)=NUM
      NUM=NUM+1
      NEW=NEW+1
      IND(NEW,7)=START
      GOTO 470
C
  510 NEW=NEW-OLD
      DO 520 I=1,NEW
       III=I+OLD
  520  IND(I,7)=IND(III,7)
      OLD=NEW
      L=L+1
      IF(L.LE.DEPTHF) GOTO 450
      IF(NUM.LE.N) GOTO 140
C
C     SECTION 9 (COMPUTATION OF KNEW)
C
      KNEW=0
      DO 670 I=1,N
       N1=NR(I)
       L1=LIST(I)
       L2=LIST(I+1)-1
       IF(L1.GT.L2) GOTO 670
       DO 660 J=L1,L2
        K=A(J)
        III=IABS(N1-NR(K))
        IF(III.GT.KNEW) KNEW=III
  660   CONTINUE
  670  CONTINUE
  680 IF(KOLD.GT.KNEW) GOTO 700
      KNEW=KOLD
      DO 690 I=1,N
  690  A(I)=I
      RETURN
  700 DO 710 I=1,N
  710  A(I)=NR(I)
      RETURN
      END
```

```
      SUBROUTINE LEVEL(START,NN,DEPTH,K3,WIDTH)
C
C     GENERATION OF THE LEVELS R(START)
C     DEPTH  DEPTH OF THE LEVELS
C     K3     NUMBER OF NODES IN THE LAST LEVEL
C     WIDTH  WIDTH OF THE LEVELS
C     NN     NUMBER OF ASSOCIATED NODES
C
      INTEGER START,NN,DEPTH,K3,WIDTH
      INTEGER A(4096),VEC(650)
      INTEGER IND(651,8),LIST(651),GRAD(650),NR(650)
      COMMON A,VEC,IND
      EQUIVALENCE (LIST(1),IND(1,1)),(GRAD(1),IND(1,2)),
     ,            (NR(1),IND(1,3))
      INTEGER J,I,BEG,END,N1,K,K2,LBR,STARTN,AI,L1,L2
      J=NN
      DO 1 I=1,J
    1  IND(I,8)=0
      BEG=1
      N1=1
      K=1
      LBR=1
      K2=1
      IND(1,6)=START
      IND(START,8)=1
C
    3 K=K+1
      END=N1
      DO 10 J=BEG,END
       STARTN=IND(J,6)
       L1=IND(STARTN,1)
       L2=IND(STARTN+1,1)-1
       DO 5 I=L1,L2
        AI=A(I)
        IF(IND(AI,8).NE.0) GOTO 5
        IND(AI,8)=K
        N1=N1+1
        IND(N1,6)=AI
    5   CONTINUE
   10  CONTINUE
      K3=K2
      K2=N1-END
      IF(LBR.LT.K2) LBR=K2
      BEG=END+1
      IF(K2.GT.0) GOTO 3
C
      DEPTH=K-1
      WIDTH=LBR
      NN=N1
      RETURN
      END
```

```
      SUBROUTINE KOMPON(START,N1)
C
C     COMPUTATION OF THE COMPONENT V(I) CONTAINING "START"
C     N1          NUMBER OF INVOLVED NODES
C     IND(I,6)    ALL NODES V(I)
C
      INTEGER START,N1
      INTEGER A(4096),VEC(650)
      INTEGER IND(651,8),LIST(651),GRAD(650),NR(650)
      COMMON A,VEC,IND
      EQUIVALENCE (LIST(1),IND(1,1)),(GRAD(1),IND(1,2)),
     *            (NR(1),IND(1,3))
      INTEGER AI,I,K2,L1,L2,STARTN,J,BEG,END,C,C2
      DATA C/10000/
      C2=2*C
      BEG=1
      N1=1
      IND(START,5)=IND(START,5)-C2
      IND(1,6)=START
C
    3 END=N1
      DO 10 J=BEG,END
       STARTN=IND(J,6)
       L1=IND(STARTN,1)
       L2=IND(STARTN+1,1)-1
       DO 5 I=L1,L2
        AI=A(I)
        IF(IND(AI,5).LT.0) GOTO 5
        IND(AI,5)=IND(AI,5)-C2
        N1=N1+1
        IND(N1,6)=AI
    5   CONTINUE
   10  CONTINUE
      K2=N1-END
      BEG=END+1
      IF(K2.GT.0) GOTO 3
      RETURN
      END
```

```
      SUBROUTINE SSORT1(F,L)
C     SORTING OF A FROM A(F) TO A(L)
      INTEGER F,L
      INTEGER A(4096),VEC(650)
      INTEGER IND(651,8),LIST(651),GRAD(650),NR(650)
      COMMON A,VEC,IND
      EQUIVALENCE (LIST(1),IND(1,1)),(GRAD(1),IND(1,2)),
     *            (NR(1),IND(1,3))
      INTEGER N2,S,T,LS,IS,JS,AH,I,J
      IF(L.LE.F) RETURN
      N2=(L-F+1)/2
      S=1023
      DO 100 T=1,10
       IF(S.GT.N2) GOTO 90
       LS=L-S
       DO 20 I=F,LS
        IS=I+S
        AH=A(IS)
        J=I
        JS=IS
    5   IF(AH.GE.A(J)) GOTO 10
        A(JS)=A(J)
        JS=J
        J=J-S
        IF(J.GE.F) GOTO 5
   10   A(JS)=AH
   20   CONTINUE
   90  S=S/2
  100  CONTINUE
      RETURN
      END
```

```
      SUBROUTINE SSORT2(VEC,K,F,L)
C     SORTING OF VEC FROM I=F TO I=L SO THAT
C     IND(VEC(I),K) IS WEAKLY INCREASING
      INTEGER F,L,VEC(650),K
      INTEGER A(4096),VECC(650)
      INTEGER IND(651,8),LIST(651),GRAD(650),NR(650)
      COMMON A,VECC,IND
      EQUIVALENCE (LIST(1),IND(1,1)),(GRAD(1),IND(1,2)),
     *            (NR(1),IND(1,3))
      INTEGER N2,S,T,LS,IS,JS,AH,GH,AJ,I,J
      IF(L.LE.F) RETURN
      N2=(L-F+1)/2
      S=63
      DO 100 T=1,6
       IF(S.GT.N2) GOTO 90
       LS=L-S
       DO 20 I=F,LS
        IS=I+S
        AH=VEC(IS)
        GH=IND(AH,K)
        J=I
        JS=IS
    6   AJ=VEC(J)
    5   IF(GH.GE.IND(AJ,K)) GOTO 10
        VEC(JS)=VEC(J)
        JS=J
        J=J-S
        IF(J.GE.F) GOTO 6
   10   VEC(JS)=AH
   20   CONTINUE
   90  S=S/2
  100  CONTINUE
      RETURN
      END
```

Appendix 6: The Buneman algorithm for solving the Poisson equation.

Let $G = (-1,1) \times (-1,1)$. We want to find a solution for the problem

$$\Delta u(x,y) = q(x,y), \quad (x,y) \in G$$
$$u(x,y) = \psi(x,y), \quad (x,y) \in \partial G.$$

Depending on the concrete problem, two REAL FUNCTIONs, QUELL and RAND, have to be constructed to describe q and ψ (see the example at the end below).

The parameter list for the subroutine BUNEMA includes only the parameter K in addition to the names for QUELL and RAND (which require an EXTERNAL declaration). We have H = 1./2**K, where H denotes the distance separating the lattice points. After the program has run, the COMMON domain will contain the computed approximations. All further details can be discerned from the program comments.

BUNEMA calls the subroutine GLSAR. This solves the system of equations

$$A^{(r)}x = b$$

by means of a factorization of $A^{(r)}$ (cf. Sec. 21). When first used, GLSAR calls the subroutine COSVEC, which computes a series of cosine values via a recursion formula. GLSAR also calls on the routine GAUBDS. The latter solves special tridiagonal systems of equations via a modified Gaussian algorithm (LU-splitting).

Program BUNEMA contains 50 executable FORTRAN instructions, and the other subroutines another 42 instructions. In order to make the program easy to read we have restricted ourselves to the case $G = (-1,1) \times (-1,1)$. However, it can be rewritten for a rectangular region without any difficulties.

We want to discuss the numerical results by means of the following examples:

$$\begin{aligned} &\Delta u(x,y) = 0, && (x,y) \in G \\ &u(x,y) = 1, && (x,y) \in \partial G \\ &\text{exact solution: } u(x,y) = 1 \end{aligned} \tag{1}$$

$$\begin{aligned} &\Delta u(x,y) = -2\pi^2 \sin(\pi x)\sin(\pi y), && (x,y) \in G \\ &u(x,y) = 0, && (x,y) \in \partial G \\ &\text{exact solution: } u(x,y) = \sin(\pi x)\sin(\pi y). \end{aligned} \tag{2}$$

The first example is particularly well suited to an examination of the numerical stability of the method, since the discretization error is zero. Thus the errors measured are exclusively those arising from the solving of the system of equations, either from the method or from rounding. The computation was carried out on a CDC-CYBER 76 (mantissa length 48 bits for REAL) and on an IBM 370/168 (mantissa length 21-24 bits for REAL*4 and 53-56 bits for REAL*8).

Table 1 contains the maximal absolute error for the computed approximations. Here

H = 1/2**K = 2/(N+1)

N : number of lattice points in one direction

N^2 : dimension of the system of equations

Since the extent of the system of equations grows as N^2, one would expect a doubling of N in Example 1 to lead to a fourfold increase in the rounding error. This is almost exactly

Example	N	CYBER 76 REAL	370/168 REAL*4	370/168 REAL*8
(1)	3	0.71E-14	0.77E-6	0.11E-15
	7	0.43E-13	0.77E-6	0.54E-15
	15	0.23E-12	0.10E-4	0.72E-15
	31	0.58E-12	0.55E-4	0.15E-13
	63	0.15E-11	0.45E-3	0.85E-13
	127	0.68E-11	0.14E-2	0.30E-12
(2)	3	0.23E0	0.23E0	0.23E0
	7	0.53E-1	0.53E-1	0.53E-1
	15	0.13E-1	0.13E-1	0.13E-1
	31	0.32E-2	0.32E-2	0.32E-2
	63	0.80E-3	0.69E-3	0.80E-3
	127	0.20E-3	0.31E-3	0.20E-3

Table 1. Absolute error

what happened in the computation on the CYBER 76, averaged over all N. On the IBM, the mean increase in error per step was somewhat greater.

The values for Example (1) and Example (2) show that for a REAL*4 computation on the IBM 370, N should in no case be chosen larger than 63. For greater mantissa lengths, there is no practical stability bound on either machine.

Table 2 contains the required computing times, exclusive of the time required to prepare the right side of the system of equations. In the compilation (with the exception of the G1 compiler), the parameter OPT = 2 was used.

Machine computation compiler	CYBER 76 REAL FTN	370/168 REAL*4 H-Extended	370/168 REAL*8 H-Extended	370/168 REAL*4 G1
N				
31	0.03	0.04	0.04	0.07
63	0.13	0.19	0.22	0.31
127	0.55	0.85	1.04	1.44

Table 2. Computing times in seconds

```
      SUBROUTINE BUNEMA(RAND,QUELL,K)
C
C     PARAMETERS
C
C     FUNCTIONS RAND, QUELL
      INTEGER K
C
C     VARIABLES OF THE COMMON BLOCKS
C
      REAL W(63,65),P(63,63),Q(63,65)
      EQUIVALENCE(W(1,1),Q(1,1))
      COMMON W,P
C
C     LOCAL VARIABLES
C
      INTEGER I,J,J1,J2,J3,J4,J5,K1,K2,KMAX,R,R1,R2
      REAL B(63),X,Y,H,H2
C
C     MEANING OF THE VARIABLES
C
C     H  :  DISTANCES OF THE LATTICE POINTS
C     K  :  H=1/2**K. THE PROGRAMME IS TERMINATED IN CASE
C           OF K.LT.1.
C     H2 :  =H**2
C     K1 :  =2**(K+1)-1
C     K2 :  =2**(K+1)
C     P,Q:  COMPARE DESCRIPTION OF THE METHOD. AS STARTING
C           VALUES ZEROS ARE ASSIGNED TO P AND THE RIGHT-HAND
C           SIDE OF THE SYSTEM IS ASSIGNED TO Q.
C     W  :  AFTER PERFORMING THE PROGRAMME W(I,J) CONTAINS
C           AN APPROXIMATION TO THE SOLUTION W(X,Y) AT THE
C           INTERIOR GRID POINTS. (I,J) AND (X,Y) ARE
C           RELATED BY
C                  X=(I-2**K)*H,     I=1(1)K1
C                  Y=(J-1-2**K)*H, J=2(1)K2.
C           TO SIMPLIFY THE SOLUTION PHASE OF THE PROGRAMME,
C           W(*,1) AND W(*,K2+1) ARE INITIALIZED BY ZEROS.
C           W AND Q ARE EQUIVALENCED. THERE IS NO IMPLICIT
C           USE OF THIS IDENTITY. DURING THE SOLUTION PHASE
C           THOSE COMPONENTS OF W ARE DEFINED SUCCESSIVELY,
C           THAT ARE NO LONGER NEEDED IN Q.
C     B  :  AUXILIARY STORAGE
C
      KMAX=5
C
C     ARRAY BOUNDS
C
C     THE LENGTHS OF THE ARRAYS W(DIM1,DIM2), P(DIM1,DIM1),
C     Q(DIM1,DIM2), AND B(DIM1) CAN BE CHANGED TOGETHER WITH
C     KMAX. IT IS
C     DIM1=2**(KMAX+1)-1
C     DIM2=2**(KMAX+1)+1.
```

```
C     ACCORDINGLY, THE LENGTHS OF THE ARRAYS COS2(DIM1) IN
C     SUBROUTINE GLSAR AND A(DIM1) IN SUBROUTINE GAUBDS MUST
C     BE CHANGED. THE PROGRAMME TERMINATES IF K.GT.KMAX.
C
      IF((K.LT.1).OR.(K.GT.KMAX))STOP
      K2=2**(K+1)
      K1=K2-1
      H=1.0/FLOAT(2**K)
      H2=H**2
C
C     ASSIGN ZEROS TO P AND TO PARTS OF W
C
      DO 10 J=1,K1
       W(J,1)=0.0
       W(J,K2+1)=0.0
       DO 10 I=1,K1
   10   P(I,J)=0.0
C
C     STORE THE RIGHT-HAND SIDE OF THE SYSTEM ON Q
C
      Y=-1.0
      DO 120 J=2,K2
       Y=Y+H
       X=-1.0
       DO 110 I=1,K1
        X=X+H
  110   Q(I,J)=H2*QUELL(X,Y)
       Q(1,J)=Q(1,J)-RAND(-1.0,Y)
  120  Q(K1,J)=Q(K1,J)-RAND(1.0,Y)
      X=-1.0
      DO 130 I=1,K1
       X=X+H
       Q(I,2)=Q(I,2)-RAND(X,-1.0)
  130  Q(I,K2)=Q(I,K2)-RAND(X,1.0)
C
C     REDUCTION PHASE, COMPARE EQ. (21.9) OF THE DESCRIPTION
C
      DO 230 R=1,K
       J1=2**R
       J2=K2-J1
       DO 230 J=J1,J2,J1
        J3=J-J1/2
        J4=J+J1/2
        DO 210 I=1,K1
  210    B(I)=P(I,J3)+P(I,J4)-Q(I,J+1)
        CALL GLSAR(R-1,B,K1,K)
        DO 220 I=1,K1
         P(I,J)=P(I,J)-B(I)
  220    Q(I,J+1)=Q(I,J3+1)+Q(I,J4+1)-2.0*P(I,J)
  230   CONTINUE
```

```
C
C     SOLUTION PHASE, COMPARE EQ. (21.10) OF THE DESCRIPTION
C
      R2=K+1
      DO 330 R1=1,R2
       R=R2-R1
       J1=2**R
       J2=K2-J1
       J3=2*J1
       DO 330 J=J1,J2,J3
        J4=J+1+J1
        J5=J+1-J1
        DO 310 I=1,K1
  310    B(I)=Q(I,J+1)-W(I,J4)-W(I,J5)
        CALL GLSAR(R,B,K1,K)
        DO 320 I=1,K1
  320    W(I,J+1)=P(I,J)+B(I)
  330   CONTINUE
      RETURN
      END
```

```
      SUBROUTINE GLSAR(R,B,N,K)
      INTEGER R,N,K
      REAL B(N)
C
C     SOLUTION OF THE SYSTEM A(R)*X=B BY FACTORING OF A(R)
C     (COMPARE EQ. (21.6) OF THE DESCRIPTION).
C     A(R) IS DEFINED RECURSIVELY BY
C     A(R)=2I-(A(R-1))**2, R=1(1)K WITH A(0)=A=(AIJ), I,J=1(1)N
C     AND
C         -4  IF I=J
C     AIJ= 1  IF I=J+1 OR I=J-1
C          0  OTHERWISE
C     THE PROGRAMME FAILS IF N.LT.2; OTHERWISE IT
C     TERMINATES WITH B=X.
C
      INTEGER FICALL,J,J1,J2,JS
      REAL COS2(63)
      DATA FICALL/0/
C
      IF(R.EQ.0)GOTO 30
C
C     THE SUBROUTINE COSVEC IS ONLY CALLED IF THE VECTOR COS2
C     IS NOT YET COMPUTED FOR THE ACTUAL VALUE OF K.
C
      IF(FICALL.EQ.K)GOTO 1
      CALL COSVEC(K,COS2)
      FICALL=K
C
    1 DO 10 J=1,N
   10  B(J)=-B(J)
      J1=2**(K-R)
      J2=(2**(R+1)-1)*J1
      JS=2*J1
C
C     BECAUSE OF COS2(J)=2*COS(J*PI/2**(K+1))=2*COS(I*PI)
C     THE DOMAIN OF THE INDEX I IS
C     I=2**(-R-1) (2**(-R)) 1-2**(-R-1).
C
      DO 20 J=J1,J2,JS
   20  CALL GAUBDS(COS2(J),B,N)
      GOTO 40
C
   30 CALL GAUBDS(0.0,B,N)
C
   40 RETURN
      END
```

```
      SUBROUTINE GAUBDS(C,B,N)
      INTEGER N
      REAL C,B(N)
C
C     SOLUTION OF A LINEAR SYSTEM WITH SPECIAL
C     TRIDIAGONAL MATRIX:
C     X(I-1)+(-4+C)*X(I)+X(I+1)=B(I),  I=1(1)N,
C     WHERE X(0) AND X(N+1) ARE VANISHING.
C     THE PROGRAMME TERMINATES WITH B=X. IN CASE OF N.LT.2
C     THIS SUBROUTINE FAILS. IF N.GT.63 THE ARRAY BOUNDS
C     MUST BE ENLARGED.
      INTEGER I,I1
      REAL Q,C4,A(63)
C
      C4=C-4.0
      A(1)=C4
      DO 10 I=2,N
       Q=1.0/A(I-1)
       A(I)=C4-Q
   10  B(I)=B(I)-B(I-1)*Q
      Q=B(N)/A(N)
      B(N)=Q
      I1=N-1
      DO 20 I=2,N
       Q=(B(I1)-Q)/A(I1)
       B(I1)=Q
   20  I1=I1-1
      RETURN
      END
```

```
      SUBROUTINE COSVEC(K,COS2)
      INTEGER K
      REAL COS2(1)
C
C     COMPUTATION OF
C     COS2(J)=2*COS(J*PI/2**(K+1)), J=1(1)2**(K+1)-1
C     BY MEANS OF RECURSION AND REFLECTION.
C     THE PROGRAMME FAILS FOR K.LT.1.
C
      INTEGER K2,J,J1
      REAL DC,T,CV,PI4
C
      K2=2**K
      J1=2*K2-1
      PI4=ATAN(1.0)
      COS2(K2)=0.0
      DC=-4.0*SIN(PI4/FLOAT(K2))**2
      K2=K2-1
      T=DC
      CV=2.0+DC
      COS2(1)=CV
      COS2(J1)=-CV
      DO 10 J=2,K2
       J1=J1-1
       DC=T*CV+DC
       CV=CV+DC
       COS2(J)=CV
   10  COS2(J1)=-CV
      RETURN
      END
```

EXAMPLE (MENTIONED IN THE TEXT):

```
      REAL FUNCTION QUELL(X,Y)
      REAL X,Y
      DATA PI/3.14159265358979/
      QUELL=-2.0*PI*PI*SIN(PI*X)*SIN(PI*Y)
      RETURN
      END
```

```
      REAL FUNCTION RAND(X,Y)
      REAL X,Y
      RAND=0.0
      RETURN
      END
```

BIBLIOGRAPHY

Abramowitz, M., Stegun, I. A.: Handbook of Mathematical Functions, New York: Dover Publications 1965.

Ahlfors, L. V.: Complex Analysis, New York-Toronto-London: McGraw-Hill 1966.

Ansorge, R., Hass, R.: Konvergenz von Differenzenverfahren für Lineare und Nichtlineare Anfangswertaufgaben, Lecture Notes in Mathematics, Vol. 159, Berlin-Heidelberg-New York: Springer 1970.

Beckenbach, E. F., Bellman, R.: Inequalities, Berlin-Heidelberg-New York: Springer 1971.

Birkhoff, G., Schultz, M. H., Varga, R. S.: "Piecewise hermite interpolation in one and two variables with applications to partial differential equations," Numer. Math., 11, 232-256 (1968).

Busch, W., Esser, R., Hackbusch, W., Herrmann, U.: "Extrapolation applied to the method of characteristics for a first order system of two partial differential equations," Numer. Math., 24, 331-353 (1975).

Buzbee, B. L., Golub, G. H., Nielson, C. W.: "On direct methods for solving Poisson's equations," SIAM J. Numer. Anal., 7, No. 4, 627-656 (1970).

Coddington, E. A., Levinson, N.: Theory of Ordinary Differential Equations, New York-Toronto-London: McGraw-Hill 1955.

Collatz, L.: Funktionalanalysis und Numerische Mathematik, Berlin-Göttingen-Heidelberg: Springer 1964.

Collatz, L.: The Numerical Treatment of Differential Equations, Berlin-Heidelberg-New York: Springer 1966.

Courant, R., Friedrichs, K. O., Lewy, H.: "Über die partiellen differenzengleichungen der mathematischen physik," Math. Ann., 100, 32-74 (1928).

Cuthill, E., McKee, J.: "Reducing the bandwidth of sparse symmetric matrices," Proc. 24th ACM National Conference, 157-172 (1969).

Dieudonné, J.: Foundations of Modern Analysis, New York-London: Academic Press 1960.

Dorr, F. W.: "The direct solution of the discrete Poisson equation on a rectangle," SIAM Rev., 12, No. 2, 248-263 (1970).

Fox, L. (Editor): Numerical Solution of Ordinary and Partial Differential Equations, London: Pergamon Press 1962.

Friedman, A.: Partial Differential Equations, New York: Holt, Rinehart and Winston 1969.

Friedrichs, K. O.: "Symmetric hyperbolic linear differential equations," Comm. Pure Appl. Math., 7, 345-392 (1954).

Gerschgorin, S.: "Fehlerabschätzungen für das differenzenverfahren zur lösung partieller differentialgleichungen," ZAMM, 10, 373-382 (1930).

Gibbs, N. E., Poole, W. G., Stockmeyer, P. K.: "An algorithm for reducing the bandwidth and profile of a sparse matrix," SIAM J. Numer. Anal., 13, 236-250 (1976).

Gilbarg, D., Trudinger, N. S.: Elliptic Partial Differential Equations of Second Order, Berlin-Heidelberg-New York: Springer 1977.

Gorenflo, R.: "Über S. Gerschgorins Methode der Fehlerabschätzung bei Differenzenverfahren." In: Numerische, Insbesondere Approximations-Theoretische Behandlung von Funktionalgleichungen, Lecture Notes in Mathematics, Vol. 333, 128-143, Berlin-Heidelberg-New York: Springer 1973.

Grigorieff, R. D.: Numerik Gewöhnlicher Differentialgleichungen, Studienbücher, Bd. 1. Stuttgart: Teubner 1972.

Hackbusch, W.: Die Verwendung der Extrapolationsmethode zur Numerischen Lösung Hyperbolischer Differentialgleichungen, Universität zu Köln: Dissertation 1973.

Hackbusch, W.: "Extrapolation to the limit for numerical solutions of hyperbolic equations," Numer. Math., 28, 455-474 (1977).

Hellwig, G.: Partial Differential Equations, Stuttgart: Teubner 1977.

Householder, A. S.: The Theory of Matrices in Numerical Analysis, New York: Blaisdell 1964.

Janenko, N. N.: The Method of Fractional Steps, Berlin-Heidelberg-New York: Springer 1971.

Jawson, M. A.: "Integral equation methods in potential theory I," Proc. R. Soc., Vol. A 275, 23-32 (1963).

Kellog, O. D.: Foundations of Potential Theory, Berlin: Springer 1929.

Knuth, D. E.: The Art of Computer Programming, Reading, Massachusetts: Addison-Wesley 1973.

Kreiss, H. O.: "On difference approximations of the dissipative type for hyperbolic differential equations," Comm. Pure Appl. Math., 17, 335-353 (1964).

Lax, P. D.: "On the stability of difference approximations to solutions of hyperbolic equations with variable coefficients," Comm. Pure Appl. Math., 14, 497-520 (1961).

Lax, P. D., Wendroff, B.: "On the stability of difference schemes," Comm. Pure Appl. Math., 15, 363-371 (1962).

Lax, P. D., Nirenberg, L.: "On stability for difference schemes; a sharp form of Garding's inequality," Comm. Pure Appl. Math., 19, 473-492 (1966).

Lehmann, H.: Fehlerabschätzungen in Verschiedenen Normen bei Eindimensionaler und Zweidimensionaler Hermite-Interpolation, Universität zu Köln: Diplomarbeit 1975.

Lierz, W.: Lösung von großen Gleichungssystemen mit Symmetrischer Schwach Besetzter Matrix, Universität zu Köln: Diplomarbeit 1975.

Loomis, L. H., Sternberg, S.: Advanced Calculus, Reading, Massachusetts: Addison-Wesley 1968.

Magnus, W., Oberhettinger, F., Soni, R. P.: Formulas and Theorems for the Special Functions of Mathematical Physics, New York: Springer 1966.

Meis, Th.: "Zur diskretisierung nichtlinearer elliptischer differentialgleichungen," Computing, 7, 344-352 (1971).

Meis, Th., Törnig, W.: "Diskretisierungen des Dirichletproblems nichtlinearer elliptischer differentialgleichungen. In: Methoden und Verfahren der Mathematischen Physik, Bd. 8. Herausgeber: B. Brosowski und E. Martensen, Mannheim-Wien-Zürich: Bibliographisches Institut 1973.

Meuer, H. W.: Zur Numerischen Behandlung von Systemen Hyperbolischer Anfangswertprobleme in Beliebig Vielen Ortsveränderlichen mit Hilfe von Differenzenverfahren, Technische Hochschule Aachen: Dissertation 1972.

Mizohata, S.: The Theory of Partial Differential Equations, Cambridge: University Press 1973.

Natanson, I. P.: Theorie der Funktionen einer Reellen Veränderlichen, Berlin: Akademie-Verlag 1961.

Ortega, J. M., Rheinboldt, W. C.: Iterative Solution of Nonlinear Equations in Several Variables, New York-London: Academic Press 1970.

Ostrowski, A. M.: "On the linear iteration procedures for symmetric matrices," Rend. Math. e. Appl., 14, 140-163 (1954).

Peaceman, D. W., Rachford, H. H.: "The numerical solution of parabolic and elliptic differential equations," J. Soc. Indust. Appl. Math., 3, 28-41 (1955).

Perron, O.: "Über existenz und nichtexistenz von integralen partieller differentialgleichungssysteme im reellen Gebiet," Mathem. Zeitschrift, 27, 549-564 (1928).

Petrovsky, I. G.: Lectures on Partial Differential Equations, New York-London: Interscience Publishers 1954.

Reid, J. K.: "Solution of linear systems of equations: direct methods (general). In: Sparse Matrix Techniques, Lecture Notes in Mathematics, Vol. 572, Editor V. A. Barker, Berlin-Heidelberg-New York: Springer 1977.

Richtmyer, R. D., Morton, K. W.: Difference Methods for Initial Value Problems, New York-London-Sydney: Interscience Publishers 1967.

Rosen, R.: Matrix Bandwidth Minimization, Proc. 23rd ACM National Conf., 585-595 (1968).

Sauer, R. S.: Anfangswertprobleme bei Partiellen Differentialgleichungen, Berlin-Göttingen-Heidelberg: Springer 1958.

Schechter, S.: "Iteration methods for nonlinear problems," Trans. Amer. Math. Soc., 104, 179-189 (1962).

Schröder, J., Trottenberg, U.: "Reduktionsverfahren für differenzengleichungen bei randwertaufgaben I," Numer. Math., 22, 37-68 (1973).

Schröder, J., Trottenberg, U., Reutersberg, H.: "Reduktionsverfahren für differenzengleichungen bei randwertaufgaben II," Numer. Math., 26, 429-459 (1976).

Shell, D. L.: "A highspeed sorting procedure," Comm. ACM, 2, No. 7, 30-32 (1959).

Simonson, W.: "On numerical differentiation of functions of several variables," Skand. Aktuarietidskr., 42, 73-89 (1959).

Stancu, D. D.: "The remainder of certain linear approximation formulas in two variables," SIAM J. Numer. Anal., Ser. B.1, 137-163 (1964).

Stoer, J., Bulirsch, R.: Introduction to Numerical Analysis. New York-Heidelberg-Berlin: Springer 1980.

Swartz, B. K., Varga, R. S.: "Error bounds for spline and L-spline interpolation," J. of Appr. Th., 6, 6-49 (1972).

Thomée, V.: "Spline approximation and difference schemes for the heat equation. In: The Mathematical Foundations of the Finite Element Method with Applications to Partial Differential Equations, Edited by K. Aziz, New York-London: Academic Press 1972.

Törnig, W., Ziegler, M.: "Bemerkungen zur konvergenz von differenzenapproximationen für quasilineare hyperbolische anfangswertprobleme in zwei unabhängigen veränderlichen." ZAMM, 46, 201-210 (1966).

Varga, R. S.: Matrix Iterative Analysis, Englewood Cliffs: Prentice-Hall 1962.

Walter, W.: Differential and Integral Inequalities, Berlin-Heidelberg-New York: Springer 1970.

Whiteman, J. R. (Editor): The Mathematics of Finite Elements and Applications, London-New York: Academic Press 1973.

Whiteman, J. R. (Editor): The Mathematics of Finite Elements and Applications II, London-New York-San Francisco: Academic Press 1976.

Widlund, O. B.: "On the stability of parabolic difference schemes," Math. Comp., 19, 1-13 (1965).

Witsch, K.: "Numerische quadratur bei projektionsverfahren," Numer. Math., 30, 185-206 (1978).

Yosida, K.: Functional Analysis, Berlin-Heidelberg-New York: Springer 1968.

Young, D. M.: Iterative Methods for Solving Partial Differential Equations of Elliptic Type, Harvard University, Thesis 1950.

Zlamal, M.: "On the finite element method," Numer. Math., 12, 394-409 (1968).

Textbooks

Ames, W. F.: Numerical Methods for Partial Differential Equations, New York- San Francisco: Academic Press 1977.

Ansorge, R.: Differenzenapproximationen Partieller Anfangswertaufgaben, Stuttgart: Teubner 1978.

Ciarlet, Ph. G.: The Finite Element Method for Elliptic Problems, Amsterdam: North-Holland 1978.

Collatz, L.: The Numerical Treatment of Differential Equations, Berlin-Heidelberg-New York: Springer 1966.

Forsythe, G. E., Rosenbloom, P. C.: Numerical Analysis and Partial Differential Equations, New York: John Wiley 1958.

Forsythe, G. E., Wasow, W. R.: Finite Difference Methods for Partial Differential Equations, New York-London: John Wiley 1960.

John, F.: Lectures on Advanced Numerical Analysis, London: Gordon and Breach 1967.

Marchuk, G. I.: Methods of Numerical Mathematics, Vol. 2, New York-Heidelberg-Berlin: Springer 1975.

Marsal, D.: Die Numerisch Lösung Partieller Differentialgleichungen in Wissenschaft und Technik, Mannheim-Wien-Zürich: Bibliographisches Institut 1976.

Mitchell, A. R.: Computational Methods in Partial Differential Equations, London-New York-Sydney-Toronto: John Wiley 1969.

Ortega, J. M., Rheinboldt, W. C.: Iterative Solution of Nonlinear Equations in Several Variables, New York-London: Academic Press 1970.

Ralston, A., Wilf, H. S.; Mathematical Methods for Digital Computers, New York: John Wiley 1960.

Richtmyer, R. D., Morton, K. W.: Difference Methods for Initial Value Problems, New York-London-Sydney: Interscience Publishers 1967.

Schwarz, H. R.: Methode der Finiten Elemente, Stuttgart: Teubner 1980.

Smith, G. D.: Numerical Solution of Partial Differential Equations, London: Oxford University Press 1969.

Stoer, J., Bulirsch, R.: Introduction to Numerical Analysis, New York-Heidelberg-Berlin: Springer 1980.

Strang, G., Fix, G. J.: An Analysis of the Finite Element Method, Englewood Cliffs: Prentice-Hall 1973.

Varga, R. S.: Matrix Iterative Analysis, Englewood Cliffs: Prentice-Hall 1962.

Wachspress, E. L.: Iterative Solution of Elliptic Systems, Englewood Cliffs: Prentice-Hall 1966.

Young, D. M.: Iterative Solution of Large Linear Systems, New York-London: Academic Press 1971.

Zienkiewicz, O. C.: The Finite Element Method in Engineering Science, London: McGraw-Hill 1971.

INDEX

Applied Mathematical Sciences

1. John: **Partial Differential Equations,** 3rd ed.
2. Sirovich: **Techniques of Asymptotic Analysis.**
3. Hale: **Theory of Functional Differential Equations,** 2nd ed. (cloth)
4. Percus: **Combinatorial Methods.**
5. von Mises/Friedrichs: **Fluid Dynamics.**
6. Freiberger/Grenander: **A Course in Computational Probability and Statistics,** Rev. Ptg.
7. Pipkin: **Lectures on Viscoelasticity Theory.**
8. Giacaglia: **Perturbation Methods in Non-Linear Systems.**
9. Friedrichs: **Spectral Theory of Operators in Hilbert Space.**
10. Stroud: **Numerical Quadrature and Solutions of Ordinary Differential Equations.**
11. Wolovich: **Linear Multivariable Systems.**
12. Berkovitz: **Optimal Control Theory.**
13. Bluman/Cole: **Similarity Methods for Differential Equations.**
14. Yoshizawa: **Stability Theory and the Existence of Periodic Solutions and Almost Periodic Solutions.**
15. Braun: **Differential Equations and Their Applications,** 2nd ed.
16. Lefschetz: **Applications of Algebraic Topology.**
17. Collatz/Wetterling: **Optimization Problems.**
18. Grenander: **Pattern Synthesis. Lectures in Pattern Theory,** vol. I.
19. Marsden/McCracken: **The Hopf Bifurcation and Its Applications.**
20. Driver: **Ordinary and Delay Differential Equations.**
21. Courant/Friedrichs: **Supersonic Flow and Shock Waves.** (cloth)
22. Rouche/Habets/Laloy: **Stability Theory by Liapunov's Direct Method.**
23. Lamperti: **Stochastic Processes. A Survey of the Mathematical Theory.**
24. Grenander: **Pattern Analysis. Lectures in Pattern Theory,** vol. II.
25. Davies: **Integral Transforms and Their Applications.**
26. Kushner/Clark: **Stochastic Approximation Methods for Constrained and Unconstrained Systems.**
27. de Boor: **A Practical Guide to Splines.**
28. Keilson: **Markov Chain Models—Rarity and Exponentiality.**
29. de Veubeke: **A Course in Elasticity.**
30. Śniatycki: **Geometric Quantization and Quantum Mechanics.**
31. Reid: **Sturmian Theory for Ordinary Differential Equations.**
32. Meis/Marcowitz: **Numerical Solution of Partial Differential Equations.**
33. Grenander: **Lectures in Pattern Theory,** vol. III.
34. Cole/Kevorkian: **Pertubation Methods in Applied Mathematics.**

Applied Mathematical Sciences

1. John: Partial Differential Equations, 4th ed.
2. Sirovich: Techniques of Asymptotic Analysis.
3. Hale: Theory of Functional Differential Equations, 2nd ed. (cloth)
4. Percus: Combinatorial Methods.
5. von Mises/Friedrichs: Fluid Dynamics.
6. Freiberger/Grenander: A Short Course in Computational Probability and Statistics.
7. Pipkin: Lectures on Viscoelasticity Theory.
8. Giacaglia: Perturbation Methods in Non-linear Systems.
9. Friedrichs: Spectral Theory of Operators in Hilbert Space.
10. Stroud: Numerical Quadrature and Solution of Ordinary Differential Equations.
11. Wolovich: Linear Multivariable Systems.
12. Berkovitz: Optimal Control Theory.
13. Bluman/Cole: Similarity Methods for Differential Equations.
14. Yoshizawa: Stability Theory and the Existence of Periodic Solutions and Almost Periodic Solutions.
15. Braun: Differential Equations and Their Applications, 3rd ed.
16. Lefschetz: Applications of Algebraic Topology.
17. Collatz/Wetterling: Optimization Problems.
18. Grenander: Pattern Synthesis: Lectures in Pattern Theory, Vol. I.
19. Marsden/McCracken: Hopf Bifurcation and Its Applications.
20. Driver: Ordinary and Delay Differential Equations.
21. Courant/Friedrichs: Supersonic Flow and Shock Waves.
22. Rouche/Habets/Laloy: Stability Theory by Liapunov's Direct Method.
23. Lamperti: Stochastic Processes: A Survey of the Mathematical Theory.
24. Grenander: Pattern Analysis: Lectures in Pattern Theory, Vol. II.
25. Davies: Integral Transforms and Their Applications.
26. Kushner/Clark: Stochastic Approximation Methods for Constrained and Unconstrained Systems.
27. de Boor: A Practical Guide to Splines.
28. Keilson: Markov Chain Models—Rarity and Exponentiality.
29. de Veubeke: A Course in Elasticity.
30. Sniatycki: Geometric Quantization and Quantum Mechanics.
31. Reid: Sturmian Theory for Ordinary Differential Equations.
32. Meis/Markowitz: Numerical Solution of Partial Differential Equations.
33. Grenander: Regular Structures: Lectures in Pattern Theory, Vol. III.
34. Kevorkian/Cole: Perturbation Methods in Applied Mathematics.